"Professor Oakley has done that rare thing: writte[illegible] is at once informative and eminently readable. She has taken 'evil' out of the realm of the religious and metaphysical, placing it instead where it belongs—inside ourselves, as in the famous Pogo cartoon: 'we have met the enemy, and he is us!' Her book is filled with many examples, some drawn from close personal experience, of the complicated ways genetics and environment interact to predispose toward evil. The genetic side of the story has been neglected far too long. *Evil Genes* makes an important and timely addition to the literature on this most fascinating topic. Professor Oakley's book aims at, and deserves, a wide readership."

—Michael H. Stone, MD
Professor of clinical psychiatry, Columbia University

"A magnificent tour through the sociology, psychology, and biology of evil. No one should pass up the experience of stepping through the portals of this fascinating book to answer Oakley's crucial question: Why are there evil people, and why are they sometimes so successful?"

—Dr. Cliff Pickover
Author of *A Beginner's Guide to Immortality* and *The Heaven Virus*

"This book conveys an enormous amount of complex, up-to-date scientific information in an extremely 'digestible' manner. Dr. Oakley manages to illustrate how, although our genetic makeup is not our destiny, there are clearly people who have an unfortunate dose of risk genes. These people often have impoverished social and emotional experience and can cause suffering to those around them. Although firmly grounded in science, this book is also compassionate and forces the reader to examine their own beliefs and prejudices in the light of what is currently known about the nature and nurture of 'evil.'"

—Essi Viding, PhD
Department of Psychology
University College London

"A highly readable, entertaining, groundbreaking, must-read study with notable insights on the rise and fall of empires; but more importantly, it offers, perhaps for the first time, a distinctly plausible mechanism for explaining the origin and persistence of social inequality."

—Glenn Storey
President, Archaeological Institute of America, Iowa Society
Associate professor of classics and anthropology, University of Iowa
Author of *Urbanism in the Preindustrial World: Cross-Cultural Approaches*

"As a forensic psychologist who has spent much of my career delving into the darkest recesses of the criminal mind, I have often wondered what roles genes and environment play in subsequent psychopathic behavior. Barbara Oakley's outstanding *Evil Genes* provides the answer."

—Helen Smith, PhD
Author of *The Scarred Heart: Understanding and Identifying Kids Who Kill*

"Blending brisk studies of notorious evildoers with her own difficult family history, Dr. Oakley skillfully weaves together a panoramic mix of history, psychology, and the complications of human behavior to make a stimulating, provocative, and accessible read."

—Adam LeBor
Author of *Milosevic: A Biography* and
Complicity with Evil: The United Nations in the Age of Modern Genocide

"Oakley has dealt Hitler, Stalin, and all of their kind—past, present, and future—a telling blow with her perceptive exposition of their psychopathy. A courageous, groundbreaking exploration not only of evil in the modern world, but of her family's darkest secrets."

—Dmitri Nabokov, son of Vladimir Nabokov
Principal English translator of his father's Russian works

A major theme of Vladimir Nabokov's writing (in short stories such as "Tyrants Destroyed" and novels such as *Invitation to a Beheading* and *Bend Sinister*) is the psychological aberrations so brilliantly brought to light in *Evil Genes*.

"A fascinating book that manages to make the neurosciences intelligible to a layperson. Its argument that at least some of human evil is encoded on our genes is hard to refute, and this reader wasn't tempted to do so."

—Dr. Colleen McCullough, AO (Order of Australia)
Author and neuroscientist

"Einstein once said that all important new science would be found at the interstices of existing disciplines; if you need proof of that, this book is it. Starting with a background in the military, linguistics, and electrical engineering, Oakley deftly moves through psychology, functional brain imagery, and molecular biology to weave a compelling and provocative case for a genetic base for evil. 'Scientific nonfiction' and 'page turner' aren't two phrases I'd expect in the same sentence, but for the remarkable *Evil Genes*, they fit."

—William A. Wulf, President Emeritus
National Academy of Engineering

"Through a fascinating blend of state-of-the-art science, political biography, and personal catharsis, *Evil Genes* constructs a provocative blueprint for our understanding of the 'successfully sinister' among us."

—David J. Buller, Presidential Research Professor
Northern Illinois University
Author of *Adapting Minds*

"Many of us encounter people whose reactions are puzzling. They are easily hurt and offended. Even when someone is being generous, or kind to them they might react with anger, revengefulness, defensiveness, suspiciousness, or aloofness. These are difficult people to have as friends, relatives, colleagues, and even as patients. Dr. Oakley has written a comprehensive and compassionate explanation for why some people are like this that will be fascinating to anyone who has encountered this type of person and cared enough to wonder 'why?'"

—Regina Pally, MD, psychiatrist and psychoanalyst
Coauthor of *The Mind-Brain Relationship*

EVIL GENES

EVIL GENES

WHY ROME FELL, HITLER ROSE, ENRON FAILED, AND MY SISTER STOLE MY MOTHER'S BOYFRIEND

BARBARA OAKLEY

59 John Glenn Drive
Amherst, New York 14228–2119

Published 2008 by Prometheus Books

Inquiries should be addressed to
Prometheus Books
59 John Glenn Drive
Amherst, New York 14228–2119
VOICE: 716–691–0133, ext. 210
FAX: 716–691–0137
WWW.PROMETHEUSBOOKS.COM

19 18 17 16 15 8 7 6 5 4

Library of Congress Cataloging-in-Publication Data

Oakley, Barbara A., 1955–
Evil genes : why Rome fell, Hitler rose, Enron failed, and my sister stole my mother's boyfriend / by Barbara Oakley.
p. cm.
Includes bibliographical references and index.
ISBN 978–1–59102–665–5 (paperback : alk. paper)
1. Good and evil—Psychological aspects. I. Title.

BF789.E94O35 2007
155.7—dc22

2007027088

Printed in the United States of America on acid-free paper

To My Family

Contents

Foreword 15

Preface 19

Introduction 23

The Successfully Sinister 27
Sleuthing with the Sciences 31
A Revolution in Research 32
Putting the Puzzle Together 35

Chapter 1: In Search of Machiavelli 39

Richard Christie Founds a New Discipline 40
The Ubiquity of the Sinister 43
Perverse Admiration: The Development of a Test for Machiavellianism 46

CHAPTER 2: PSYCHOPATHY 49

Antisocial Personality Disorder 50
But What Is a Psychopath? 51
Sadism 52
The Role of Genes and Environment 54
The Genetics of Psychopathy 55

CHAPTER 3: EVIL GENES 59

Machiavellian Genes 65
Serotonin Receptors and Behavior 69
The Long and the Short of It—Serotonin Transporters 72
Pleiotropy—The Naughty-Nice Aspects of "Evil Genes" 75
Brain-Derived Neurotrophic Factor 77
Warrior or Worrier? The COMT Gene 78
Monoamine Oxidase A 80
Other Moody Genes 82
Emergenic Phenomena 83
Why We Can't Simply Eliminate "Evil" Genes 85

CHAPTER 4: USING MEDICAL IMAGING TO UNDERSTAND PSYCHOPATHS 89

Looking Inside the Brain of a Psychopath 89
Emotion and Language 90
Low Response to Threatening Stimuli and Empathy Impairment 92
Lack of Anxiety 93
Executive Dysfunction 94
Emotional Control—Affective and Predatory Murderers 96
Problems with Abstract Reasoning 97
Seeing the Human Conscience 100
Why Does Psychopathy Develop? 102

Empathy and Mirror Neurons 104
Successful Psychopaths 105

CHAPTER 5: INSIGHTS FROM MY SISTER'S LOVE LETTERS 109

The Early Years 110
The Missing Decade 117
Memories of Carolyn 119
The Letters 120

CHAPTER 6: THE CONNECTION BETWEEN MACHIAVELLIANISM AND PERSONALITY DISORDERS 131

The Dimensional Approach to Understanding Personality Disorders 132
McHoskey's Findings 133
Borderline Personality Disorder 136
Borderline Coping Behaviors 143
The Impact of Borderline Personality Disorder on Others 147

CHAPTER 7: SLOBODAN MILOSEVIC: THE BUTCHER OF THE BALKANS 151

The Quintessential Machiavellian: Slobodan Milosevic 153
Identity Disturbance 155
The DSM-IV Description of Borderline Personality Disorder 157
Was Milosevic a Borderline? 162
The Dimensional Approach to Describing Borderline Personality Disorder 163
Was Milosevic a Psychopath? 166
Personal Impact: Milosevic and My Family 168

CHAPTER 8: LENSES, FRAMES, AND HOW BROKEN BRAINS WORK 173

Lenses and Frames 174
Neurological Systems and How They Function to Regulate Emotion 179
The Cerebral Cortex 179
The Limbic System 182
Neural Connections 183
"Feel Good" Politics: How Machiavellians—and Altruists—Manipulate Emotions 187
Seeing Subtle Defects in the Emote Control System—Borderline Personality Disorder 193
Emotional Dysregulation: Limbic System 193
Impulsivity: Anterior Cingulate and Orbitomedial Prefrontal Systems 195
Cognitive-Perceptual Impairment: Dorsolateral, Ventromedial, and Orbitofrontal Prefrontal Systems 203
But What's the Big Picture? 205
A Link with the Immune System? 207
Overlapping Personality Disorders 208

CHAPTER 9: THE PERFECT "BORDERPATH": CHAIRMAN MAO 211

Mao as Borderpath 216
The Early Years 218
Early Antisocial Tendencies 218
Markedly Disturbed Relationships 219
The Confusing Façade—Sympathy with Little Empathy 225
Impulsivity 228
Poorly Regulated Emotions 229
Drug and Sexual Addictions 232
Cognitive-Perceptual Impairment 234

Control and the Purges 236
Sadism 239
Ability to Charm 240
Narcissism 242
Paranoia 245
Mao's Manipulation 246
Did Mao Believe in Communism? 247
Endgame and Aftermath 248

Chapter 10: Evolution and Machiavellianism 253

Tit for Tat 256
Throwing Away the Steering Wheel 260
Blink and You'll Miss It—The Quickness of Evolution 261
Baldwinian Evolution 263
Can Culture Create Machiavellians? 264
Gold Diggers, Stable Sinister Systems, and the Slow-Motion Implosion of Empires 271
Defining "Machiavellian" 280
Linda Mealey 283

Chapter 11: Shades of Gray 285

Narcissism, Deceit, Humbleness, and Conscience 287
Enron—The Power of Unchecked, Mutually Supportive Machiavellians 294
Temper, Temper, Temper 298
Cognitive Function and Dysfunction 300
Delusions 302
The Delusions of Dictators 304
The Delusions of Madmen 305
Personality Underlies Ideology 307
Ambition and Control 308

The Surprising Attributes 310
So, Are the Successfully Sinister Really Different? 314
Naivete 315
Warren Buffett—Multifaceted Genius 317
Healthy Cynicism 319

Chapter 12: The Sun Also Shines on the Wicked 323

Evil and Free Will 327
Who *Are* the Successfully Sinister? 332
How Can You Tell? 337
Carolyn 340

Afterword 345

For Pondering 357

Acknowledgments 361

Text Credits 365

Illustration Credits 367

Endnotes 371

Glossary 413

Index 429

I would like to take this opportunity to point out that the excerpts from Carolyn's diaries and letters are virtually verbatim, with only a few slight emendations to disguise the individuals in question, for brevity's sake, or to clarify personal shorthand.

There's a glossary in the back if you've set the book down for a day or two and want to refresh your memory about something. And if you're not a science type, don't worry—I'm on your side. I've written the book so you can skim over the sections that might not be your cup of tea.

—Barbara Oakley, Rochester, Michigan,
July 8, 2007

"He is a man with tens of thousands of blind followers. It is my business to make some of those blind followers see."

—Abraham Lincoln on the covertly proslavery, and amoral, Stephen Douglas[1]

Foreword

The natural world includes millions of species that evolved to survive and reproduce in different ways. These species have long been admired for the sophistication of their bodies—the shape of the bird's wing, the speed of the cheetah, the insect that looks exactly like a leaf. Only recently have we begun to appreciate the sophistication of their *behaviors*. For centuries and millennia, we have prided ourselves as being set apart from the rest of life by our intelligence. Yet, tune into any of the wonderful nature documentaries that are so widely available today, and you will see animals from insects to primates behaving amazingly like . . . us.

If we are set apart from the rest of life, it is primarily in our behavioral *flexibility*. We can adapt to new problems in ways that other species cannot. It is this ability that enabled our ancestors to spread over the globe, displacing other hominids and many other species along the way. Our cultures and individual behaviors are so successfully diverse that humans are more like an entire ecosystem than a single species.

Yet, our unique flexibility has an implication that is only just dawning upon us. If we are more like an ecosystem than a single species, *then human cultural and behavioral diversity can be understood in the same way as biological diversity*. We have not escaped evolution, as is so commonly assumed. We experience evolution in hyperdrive.

Explaining human diversity from this perspective is so new that it is like the earlier days of exploration. Instead of perilous voyages on ships and encounters with strange tribes on foreign shores, there are voyages to academic disciplines that have been largely disconnected from each other throughout their histories and whose members are often as fierce toward outsiders as any "primitive" tribe. Evolutionary theory can transcend disciplinary boundaries for the study of our own species, as it already has for the biological sciences.

In *Evil Genes*, our evolutionary explorer is Barbara Oakley, as colorful a character, in her own way, as Indiana Jones. An Air Force brat, she moved ten times by the tenth grade and enlisted as a private in the army before entering college on an ROTC scholarship to study Slavic languages and literature. After four years as a communications officer in Germany, she moved to Seattle and alternated between getting a second college degree in electrical engineering and serving as a translator on Russian fishing trawlers, which led to her first book, *Hair of the Dog: Tales from Aboard a Russian Trawler*. She met her husband during a stint as a radio operator in Antarctica and settled down to a more normal life, working in industry, having two children, and inventing a popular board game called Herd Your Horses along the way. After her children were old enough, she earned her PhD in engineering and as a professor has won teaching awards and performed research on noninvasive pressure sensing and the effects of electrical fields upon cells. As she recounts, Oakley's inquiry that led to *Evil Genes* was initially sparked by her own family history. If her academic pedigree appears unorthodox, the new evolutionary explorers come from every conceivable background and their discoveries must be evaluated on their own merits. I would not underestimate the abilities of an engineering professor/army officer/Russian translator/game inventor/author/wife/mother. In fact, it's doubtful that someone coming from a standard academic perch could have crossed so many disciplines—and perspectives—to develop such an encompassing, thought-provoking thesis.

Biological ecosystems include the full spectrum of relationships among species, from ruthless exploitation to obligate mutualisms. The human ecosystem is much the same. My own work on altruism, morality, and religion emphasizes the cooperative end of the spectrum. It is gratifying how often goodness succeeds as an evolutionary strategy. It does not always succeed, however, leaving room for the human equivalent of predators, parasites, and competitors. The more ruthless strategies of this segment of the population might succeed for the individuals and groups that employ them, at least over the short term, but at a massive cost to others and society as a whole over the long term. It is on this end of the spectrum that Barbara Oakley concentrates in *Evil Genes*.

As the humorous subtitle implies, exploitation can take place on any scale, from a single family (Barbara's sister, who stole her mother's boyfriend), to major corporations (Enron) and whole nations (the Fall of Rome and Rise of Hitler). In addition to the scientific merits of her thesis, you will be entertained by the many stories from her personal life and academic explorations. The main title, *Evil Genes*, might lead you to expect a simple scientific story about genes that directly code for evil behaviors. Nothing could be further from the truth. Oakley is remarkable for the degree to which she appreciates the complexity of the story—replete with genetic, developmental, and environmental interactions—and conveys the complexities in a way that remains entertaining and enlightening. Just as Darwin's books were read by all sectors of the population, not just other scientists, *Evil Genes* deserves to be read by everyone from high school students to the most distinguished professors. Through this book, you, too, can ponder the big questions about nature and human nature.

David Sloan Wilson
Author of *Evolution for Everyone: How Darwin's Theory Can Change the Way We Think about Our Lives*

PREFACE

The Wizard of Oz said loudly to Dorothy and her friends:

> "I am Oz, The Great and Terrible! Pay no attention to the small man behind the curtain."

Imagine what would have happened if Dorothy had obeyed the wizard. Would she ever have made it back to Kansas? Instead of obeying, she continued to think and observe, leading herself and her friends toward their goals.

You face a similar challenge in reading this book: you'll have to be courageous. Go ahead, buy it; it's important. It's also quite entertaining.

Why is a psychiatrist writing this preface? Well, you see, although many people think psychiatrists can "read people's minds," we cannot. Instead, the principal value of a good psychotherapist is to enable people to ask dangerous questions and *tolerate the answers,* while maintaining a sense of hope. Since Dr. Oakley's book asks some very dangerous questions, my goal in the next few paragraphs is to prepare you to tolerate the information she presents and the implications, while maintaining your own sense of hope.

First, as you read this book, you must remember that on the whole, humans are remarkably good. This is not just a wishful statement. Sophisticated brain science (the type Dr. Oakley so superbly displays in this book) shows how deeply our tendency to trust and cooperate with others is rooted in brain anatomy and function—even when such behaviors place us at direct personal risk. For example, a research team at the Center for Neural Science at New York University recently conducted a remarkable experiment examining the natural willingness many people have to trust others. They showed that this tendency is rooted in the brain circuits we use for learning through trial and error, a region called the caudate. Even when subjects were repeatedly being taken advantage of, their caudate continued to respond in a trusting manner, *if* they had been led to believe they were working with a "good" person. (If they thought they were working with someone neutral or "bad," they figured out what was going on quickly.) This caudate response, and their trusting behavior, persisted even when the subjects *understood* the error they were making! Their brains appeared to be *wired* for cooperation. At times you may find yourself quite disturbed by the implications of this book, which describes some very opposite research results, applying to a minority of humans. So please remember: most of us are surprisingly good.

A second source of hope is especially important when "belief"—certainty in one's knowledge, often wearing the robes of religious faith or walking the halls of politics—has so actively challenged scientific thought. This hope springs from the value of true understanding. Accurate information, on its own, is never dangerous. What we think of it, and what we do with it, can indeed be potentially very harmful. But obtaining accurate information, and thinking clearly about it, can only be helpful to us. The more we avoid trying to understand what is really happening around us, because it makes us uncomfortable or does not fit with our existing beliefs, the worse off we will be. Instead, when we use a rigorous scientific method to ask important questions and then think clearly about the implications of the research results, we are more likely to live free and prosperous lives. This book is a testimony to that process.

Some psychiatrists could quibble about the way in which borderline, antisocial, and bipolar disorders are treated herein, but these conditions are highly related and overlapping and not inaccurately treated here. However, I suspect there will indeed be screaming about this book. "You want to label people, even before they are born!" "This is the beginning of a slippery slope, which leads to genetic screening and sterilization; this is a return to eugenics!" True, there is a great risk that we will use genetic information wrongly, driven by our "us versus them" tendencies, our inclination to stereotype people who are not like us and make them "enemies." You can almost see it coming: "evil genes" will be thought to be more common among groups whom we are inclined to hate and used as a justification for striking out at such peoples. As you know, this kind of thinking has even led to efforts to completely eradicate some ethnic groups. And yet, as Dr. Oakley shows in this book, such societal insanity is more likely when the leaders themselves possess antisocial traits. Perhaps the world will be safer if we watch for these traits among leaders and anticipate their behaviors (although trying to solve that problem by invading their country has not recently seemed to be the best way to approach things, one might argue).

We should not fear this book. We should fear the implications of what Dr. Oakley describes, but the more we understand about "bad behaviors," the better equipped we will be to deal with them. Therefore you will find it useful, as well as important, to read all the way through this book. Fortunately, Dr. Oakley has written it as a combination of personal narrative and science story. The result is not unlike *The Wizard of Oz*: an entertaining story in which one encounters some very nasty characters. Remember to take with you the Lion and the Tin

Man; that is, remember that *most* humans are remarkably kind and generous and brave. One simply has to learn one's way around the kingdom, with eyes wide open.

Jim Phelps, MD
Author of *Why Am I Still Depressed?: Recognizing and Managing the Ups and Downs of Bipolar II and Soft Bipolar Disorder*

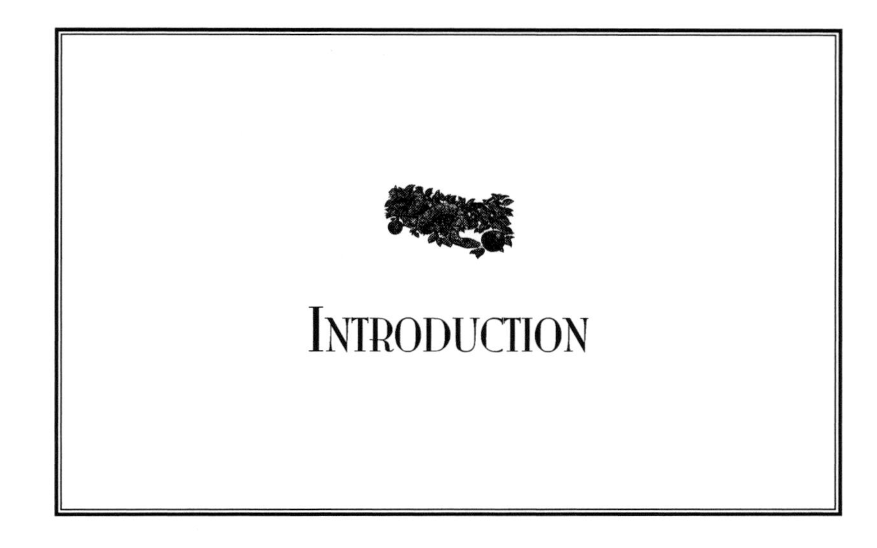

INTRODUCTION

"My mother's obsession with the *good* scissors always scared me a bit. It implied that somewhere in the house there lurked: the evil scissors."

—Tony Martin, *The Late Show*

"*Back to the real world after panic attack. Must ease Jack out. Can't tolerate the smoke or the late night 'sloppies.' He is still a good friend to have.*"

There they are: Carolyn's last written words, directly from the diary found lying on her bed stand after her death. Words mattered to Carolyn. Did she suspect what was about to happen?

I knew who Jack*—the "good friend"—was. He and my sister Carolyn had lived together on and off for years, beginning soon after Carolyn had decided to move up from southern California to Sequim, Washington. Sequim rhymes with "swim." As the T-shirt says: "Sink or Swim in Sunny Sequim."

---

*Names and identifying details of Carolyn's friends and acquaintances have been changed, as have similar details of several other individuals for the sake of privacy.

Sequim is an oddly bipolar town, crouched in the rain shadow of the Olympic Mountains. Mount Olympus, less than thirty miles west near the coast, gets nearly seventeen *feet* of rain a year. But by the time the air travels over the mountains to Sequim, a paltry seventeen inches a year is all that's left—not much more than what Tucson gets. The same air then continues over Puget Sound, picking up moisture again to drizzle up to a yard a year in Seattle. Wedged between two mother lodes of rain, Sequim is actually a postage stamp desert weirdly laced with irrigation ditches, streams, and rivers, all of which funnel rainwater and snowmelt from the Olympic Mountains down toward the sea. Living in Sequim is like living in a vortex. You can look up nearly every day and see bright blue skies overhead. But often as not, the town is surrounded by a ring of ominous storm clouds, kept at bay as if by some awesome force.

Retirees often move to Sequim, attracted by the unusual juxtaposition of mountains, sea, sunshine, and mild winters. They build huge homes on bluffs and hillsides overlooking the ocean—virtual villas with lovely gardens and masses of windows to take in the scenic vista. But after a year or so, some retirees discover that their chirpy realtor had inflated the average temperatures for the area by ten degrees. Newcomers expecting the climate of Tucson often find temperatures closer to those of Juneau. It's never exactly cold in Sequim, but it's never exactly warm, either. Year-old villas are often for sale in Sequim.

Both my parents were living separately in Sequim, each for different reasons, at the time my sister moved to town. My father reveled in the outdoors. After being a veterinarian for thirty years, he spent his retirement days happily felling trees in the National Forest and hauling them home to peel and insert in his ever-growing complex of log structures: log cabin, log storage shed, log garage, log guesthouse, log pantry buildings, log woodsheds, log sauna, log bridges, log greenhouse, and even, for reasons only he could explain, a three-story log water tower with a thousand-gallon tank and a panoramic view of the jutting peaks of the Olympics. The entire complex was nestled in a private forest of soaring Douglas firs, graceful hemlock, and spidery madrone.

Fig. Intro.1. My father at play setting the corner post for a covered log bridge. A few weeks after this picture was taken, he fell twenty feet from the top of the bridge onto the stone creek bed. Swelling from the resulting concussion, egged on by his genetic predisposition, led to the peculiar gene-environment mix that probably caused his death.

My mother, on the other hand, reveled only in my father, which was a little creepy, all things considered. Long after their divorce, she had tracked him down and moved about three miles away to a small apartment near Sequim's slightly bedraggled downtown, across the street from a dusty field of quackgrass, wild barley, and a gaggle of Garry oaks. Despite the proximity to my father, it took years before my mother finally realized that merely extricating herself from alcohol and acting nice wouldn't resurrect her ruptured thirty-year marriage. So she lived a lonely life serving as a hostess in a Mexican restaurant. Gradually, as the owners realized her skills, she became the restaurant's bookkeeper and, ultimately, a manager.

Carolyn, like my mother, had moved to Sequim for a reason.

About a decade after my parents' divorce, my mother had begun dating a wealthy emphysemic—a wheezy fellow named Ted, who planned to take her and his oxygen tanks on an extensive trip through Europe. Ted and his breathing apparatus were perhaps no great catch, but he was good company for my mother, as she probably was for him. And in all the years of our family's moves around the United States, my mother had never before been overseas. I remember listening to her talk about her upcoming trip to Europe with Ted, her breath coming quick and eyes sparkling as she wondered about the food in France—was it as good as they say?—the cathedrals, and even the width of the streets. It was the first time in years I'd seen her show any real animation or enthusiasm.

When I saw my mother again several months later, rather than discussing baguettes and béchamel, she told me how her Mexican restaurant made vegetables look greener by taking the lid off the steaming pot. She was mum about her social life, so only much later did I find out the particulars. Apparently, about a month before departure, my mother had mentioned her pending European adventure during a rare, probing telephone call from Carolyn. A few days after the phone call, Carolyn had pulled up her Southern California stakes and abruptly moved to Sequim. One leg still limp from her childhood bout with polio, my sister tucked herself and her crutches up beside the less-than-active Ted while my mother pitched in to get her situated in her new apartment. Carolyn had a dazzling knowledge of French food and wine. She paused frequently in her connoisseur's conversation—each pause just long enough to catch Ted's eye.

Soon Carolyn was comfortably ensconced beside Ted's oxygen bottle on a flight to Paris. Just another underhanded episode in a lifetime of such episodes. My mother never did get to see Europe.

The Successfully Sinister

Prompted by my sister, even as a child I used to wonder about subtly nasty characters—the ones who get really close to you so the knife goes deeper. I read about the alluring but sometimes sinister wives and concubines of the Roman emperors and Ottoman sultans. Were these women perhaps like my sister? I learned of the evil machinations of Count Romulus, some two thousand years ago in North Africa; his legendarily malevolent nature has passed down to modern culture in the naming of *Star Trek* Romulans. I shivered over stories of China's dowager empress Cixi, famed for her beauty, charm, love of power, and utter ruthlessness.[1] She was accused of killing her own grandchild to retain her hold on the throne—her narrow-minded policies undoubtedly set the stage for China's gruesome self-immolation during the twentieth century. Like tens of thousands of other children my age, I read *The Diary of Anne Frank* and wondered at how the horrific policies of a single demented leader could resonate in an echo chamber of banal evildoers and result in the deaths of millions.

As I grew older, I noticed literature, movies, TV shows, video games, and comics that brimmed with quirky, evil antagonists: Shakespeare's sinister Iago; *David Copperfield*'s servile Uriah Heep; *The Lovely Bones*' pitiful serial killer, George Harvey; Glenn Close's psychotic book editor Alex in *Fatal Attraction*; *Silence of the Lambs*' Hannibal Lector; Nurse Ratched of *One Flew over the Cuckoo's Nest*; Captain Hook; manga's twisted gang girl leader Mitsuko Souma;

Batman's Joker; Seinfeld's Soup Nazi. Entertainment, it seemed, could hardly be entertainment without a bad guy (or gal) smiling through the duplicity.

Real life, of course, is much worse. Hitler, Mussolini, Pol Pot, Nicolae Ceausescu, Rafael Trujillo, Slobodan Milosevic, Anastasio Somoza, Saddam Hussein, Idi Amin—the twentieth century's list of monstrous leaders grows longer the more you look. Altogether these bloody despots were responsible for policies that caused the murders of well over one hundred and fifty million people during the twentieth century alone—that's about fifteen hundred families for each word in this book.

And each of these ruthless dictators shared a similar devious expertise in manipulation and control. The subtly deceitful Stalin, for example, was prone to tricks such as having newly promoted field marshal Grigory Kulik entertained in the office above the cell where the marshal's wife, the beautiful Countess Kira, mother of an eight-year-old daughter, was being tortured after having been kidnapped on her way to the dentist. This was Stalin's way of ensuring that his people toed the line. (One month later, the countess was coldly executed with a bullet in the head; Kulik himself was quietly shot a decade later.) Millions died during Stalin's grisly purges, which assigned quotas for executions by the thousands; millions more died during his enforced starvation policies in the Ukraine.*[2]

In China, Mao followed Stalin's lead. Rather than sending hapless millions to the gulags to suffer and—if they were lucky—die, Mao turned all of China into an über-gulag. Using gullible Western writers, Mao created a legend for himself as a Chinese Robin Hood who won the respect of all those he led. The reality was that he ruled by savage

*Robert Conquest's monumental work on Stalinist horrors, *The Great Terror*, earned enormous animosity upon its initial release in 1968—its graphic descriptions of the horrors perpetrated in the Soviet Union under Stalin's direction were felt by many to be false in virtually every particular. The opening of the Soviet archives and later verification by a host of Russian historians not only supported Conquest's findings, but showed the Stalin's "model state" had been even worse than Conquest had originally outlined. When *The Great Terror* was rereleased in a post-glasnost 1992 edition, Conquest was asked if he would like to give it a new title. His terse response was: "How about, *I Told You So, You Fucking Fools.*"

caprice, willful incompetence, and messianic egotism. Ultimately, he was responsible for the deaths of over seventy million Chinese during peacetime alone. Mao also set the example for Pol Pot, who espoused a radically revised Maoism that resulted in the "killing fields" that depopulated much of Cambodia.

When I read accounts of these tyrants, I shuddered. But something always baffled me, just as it baffled all of the tyrants' careful biographers and their readers, and just as it must have baffled those victims cognizant enough to know the ultimate source of their needless suffering. How did these seeming psychopaths get to the top?*[3] Shouldn't people have noticed these tyrants were a little, well, odd before they ascended to power? How could they fool and manipulate people so easily? And, in this new era where dictators like Slobodan Milosevic and Saddam Hussein are caught and tried rather than butchered on the spot, it's disconcerting to watch these caged mini-Hitlers face the overwhelming evidence of their atrocities and—*poof*—pretend it didn't happen, or wasn't that serious, or was someone else's fault. I can't help but wonder, what is going on in these people's minds?

Compared to people like Mao, Stalin, and Milosevic, my sister's many devious manipulations and deceits were small potatoes indeed. But for my parents and the many people she affected through her mysteriously foreshortened life, the pain of her purposefully malign actions was real and devastating. I thought a lot about my sister as I grew older and learned more about her ability to deceive. I thought about her wit, her intelligence, and her uncanny ability to charm.

While working as a Russian translator on Soviet trawlers during the cold war of the early 1980s, I studied the avuncular picture of "Uncle Joe" Stalin on the captain's wall. I knew all about Stalin's loathsome policies and personality, yet I still couldn't help but speculate—would I have known what Stalin was really like if I'd naively met him at his wedding cake–shaped mansion, or at a well-lubricated

---

*Psychopaths might best be described as "predators who use charm, manipulation, intimidation, and violence to control others and to satisfy their own selfish needs. Lacking in conscience and in feelings for others, they cold-bloodedly take what they want and do as they please, violating social norms and expectations without the slightest sense of guilt or regret."

Kremlin dinner party? After all, even Winston Churchill, a man who clearly had Hitler's number, was fooled. I read with interest how Churchill was charmed and dazzled by Stalin, "that great man," with whom he shared cheerful drinking bouts and similar paternal adoration of their redheaded daughters.[4] Stalin was a gifted organizer who was capable of working prodigious hours. But, as Stalin's most perspicacious biographer, Simon Sebag Montefiore, discovered, the "archives confirm that his real genius was something different—and surprising: 'he could charm people.' He was what is now known as a 'people person.' While incapable of true empathy on the one hand, he was a master of friendships on the other. He constantly lost his temper, but when he set his mind to charming a man, he was irresistible."[5]

Near the little town of Rochester, Michigan, not far from Detroit, where I now live with my family, I often drive by the site of the old Machus Red Fox Restaurant—the last place Jimmy Hoffa was seen alive. Hoffa was the dark mastermind who created an enduring image of Teamsters as bullies who achieved the same thuggish levels as their managerial opponents; Hoffa's unresolved disappearance that lazy summer day in 1975 fixed a stain on his legacy that will never be erased.

What is it about some men that makes them willing to sign a pact with the devil?

At universities, I've watched the machinations and manipulations of a small pool of academicians—strange, deeply power-hungry professors who terrorize students and drive the staff insane but who earn kid-glove treatment from administrations. Scanning the news, I read about the widespread pedophilia of the Catholic Church and how it was condoned by leaders who set the perpetrating clergy loose again and again to molest tens of thousands of children. I watched as Enron became a buzzword for executive skullduggery, and read of the horrific private lives of business executives like Chainsaw Al Dunlap, who liked to liven things up by telling his soon-to-be ex-wife how he liked to torture children.[6] One image consultant Dunlap tried to hire noted: "He was the most unpleasant, personally repulsive businessman I ever met in my life."[7]

"There are two ways to rise to the top," says my business executive husband, with his hypersensitive bullshit detector. "One is to be the cream. The other is to be the scum."

Ever since the early fascination with my sister's many devious successes, it is the scum who have long held my interest.

Sleuthing with the Sciences

My early engineering studies led in a spiraling path toward research in bioengineering—a relatively new discipline that integrates biology and medicine with engineering to solve problems related to living systems. Indeed, the scope of bioengineering is immense, covering many disciplines. One such discipline is genomics, which involves figuring out exactly how the molecular building blocks of DNA have been stacked to build genes, what each gene does, as well as where each gene is placed among an organism's chromosomes. Understanding how both normal and rogue genes work can lead to improved detection, diagnosis, and treatment of disease—potentially including the intractable conditions underpinning personality disorders. A related field is proteomics—a sort of molecular geography describing the location, interactions, structure, and function of proteins. Advances in proteomics have encompassed discoveries about a variety of cellular processes, including those crucial for understanding the cascade of molecular relay systems underlying human thought and emotion. Bioengineering also explores imaging and image processing that allow us to "see" inside the human body. And bioengineering involves understanding the warp and woof of neural structures in human beings, allowing us to build devices that can help the paralyzed to walk, the deaf to hear, and the blind to see.

How do you apply all this knowledge to the study of sinister people? How would you begin a search of the literature for people who have tried? What keywords would you even use to do a search?

<Manipulative>?
<Duplicitous>?
<Evil>?

Medline, a standard medical search engine, doesn't provide any relevant answers.

How about <Machiavellian>? It means charming on the surface, a genius at sucking up to power, but capable of mind-boggling acts of deceit for control or personal gain.

Machiavellian, in fact, hits the jackpot for keywords. It turns out to describe an entire field of study—one that takes the ideas of Renaissance Florentine statesman Niccolo Machiavelli, author of *The Prince*, and builds them into a sweeping—and often unsavory—portrait of humankind.

A Revolution in Research

Over the years, I've learned that much of what's known from psychological, psychiatric, and neuroscientific studies related to Machiavellians or deceitful, manipulative people isn't easily accessible. Often, it's because the findings are relatively new—they haven't had time to chew their way through academic leather straps and into the public domain. Occasionally, it is because the implications of the research findings are controversial and in conflict with other long-held beliefs. The public sometimes catches glimpses of important study results—a hint of information about offbeat neural images related to psychopaths in this science magazine, another tidbit about the effects of genes on impulsivity in that newspaper article. It's as if all the growing pieces of information about the "successfully sinister" (which I use as a synonym for *Machiavellian*), are lying about disassembled.* These pieces

*Niccolo Machiavelli's original writings described cunning, scheming, and unscrupulous behavior that any pragmatic leader would have to use occasionally to achieve overall good government. I should make it clear here that I use the modern definition of the term *Machiavellian*, which involves cunning, scheming, and unscrupulous behavior that can foment evil. The term's modern negative connotations arose in part because Machiavelli's works were forbidden by the Church for placing advantage over morality.

are waiting for someone to tie them together to hold up a new model to the light—one that goes far beyond the crude label of psychopath. This new model contains surprises—among them, that evil may be unavoidable and that it can even have an unexpectedly good flip side.

Researching the topic as the years have gone by, I've also found intriguing oddities. Hitler, for example, has been diagnosed with dozens of different disorders, including narcissistic, borderline, and antisocial personality disorders; schizophrenia; psychopathy; syphilis; encephalitis; paranoia; malignant narcissism; moral cretinism; Parkinson's disease; hubris-nemesis complex; "enfeebled and unformed self"; "destructive and paranoid prophet"; "a constitutional left-side weakness that allowed his right cerebral hemisphere to exert a strong influence on his thought and behavior"; and, believe it or not, "sibling rivalry."[8] On the other hand, sifting carefully through the two main psychology databases, I find no articles at all about Pol Pot, who was responsible for the deaths of some two million Cambodians—perhaps a quarter of the Cambodian population. Current research on malevolent dictators, I discover, consists of a hodgepodge of contradictory and missing studies.

I type the subject term for one of the most commonly studied psychiatric conditions, "antisocial personality disorder," into Medline, one of the world's most comprehensive sources of life sciences and biomedical bibliographic information. It pops up, virtually instantaneously, with 5,494 hits. The subject "borderline personality disorder" gets 3,090 hits—meaningful hits, including hundreds of medical imaging studies, genetic studies, drug studies, and so on. On the other hand, if I type in *malignant narcissism*—a term used by world-class psychiatrists like Otto Kernberg and Jerrold Post, along with hundreds of thousands of Googled others, to describe the kind of malevolent, yet high-functioning people I'm researching—I get nothing. Zero hits. *No medical studies whatsoever.*[9]

It is unsettling to discover this kind of omission—like hearing that the oncologist about to operate on your father's cancerous liver actually has a fake degree from a diploma mill. Where's the science here?

The limited interaction between biology and psychology regarding the study of malevolent but high-achieving individuals is evident in the recent article on dictators "Why Tyrants Go Too Far: Malignant Narcissism and Absolute Power" in the prestigious journal *Political Psychology*. Key neurological factors such as neurotransmitters, the hippocampus, or amygdala—all of which have been profoundly implicated in the kind of impulsive, antisocial behavior often seen in despots—are not even mentioned.[10]

It turns out, however, that if you're willing to peer directly into the witches' cauldron of research results, this first decade of the new millennium is an extraordinarily lucky time to be focusing on Machiavellianism. Neuroimaging has progressed well past the point of simply determining the shape and structure of the human brain—now we can watch the molecules of emotion scurry about the cells as they complete their neurological chores. Neuroinformatics is allowing researchers to access libraries of data about thousands of different brains to see what is usual and what is unusual. Brain atlases—images of what a normal functioning brain looks like—are providing detailed roadmaps to help guide research efforts.

Underlying the images we see of the brain's structure and functioning are the genes that help serve as neural operating instructions. Since the completion of the human genome project in 2003, researchers have dived even deeper to understand not only where genes are located on the human chromosomal framework but also how those genes are structured and what they do. Microarray chips give indications about which genes are turned on or off at any given time. This provides information crucial for understanding how genes and proteins communicate within and between cells. The upshot of all this? The entire field of biology is now undergoing a genome-based revolution.

A new field is that of systems biology, the "science of everything"—everything living, at least. This new discipline looks at the piecemeal information that has been found related to genes and knits it together with other research to form a big picture describing how cells signal each other and how neurons interconnect. Ultimately, this helps us to understand how slight molecular and genetic differences can result in dramatic changes, not only in how a person looks, but in his or her temperament. This, then, is where we need to look to ultimately understand Machiavellian—unscrupulous, self-serving, often deeply malign—behavior.

Perhaps surprisingly, the more I've learned about Machiavellians, the more I've discovered how fascinated people in general are by the latest scientific breakthroughs in studying them. After all, the social achievements of clearly disturbed individuals such as rapacious pedophile priest John Geoghan, quintessentially greedy Enron CFO Andrew Fastow, or sadistic leaders such as Saddam Hussein can be mystifying. Even more mystifying are the occasional successes of malign individuals who people know personally: colleagues, supervisors, teachers, doctors, lawyers, pastors, or elected officials. Surprisingly often, a successfully malign example turns out to be a family member whose sinister characteristics, often confusingly combined with more lovable traits, have kept the family walking on eggshells for years.

And once we start wondering about the successfully sinister, more questions abound. Do hereditary aristocracies tend to attract those, like Princess Diana, with certain dysfunctional personality traits? Could a propensity for these traits be passed down from generation to generation, leading to the decay of empires? What happens when the successfully sinister take a prominent role in religions? Are Machiavellians skilled at using neurologically based tricks to disable the thinking of those who might oppose them? Should the successfully sinister be despised for what they do to others? Or, like sharks, should they be warily respected for the perversely successful roles they play?

Putting the Puzzle Together

This book describes my own attempt to outline results from various fields of research that describe why seemingly evil people exist and how they can function in and even rise to the top of many types of social structures, including government, religion, academia, industry, the everyday workplace, and the ordinary family. The book also explores interesting sociological patterns related to the rise of these malign individuals—patterns that can be observed in social structures as different from one another as Enron, Chairman Mao's China, and the Roman Empire.

For me, a touchstone of this research has always been my sister Carolyn and her life, so I share some of that knowledge. This includes insights gleaned from Carolyn's life story, as well as from her letters and diaries. Although some might expect this book to contain only a traditional objective translation of research findings, I think that just looking at academic papers can result in losing sight of the human picture of what this research means in our lives. And so I describe not only my sister and her relationship with others but additional instructive experiences—a bit of late-night partying on Soviet trawlers; the compulsive collecting of obscure teapots in the far corners of China; drama in a German *kaserne*.

As background for this book, I've interviewed and communicated with psychologists, psychiatrists, imaging specialists, geneticists, immunologists, biologists, lawyers, historians, philosophers, sociologists, and anthropologists, and have read extensively in the research literature of many of these fields. I've tried to distill these multifaceted findings and also give some sense of the struggle dedicated researchers have faced as they've tried to find a new framework for what was previously unknown about the successful sinister people among us.

My doctoral training in systems engineering—a unique field of study that provides training in understanding the big picture patterns of diverse disciplines—provides a handy vantage point. It's far enough from any of the schools of thought I'm looking at to help keep me from being blinded by the unwitting side effects of each school's perspective.[11] Yet this training underlies many of the cutting-edge techniques used now for medical imaging and genomics—techniques that are providing stunning revelations about how genes are organized and how our brains work. By putting hard-won insights gleaned by researchers in the physical and biological sciences together with extraordinary related results from the social sciences, I hope to provide a look at sinister people with a fresh perspective.

After all, the successfully sinister affect virtually everyone sooner or later. We obsess about trying to reach closure about hurts we've experienced, even if our closure is only vicarious. We have a gut-level need, even in fiction, to see slimy, smiling antagonists get what they deserve. It's why we can't help but follow newspaper, Internet, and—heaven help us—even *National Enquirer* articles about the latest senator caught in a payoff scandal or the most recent holier-than-thou televangelist caught in bed with a prostitute.

But if there's one thing we are even more fascinated with, it's wanting to know *why*. Why would anyone spread such malicious gossip? Why would anyone ever use a publicly owned company as a private piggy bank? Why would anyone knowingly starve millions of his own people? Psychology, with explanations founded on "defense mechanisms," "countertransference," and "acting out," can go only so far. Neuroscience is fleshing out the field nicely, but unfortunately, popular hard science–based books about the successfully sinister are rarer than frequent flyer mile seats to Hawaii at Christmas. It appears we'll have some time to wait before we start seeing popular books with titles like *He Really Is Driving You Crazy: Understanding Theta Wave Activity*, or *Bitch: The Science behind the Savagery*.

But genetics is as important as neuroscience in understanding the successfully sinister. Groundbreaking books such as Judith Rich Harris's *The Nurture Assumption* and Harvard psychologist Steven Pinker's *The Blank Slate* have served as fulcrums to help swing researchers off their centuries-long love affair with the idea that people are naturally good.[12] Under this well-intentioned ideology, "evil" people were believed to be created and shaped solely by their environment. The advantage of this belief is that it gives researchers the comfort of thinking that humans have direct control over evil—

that by somehow reengineering the social environment, human evil can be eliminated. The disadvantage of this belief is that it is wrong—there are rafts of studies supporting the conclusion that human personalities are shaped as much by their genes as by their environment. But despite the overwhelming evidence, many academicians today have held so close the belief that people are naturally noble creatures who go astray only because of poor early nurturing that it is sometimes difficult for them to come to grips with the implications of modern research findings.* All of this has meant that over the last decade, ever since neuroscience hit its stride and the human genome has been sequenced, there has been a gap in communicating the implications of scientific findings about the successfully sinister. People have been left largely unaware of how science is beginning to provide answers to some of their most compelling questions: Why are there evil people, and why are they sometimes so successful?

Evil Genes was written to help answer those questions.

---

*One brilliant psychology professor I know provided feedback after she had read the first few chapters of an early draft of this manuscript. The paraphrased essence of her thoughts was: Even if you're right about genes influencing behavior, it's impossible to change, so what good can come of telling people about it? While you cannot control the whole environment, you can put people on notice with regard to their behavior.

In responding, I kept myself from pointing out the obvious ineffectiveness of, for example, putting career criminals on notice about their behavior. Instead, I pointed out that good parents who receive an unlucky shake of the genetic dice and happen to have a psychopathic child might want to hear that their child's behavior isn't directly their fault. Certainly mothers of autistic children, told for decades that their child's autism was directly due to their cold parenting style, have benefited from recent research revealing the strong genetic component involved in the disorder.

But my friend didn't respond to my comments—I'd already lost her. Like a fundamentalist discovering she'd been roped into reading an evolutionary screed, she read no further.

CHAPTER 1

IN SEARCH OF MACHIAVELLI

"They're beautiful, they're elegant, they're vicious as hell . . . there's a real life lesson here somewhere."

—Professor Ralph Noble, Rensselaer Polytechnic Institute,
Psychology of Motivation class, Fall 1991,
on Siamese fighting fish

One might easily argue that the modern study of sinister people began in 1954—nine years after my sister Carolyn's birth. It was an oddly tenuous time to begin work in this area. The first clunky medical imaging machines wouldn't come online for another quarter century. Genetic analysis was an impossibility; the structure of DNA had only just been determined, and sophisticated gene sequencing methods lay decades in the future. In the social sciences, researchers not only didn't have the tools they needed to climb to a new level in penetrating the human psyche, they often didn't know they needed them. The successfully sinister, if thought of at all, were felt to be people who consciously chose, or were conditioned by their early environment, to be nasty.

Nobel Prize–winning biologist Albert Szent-Gyorgyi said: "Research is to see what everybody else has seen, and to think what nobody else has thought."[1] In the conformist 1950s, it would take a special person to see that duplicitous, manipulative, yet not necessarily criminal individuals were an interesting problem, worth defining and studying.

Psychologist Richard Christie was just such a person.

The groping progress of this brilliant researcher is worth following, not only because it underpins the modern study of devious behavior, but also because it gives an idea of the difficulties that investigators in this area faced in an era before imaging or genetic studies became available. After all, even now most people face the same difficulty Christie did in trying to understand the deceitful, conniving motivations and actions of the successfully sinister.

Richard Christie Founds a New Discipline

In 1954, Richard Christie had been given a unique opportunity: a year off as a fellow of the prodigiously endowed Center for Advanced Study in the Behavioral Sciences, which overlooks Stanford University. The Center was a low-key, almost deliberately secretive organization that pampered its fellows with privacy amid a tranquil setting of blue skies, kelly green lawns, palms, live oaks, and camellias. (A quarter century later, this same landscape elixir would begin fueling the creativity of Silicon Valley.) For Christie, the hilltop Eden signaled both an expectation and a problem. The expectation was that Christie was to do something extraordinary. The dilemma was, however, that Christie had no idea what to do.

Christie soon discovered that fellows from other disciplines shared his dilemma. He wrote, "Most of us had never had a year without any outside commitments, and we were enjoying it in idyllic surroundings with others who were equally overwhelmed by their good fortune. After the initial shock had worn off, a period of contagious anxiety began to develop. What could we possibly do to justify this opportunity? Few of us were particularly modest about our scholarly abilities, but the perceived necessity of producing a Great Work was awesome."[2]

What Christie *wanted* to work on was psychology's equivalent of Einstein's vision of a unified field theory. The problem was that unlike physics—where the equations themselves cried out for a unifying theory relating existing models of nuclear forces, electrical charge, and gravitational pull of the stars—psychology's science was a patchwork quilt of observations. What was important in psychology was often a matter of opinion. And opinion at the time was just beginning the long slide into the endorsement of behaviorism, where humans were thought to be capable of being shaped into anything—"doctor, lawyer, or Indian chief"—by simple reinforcement methods. The mind was a "black box"—the strict rules of behaviorism forbade hypothesizing neurological mechanisms for such psychological concepts as fear, desire, and consciousness. Only overt, observable behavior was thought to be the proper avenue for psychological inquiry. This type of black box thinking seems laughably wrong today, but it had an incalculably restrictive effect on a generation of psychological research.[3]

Itchy to find their Great Work, the fellows formed groups centered on areas of interest. Cynics observed that these groups appeared to be formed more for mutual alleviation of anxiety than any real scientific function. But still, conversation helped, and the supporting money and residence of the fellowship—similar to what Einstein had been given at Princeton—allowed Christie to sit back and put disparate strands of psychology research into perspective. Gradually, following his own instincts while talking with his new colleagues, Christie began to realize that *the* most important problem in his field was one that had, in the recent hellish years of World War II, profoundly affected virtually everyone on the planet: the problem of manipulative, deceitful leaders.

Machiavellianism.

The Prince, written in 1532 by Niccolo Machiavelli, was the first and is still among the best of resources at providing an in-depth description of the type of nasty behavior that Christie was trying to understand. As Annie Paul, a former *Psychology Today* senior editor, writes: "Machiavelli's matter-of-fact instruction that rulers must be prepared to lie, cheat and steal to hang on to their thrones—all the while acting the part of the benevolent leader—has not lost its razor edge. Even in this era of cynicism, Machiavelli's view of humanity as greedy and self-seeking or stupid and easily tricked still seems remarkably dark—and to some, remarkably relevant."[4]

The real problem was to figure out what a modern, scientific understanding of a prototypical Machiavellian might consist of. Although there was a boatload of then-current literature on the duped followers of Machiavellian leaders, as well as on leadership in general, there were very few studies of unethical leaders themselves.

Christie and his colleagues would have to fashion a new discipline, one studying manipulative, duplicitous individuals from the perspective of social psychology. Only later would researchers notice parallels between Christie's Machiavellians and the more criminal-prone psychopaths outlined in the traditions of clinical psychology. By then, the benefits of viewing duplicitous individuals from Christie's nonclinical, more broadly sociological perspective would be too valuable to ignore.

Christie suggested that four sinister and manipulative, but not necessarily criminal, personality traits underpinned this new research area. In Christie's view then, Machiavellian personalities:[5]

- View others as objects to be manipulated rather than as individuals with whom to empathize.
- Lack concern with conventional morality. Lying, cheating, and other forms of deceit are acceptable forms of behavior.
- Lack obvious psychopathology. Although perhaps not the epitome of mental health, contact with the more objective parts of reality would be within normal range.
- Have low ideological commitment. They are more concerned with tactics for achieving possible ends than in an inflexible striving for an ultimate idealistic goal.

Christie's listing of sinister traits—predicated on the manipulator's lack of obvious psychopathology—eerily presaged characteristics that would resurface nearly half a century later in sophisticated technologically based studies. Faintly anomalous brain scans would reveal hardwired reasons for deception and manipulation, while peculiar patterns of cellular signaling, along with inheritance studies, would show the surprising involvement of genetics. In his own intuitive way, Christie was groping toward something innate but covert—something even more subtle than the thin mask of normalcy worn by the psychopath. By setting out a listing of key characteristics for Machiavellian personalities, Christie was sending up a flare to the technically fortified future: Here lies a problem. Here lie the clues.

The Ubiquity of the Sinister

At the time, Christie and interested colleagues were unsure whether the hypothetical four-trait Machiavellian model they had created even existed outside the textbook world. And if such Machiavellians did exist, there was no clue as to how common they might be or where they might flourish. So the small band of researchers took an unusual next step. Rather than study the pathological leaders of history books—history would come later—they decided to first tap a unique resource: the fellows surrounding them at the Center for Advanced Study. After all, virtually every fellow was either a luminary in the behavioral sciences or the protégé of one. Christie and his colleagues reasoned that interviews with junior fellows would give a great deal of information about the fellows' sponsors, that is, the highly regarded research leaders who were responsible for the fellows' training and for their being at the center.

Surprisingly, this seemingly simple interview step was a brilliant idea that would shape the whole of research in the subject for the next half of the century. The key was that Christie had begun to suspect that Machiavellian leaders were not best characterized by the rare Hitlers and Stalins and Maos that had all played such crucial and destructive

Fig. 1.1.

DOUBTFUL FRIENDS

roles in Christie's lifetime. Or, rather, they were not *only* those types of characters. Machiavellians might also appear as leaders in other areas, such as science, religion, business, or even charitable work. In fact, *Machiavellian personalities might be present in small percentages of any group of people.* In some sense, Christie was connecting the *Punch* cartoon characterization of the duplicitous Stalin and Hitler with equally devious everyday behavior, as depicted by cartoonist Matthew Henry Hall.

Christie burrowed into the interviews with junior fellows with gusto, uncovering apparently juicy material that he noted was "difficult (and possibly grounds for libel) to quote."[6] Most of the fellows reported respect for their sponsors' intellectual abilities but, at the same time, felt little personal closeness or empathy from them. The

Fig. 1.2.

"You're the only person in this department I trust."

sponsors' cool, utilitarian, somewhat manipulative relationship with the fellows jibed with the first of the four Machiavellian role model traits. Sponsors also appeared to have, as Christie wrote with his usual deliberate understatement, "relatively little concern with middle-class conventionality." Whatever their differing morality, however, sponsors showed no blatant symptoms of psychopathology—at least, nothing that was obvious to their students. That covered traits two and three. Ideological commitment, the last of the four traits, was harder to quantify. But it appeared that sponsors who were active in power positions generally didn't involve themselves in anything time-consuming. No tedious organizing petitions or rallies for them—their job was to tell others what to do.

All in all, the interviews were encouraging. Christie and his colleagues' four key Machiavellian traits seemed to dovetail neatly with the junior fellows' uncannily similar descriptions of their sponsors. But still, even if the four traits that had been devised were descriptive of a real Machiavellian personality type, the traits provided little guidance about how a person with those four characteristics would actually *act*, especially if he were in an important position.

To get a handle on how Machiavellian personality types operate, Christie and his colleagues studied both modern and historical attempts to understand people grappling for power. Their readings ranged from then-avant-garde political psychology, including Adorno's *The Authoritarian Personality* and Eysenck's *The Psychology of Politics*, to the two-thousand-year-old Chinese *Book of Lord Shang*.[7] But Christie eventually concluded that Machiavelli's *The Prince* and its companion volume, *The Discourses*, were of a different quality altogether than any other text.[8] "Unlike most power theorists," Christie would later write, "Machiavelli had a tendency to specify his underlying assumptions about the nature of man."[9] In other words, Machiavelli provided a boilerplate description of Machiavellian attitudes.

PERVERSE ADMIRATION: THE DEVELOPMENT OF A TEST FOR MACHIAVELLIANISM

Christie's group developed a series of statements based on Machiavelli's work, along with other expressions that appeared to tap into the same syndrome. These statements were along the lines of "Barnum was probably right when he said there's a sucker born every minute." Next, all fellows at the center who could be cornered were asked to go through and respond to each statement on a 1 to 7 scale that indicated the extent of their agreement or disagreement. The test was tweaked and retweaked; encouragingly, the group found that "the extent to which our respondents agreed with Machiavelli seemed to fit with our subjective estimate of their relative success in manipulating others."[10] These, Christie dubbed "high Machs." Those at the other extreme—

the people who would gladly give the shirt off their backs—Christie dubbed "low Machs." Results from the final test found that people in general spanned a spectrum in their manipulative nature, with the majority of people straddling the middle. Although the test was developed to see whether the test-taker agreed with the ideas of a sixteenth-century Italian political philosopher, Christie found that it neatly delineated differences in temperament.*[11]

Initially, Christie and his eventual collaborator, Florence Geis, had negative images of high Machs. But, after watching high Machs using their manipulative abilities—shadowy and unsavory though they might be—to trounce other players in certain games set up by psychologists, the two researchers found themselves developing a perverse admiration.

Research on Machiavellianism was to grow over the second half of the twentieth century. After all, virtually everyone has had experiences with a Machiavellian, and the new test provided the first easy-to-use tool to assess Machiavellian traits and personalities. Even working blindly, without the tools of modern technology, the fellows, through their methods and ideas, would provide a groundbreaking conceptual framework for studying the successfully sinister. Their findings would flesh out a context for the imaging and genetic studies to come.

But one major problem persisted. Christie's group was onto a new personality type—a "high Mach." This was too new a concept to have a relationship with any personality type or disorder listed in the trusty *Diagnostic and Statistical Manual of Mental Disorders* (the "DSM"—currently on its fourth, text revision edition—the DSM-IV-R). And without being associated with a personality disorder in that particular psychiatric bible, it would be impossible to relate Machiavellian behavior to the many medical studies that had already been done on those personality disorders.

---

*You can take the test online at www.salon.com/books/it/1999/09/13/machtest. As the test states: "High Machs constitute a distinct type: charming, confident and glib, but also arrogant, calculating and cynical, prone to manipulate and exploit. . . . True low Machs, however, can be kind of dependent, submissive and socially inept. So be sure to invite a high Mach or two to your next dinner party."

And so Machiavellian research stood as an eddy disconnected from much of the mainstream of psychological research. Christie himself would move on to other areas from his eventual academic perch at Columbia; a heart attack would cause this generous, decent man's passing in 1992. Conventional psychology involving duplicitous people continued to concentrate on criminal psychopaths and those with related clinically diagnosable conditions. It would not focus on Christie's high-functioning Machiavellians who could rise so artfully in various social structures. There were no real answers for those who wanted to understand the Hitlers, Maos, and bin Ladens of their day, much less the two-timing cheat of a husband who dotes on his daughter even as he emotionally abuses his long-suffering wife; or the manipulative, domineering mother-in-law; or the duplicitous supervisor who takes credit for every good idea his subordinates devise. Those answers, and the connection between Machiavellianism and clinical studies, were to come from an unlikely source.

Chapter 2

Psychopathy

"... and he's refreshingly free of emotional baggage."
—Tonya, *The Tonya Show*, apropos her pet iguana Spiro

Psychology professor John McHoskey was different. At least, that's what the evaluations suggested on a popular nationwide Web site that critiques college professors. "Looks hot with a shaved head, but he is quite an odd man," declared a student in one of the few kind reviews.

And indeed, McHoskey was a difficult man to reach, hunkered down as he was then at Eastern Michigan University in rural Ypsilanti—a small, improbably named town a lot like Peoria but without the glitter. But despite, or perhaps because, of his reclusiveness, John McHoskey is one of the world's leading experts on Machiavellianism—his research on the topic is world-class. Through nearly half a dozen meticulously researched journal papers, he has explored the relationship between Machiavellianism and such topics as narcissism, ethics, and sexuality. Perhaps his most significant paper, however, was one relating Machiavellianism to psychopathy—a tour de force of research that garnered its author an unofficial crown (princely, of course) in Machiavellian studies.[1] This paper argued convincingly that Machiavellianism and psychopathy are one and the same—any apparent differences simply arose from flukes of history. Personality psychologists and social psychologists, McHoskey pointed out, had been studying the topic under the name "Machiavellianism," while clinical psychologists had studied the same personality construct and called it "psychopathy."

ANTISOCIAL PERSONALITY DISORDER

McHoskey wasn't actually the first to see the similarity between Machiavellianism and psychopathy, as he himself noted. But McHoskey was the first to take a *close* look at the overlap between the two personality types. The puzzle for everyone is how there could be people who can do bad—even horrendous—things to others without feeling guilt. Historical attempts to explain such personality types extend as far back as the early 1800s, when Cesare Lombroso attempted to explain the biological roots of lawbreaking, and Emile Durkheim saw crime in terms of social factors.[2] Nowadays, the official DSM-IV term for the broad category into which psychopathy fits is *antisocial personality disorder*, a syndrome in which people show a pervasive pattern of disregard for and violation of the rights of others, occurring since the age of fifteen. The DSM-IV is very specific about this—a person has to have at least three of the following traits:

- fails to conform to social norms with respect to lawful behaviors as indicated by repeatedly performing acts that are grounds for arrest
- deceitfulness, as indicated by repeated lying, use of aliases, or conning others for personal profit or pleasure
- impulsivity or failure to plan ahead
- irritability and aggressiveness, as indicated by repeated physical fights or assaults
- reckless disregard for the safety of self or others

- consistent irresponsibility, as indicated by repeated failure to sustain consistent work behavior or honor financial obligations
- lack of remorse, as indicated by being indifferent to or rationalizing having hurt, mistreated, or stolen from another.

But What Is a Psychopath?

First, a bit of background. As mentioned above, the DSM-IV presently lumps antisocial dysfunctional behavior under the general term "antisocial personality disorder," and implies that the disorder is akin to psychopathy. But genetic and imaging research is beginning to provide compelling evidence that, if those with antisocial personality disorder were thought of as being at the bottom of the hole of the human race, psychopaths would form the subgroup that took out the shovel and kept digging. Psychopaths appear to be a special alien subset of humanity—glib, with shallow emotions, and completely lacking in the ability to feel guilt or remorse. They also indulge in classical antisocial behavior—stealing or violence. (Or, stealing *and* violence, as with the bank robber who shouted, "Nobody move!" and then, high on adrenaline, shot his still-moving partner.)

Both *sociopath* and *psychopath* are words used to indicate a person who is without the ability to feel remorse and other moral emotions, and who thus has no qualms about carrying out antisocial and criminal behavior. Sociologists generally use the word *sociopath* and believe that sociopathy is learned behavior, as, for example, with someone who is socialized in an antisocial subculture such as a gang. Biologists use the word *psychopath* and think that the behavior is an innate characteristic that cannot be changed. Psychology professionals disagree about which term to use.

Psychopaths know right from wrong—they just don't act that way. As preeminent neuroscientist Jorge Moll states: "A most astonishing characteristic of psychopaths resides in their ability to tell right from wrong. Their knowledge of how to behave appropriately, however, is only rhetorical and wields little, if any, impact on the guidance of their actions in real contexts. Psychopaths are the best example of the dissociation between knowing good and acting good."[3] Moll's work eerily provides two-millennia post hoc scientific validation for Aristotle's distinction between knowledge and moral virtue, where knowledge alone was felt to be insufficient for moral virtue. In short: knowing good does not entail doing good.[4]

Psychopaths are so scary that roughly a quarter of psychology professionals who meet one wind up experiencing goose bumps, hair standing up on the back of the neck, crawling skin, the "creeps" or the "willies," or a sense of hotness or coldness. Forensic psychologist J. Reid Meloy notes: "These are all signs of a very old, biologically rooted activating system within our brains that is warning us of danger. The old fear of being prey to the predator."[5]

Sadism

Often intertwined with psychopathy are notions of sadism. Sadism was slipped into the appendix of the DSM-III-R (the immediate predecessor to the current DSM-IV bible of mental disorders), but has since been excluded.[6] Sadly, the disorder was excluded in part because enshrining it within the DSM-IV—thus providing official status—could provide an excuse for psychologists and lawyers to claim that sadistic torturers of women and children were not responsible for their actions. The disorder was also ostensibly excluded because there was too little research available—but the exclusion has meant that even less research has been done. At present, then, there is little hard science research involving sadism. There does appear to be a high rate of coexistence of sadism with other personality disorders, especially antisocial and narcissistic personality disorders.[7] A small but compelling body of research shows that as evidence of psychopathy increases, evidence of sadism increases as well.[8] Sadism also appears to possess a genetic component.[9]

Sadistic Personality Disorder

The old DSM-III-R diagnostic criteria for sadistic personality disorder are as follows:[10]

1. has used physical cruelty or violence for the purpose of establishing dominance in a relationship (not merely to achieve some noninterpersonal goal, such as striking someone in order to rob him or her)
2. humiliates or demeans people in the presence of others
3. has treated or disciplined someone under his or her control unusually harshly, for example, a child, student, prisoner, or patient
4. is amused by, or takes pleasure in, the psychological or physical suffering of others (including animals)
5. has lied for the purposes of harming or inflicting pain on others (not merely to achieve some other goal)
6. gets other people to do what he or she wants by frightening them (through intimidation or even terror)
7. restricts the autonomy of people with whom he or she has a close relationship, for example, will not let spouse leave the house unaccompanied or permit teenage daughter to attend social functions
8. is fascinated by violence, weapons, martial arts, injury, or torture
9. behavior detailed above has not been directed toward only one person (e.g., spouse, one child) and has not been solely for the purpose of sexual arousal.

Psychologist Robert Hare provides a wider context for the disorder:

> It is important to emphasize that these characteristics can be expressed in differing ways in differing contexts. Perhaps the most obvious form are the tyrannical sadists who, in the workplace, choose their victims whom they abuse, intimidate, and humiliate in

> front of their colleagues; they obtain pleasure from the psychic pain and distress of those whom they subjugate. Sadists may seek and achieve social positions, which allow them to exercise control and to mete out punishment in socially sanctioned roles. They may include the judge who metes out punishment of a cruel and unusual severity, the army sergeant who brutalizes the new recruit, and the psychiatrist who misuses mental health legislation to incarcerate a patient.[11]

The Role of Genes and Environment

But what causes some people to be antisocial, or psychopathic, or sadistic? Decades of research have shown that both genes and environment work together in a complex tango of cause. Intriguing research has shown, for example, that abused boys who happen to have genes coding for low levels of an enzyme, MAO-A, which breaks down communication molecules in the brain, have a higher tendency to become violent or criminal than other abused boys. When these children are raised in a normal environment, however, the gene doesn't appear to affect behavior. The low-coding MAO-A genes provided only a *predisposition* for antisocial behavior—an example of how both nature and nurture might combine to form personality as well as personality disorders.

Some genes lead more or less directly to changes in general health and personality. Inheritance of stuttering, redundant trinucleotides on chromosome 4, for example, can cause Huntington's disease—a disorder that leads to a dramatic transformation in personality followed by a devastatingly premature death. Susceptibility to schizophrenia is associated with specific genes on chromosomes 13 and 8. Bipolar disorder is affiliated with certain genes on (at a minimum) chromosomes 1, 6, 7, 10, 18, and 21. Discussion of all these chromosomes, genes, and inherited disorders raises the question—could there be a gene or set of genes that would lead directly to the behavior associated with at least some types of antisocial personality disorder or psychopathy?

The Genetics of Psychopathy

The essentials of psychopathy can be seen even in this simple portrayal of a seven-year-old child: "Mark does not feel guilty if he has done something wrong, he does not show feelings or emotions, and he is rarely helpful if someone is hurt."[12]

This description of Mark is the opening sentence of a fundamentally important 2005 study, "Evidence for Substantial Genetic Risk for Psychopathy in 7-Year-Olds," incongruously led by a delicate-featured young blonde with guileless blue eyes and the memorable name of Essi Viding. (Researchers with unusual names are always a delight for other researchers—search engines home right in on their work.) Viding has been curious about the possible genetic origins of antisocial behavior since a stint as a junior researcher before graduate school. She noticed that despite efforts from professionals, there seemed to be some antisocial children who appeared beyond reach.

Viding's study sought to tease out how much genetics comes into play in antisocial behavior in children with outright psychopathic tendencies, as well as in those with less severe forms of antisocial behavior. To do this, Viding used results from TEDS—the Twins Early Development Study—a monumental study of over five thousand pairs of twins in the United Kingdom who were born in 1994, 1995, and 1996. The TED study was initiated to allow scientists to take advantage of the fact that while identical twins share 100 percent of their genes, fraternal twins share only about 50 percent. If the effect of genes predominates for a given characteristic—extroversion, for example—then you would expect that if one identical twin was outgoing, the other one would be too. But for fraternal twins, like my husband and his sister, it is just as likely that one could be laconic (my husband), while the other is loquacious (Jane),

The effect of environment turns out to have two components. One component has to do with the environment the twins share: a set of easygoing, loving parents, for example, or a shared diet of lard-ridden Southern cooking. You can figure out what this shared environment is by playing with the data to see if fraternal twins appear to be more similar than expected, since they share only half their genes.

The other environmental component has to do with a nonshared environment. Maybe Mom really *did* like your twin best, or maybe your stepfather decided *you* were the one he was going to beat when he had too much to drink. You can get an idea of what this nonshared environment is by looking at the data from identical twins. If the twins aren't 100 percent identical for a certain trait, the nonshared environment is suspected to be the reason.

Viding's group surveyed the teachers of 3,687 pairs of "TEDS" twins from the 1994 and 1995 births regarding their students' character traits. From this, 187 pairs of twins were found to have at least one twin who displayed extreme psychopathic traits. In another 177 pairs of twins, at least one of the twins was found to display traits of antisocial behavior but without the callous, emotionless features of psychopathy.

By careful analysis of the data for identical versus fraternal twins, what Viding found was unexpected—and striking. Antisocial behavior in those twins who were highly psychopathic was under strong genetic influence—a heritability of 81 percent.*[13] The remaining 19 percent of influence appeared to be entirely due to unshared environmental components—differences in how the two twins were raised. The twins with antisocial behavior, on the other hand, showed a moderate genetic heritability of 30 percent; the remaining influences were entirely environmental.[14]

If this seems hard to swallow, here's an analogy. Let's say that you randomly gather a group of older children of normal intelligence who

---

*This might bring some readers to wonder about identical twins pairs in which one twin has psychopathic traits and one does not. Remember—just having the genetics for a disorder does not necessarily mean the disorder is inevitable. It may mean, for example, that one twin is just over the diagnosable edge, while the other isn't. Even with identical twins growing up in the same environment, it's impossible to identically wire brains with 100 trillion synapses. And environment itself does make a difference. For example, phenylketonuria, a recessive single-gene disorder, causes severe mental retardation. However, administering a diet low on phenylalanine prevents mental retardation. Judith Rich Harris's latest book, *No Two Alike*, provides an elegant hypothesis related to the often troublesome category of "non-shared environment."

Perhaps more importantly, the idea that the variation in personalities is roughly half due to genetics and half to environment is valid, but misleading. Psychopaths, for example, sometimes appear to possess "dominant" sets of genes that override the influence of environment.

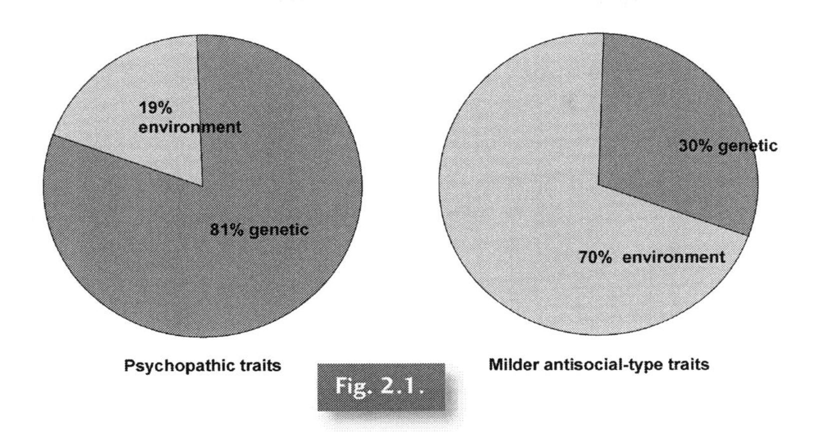

Fig. 2.1.

are poor readers. If you analyze the group carefully, you might find that some kids have a poor home environment or have disadvantaged schooling. Based on the twin studies, you would probably find that there was also evidence of a slight genetic component.

But now let's say you asked your poor readers to read and answer the simple question: "Do the words *fruit* and *boot* rhyme?" You'd find that the kids would automatically sort themselves into two groups, irrespective of schooling. One group of kids would answer the question easily, while the other group would struggle. If you then analyzed each of these groups separately by drawing on the twin studies, you would find something very different from your first analysis. The children who easily understood that *fruit* and *boot* rhyme would probably show that their reading ability was substantially influenced by their environment. But the group who had trouble seeing the rhyme would probably show considerable genetic evidence for their reading disability. This is because some or many of the children who had trouble with the unusual rhyming pattern were probably dyslexic, and dyslexia has a strong genetic component.

Let's look again at the results from the study of psychopathy. Discarding the belief in the natural innocence of children and eliminating a century of social engineering, this means that *some kids are born with a marked tendency toward evil.* Sure, traditional forms of intervening in the family environment seem to work for kids with a light set of antisocial genes. But some type of genetic vulnerability appears to cause significant differences in the neurological development of children with psychopathic tendencies. As Viding had seen at the very beginning of her academic career, these unusual children seem remarkably resistant to intervention and may be at high risk for becoming adult psychopaths. Viding's study has far-reaching implications for treatment and prevention—precisely the reason she carried out the research. For, if the cause of psychopathy is indeed genetic—that is, related to subcellular programming—an understanding of that programming can lead to methods of repair.

But what exactly is going on, genetically speaking, in those with problematic personalities? We have finally reached a new, genomic era, where we can begin to understand.

And perhaps, just perhaps, I can find some clues about Carolyn.

CHAPTER 3

EVIL GENES

"The problem with the gene pool is there aren't any lifeguards."
—Anonymous

We already know that genes have dramatic effects on virtually every aspect of the human body—height, weight, skin color, and even the ability to process oxygen. But sometimes we forget how important genes are in shaping personality. As behavioral geneticist Robert Plomin has pointed out, the answer to the question "how much does heredity affect behavior" is "a lot."[1] Indeed, as Plomin notes, genetic influence is so ubiquitous that we should not be asking what is heritable with regard to behavior. Instead, we should be asking what is *not* heritable. "So far," he notes, "the only domain that shows little or no genetic influence involves beliefs such as religiosity and political values; another possibility is creativity independent of IQ."*[2]

*Since Plomin's comments, a twin study has shown that a tendency toward being religious does indeed appear to be moderately heritable. Such a tendency, it seems, is strongly influenced by the environment during adolescence. However, by adulthood, religiousness seems to slip away somewhat unless you've got genes that predispose you toward religion.

Recent research has even undermined the long-held belief that troubled, argumentative marriages cause problematic behavior in the children brought up in those households. A recent study of adult twins and their offspring revealed that it is not the family discord that causes problematic behavior but rather the genes that troubled parents pass along. In fact, the parents' *own* genes apparently determine how often they argue with each other.[3]

Ultimately, then, since genes are so crucially important to understanding personality—including its Machiavellian manifestations—it's a good idea to take a quick review of some fundamentals. A human body is composed of about a trillion tiny, membrane-enclosed cells—bone cells, nerve cells, white blood cells, and so on. Each cell is alive and carries out its own suite of "life functions" by following the instructions encoded in its genes. The genes are portions of chromosomes, sequestered in the cell's spherical nucleus. Most human cells (there are a few exceptions) contain forty-six chromosomes—twenty-three inherited from the mother, and another twenty-three from the

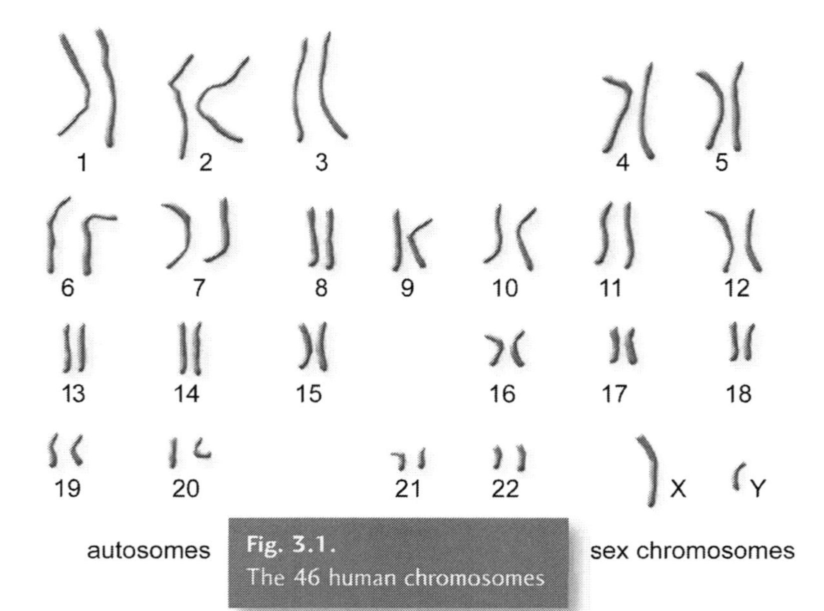

Fig. 3.1.
The 46 human chromosomes

father. Each chromosome is a long, slender DNA molecule. If all the DNA in one cell's chromosomes were stretched out, they would total about six feet in length. That means that the average chromosome is a couple of inches long—which makes a DNA strand a giant, as molecules go. But because DNA is extraordinarily slender—thousands of times thinner than a hair—all forty-six human chromosomes are easily wadded up like a molecular-sized ball of string inside the cell's nucleus.

DNA molecules are not only long and skinny but also very simple in structure. The famous DNA "double helix" consists of two parallel

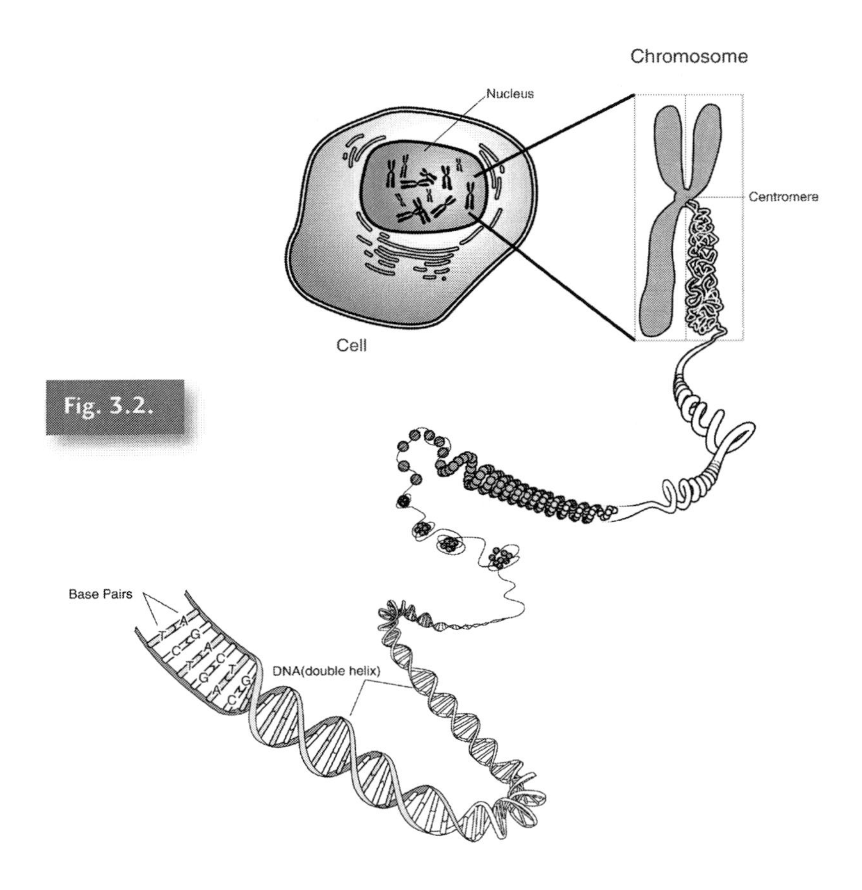

Fig. 3.2.

(but twisted) chains of molecular building blocks called nucleotides. There are only four kinds of nucleotides, and what makes the four different from one another are parts of the nucleotides called bases—adenine, guanine, thymine, and cytosine. The four DNA bases are almost always abbreviated A, G, T, and C. The DNA "code" consists of sequences of nucleotides with various bases. ATTCGACCTCC tells a cell to do one thing, TGACCTGCAG says something else. You can think of a cell's chromosomes as a set of cookbooks, with each chromosome being a volume, and each gene a recipe. There are hundreds of gene "recipes" strung along each chromosome, and roughly 70 percent of all those genes participate in the development and operation of our brain.

Along the chromosome, each gene—each recipe, if you will—is a sequence of DNA letters that tells a cell how to make a particular protein. The proteins, in turn, make up much of the structure of a cell and, hence, the body, and carry out cellular functions. Thus, the genome consists of genes—sequences of DNA bases along the chromosomes—each of which codes for the assembly of a particular protein. Depending on the information in the genes, the proteins build an ant, a pine tree, or a human—with brown eyes or blue.

We humans are estimated to have about twenty-two thousand genes sprinkled among our chromosomes—that's only about 5 percent of the total DNA.[4] The other 95 percent, sometimes called "junk DNA," has long been thought of as sitting around twiddling its molecular thumbs doing nothing in particular. However, researchers are discovering that some junk DNA is intricately involved in the regulation of which genes are turned on or off in which tissues. This in turn ultimately determines an organism's phenotype—how an organism appears. Thus, although genes are still of paramount importance, there is a whole additional layer of complexity involving control of those genes—a sort of "index"—that we are just barely beginning to understand. What makes us special among primates may not so much be the genetic recipes themselves but when and where the regulatory sequences turn the genes on and off.

Even more interesting are recent findings based on how often genes repeat themselves—"copy number variants."[5] Perhaps 10 percent of genes, it seems, are in regions that can easily find themselves doubled, tripled, quadrupled, deleted, or scrambled. These different numbers and types of genetic copies can make dramatic differences in the genetic makeup between different population groups, as well as between more closely related individuals—perhaps even between siblings. The number of copies of different genes has already been linked with a variety of medical conditions, including Alzheimer's, kidney disease, and HIV. The expectation is that these copy number variants may become very significant in personality disorder research as well.

Differing versions of a gene that can fill a slot on a chromosome are called *alleles*. Alleles are simply variants of genes—kind of like a recipe variant where egg whites are substituted for egg yolks when baking a cake. But alleles can also be thought of as competitive versions of a gene. A different, "new and improved" version of an allele, for example, might help build a better molecule for ferrying oxygen around or might help grow sturdier bones.

But how do different alleles for a gene arise? Primarily through mutation. One of the bases—A, C, T, or G—might be miscopied when reproduction is taking place. Alternatively, sometimes genes stutter when they are copied, repeating certain parts of themselves, or, like dropping a stitch, losing a section. Either way, a slightly different allele is created that is passed down to future generations. Incidentally, about 25 percent of all human genes have alternate versions available. The lowest average number of alternate versions is found in the populations of New Guinea and Australia, while the dazzlingly highest number of alternate versions is found in the Middle East, western Asia, and southern, central, and eastern Europe.[6]

Yet even one simple change of a nucleotide at a single location can lead to problems—as if a cook used a teaspoon of salt instead of yeast in a recipe for bread. Such a change in a gene (making a new allele, or "flavor" of that gene) causes a change in the protein it builds. This often means that the protein doesn't function normally. Just such single mutations are responsible, not only for cystic fibrosis but for dozens of other devastating conditions such as hemophilia and sickle-cell anemia. Diseases such as Alzheimer's or schizophrenia, on the other hand, often involve more complicated confluences of unlucky alleles. The illustration below shows a few of the common and unusual illnesses that have so far been found to be associated with genes just on chromosome 17.

Our *genotype* is the actual information contained on the long strands of DNA molecules in the nucleus of our cells—we can determine our genotype only by using molecular methods. Our *phenotype*, on the other hand, relates to our appearance, which is determined by the output of the genotype and sometimes by environment as well.

Now we're ready to return to the subject of evolution. As we have seen, traits are controlled by a mix of genes—various alleles that affect characteristics such as skin color, disease resistance, memory, and even novelty seeking. If an allele helps produce a trait that confers

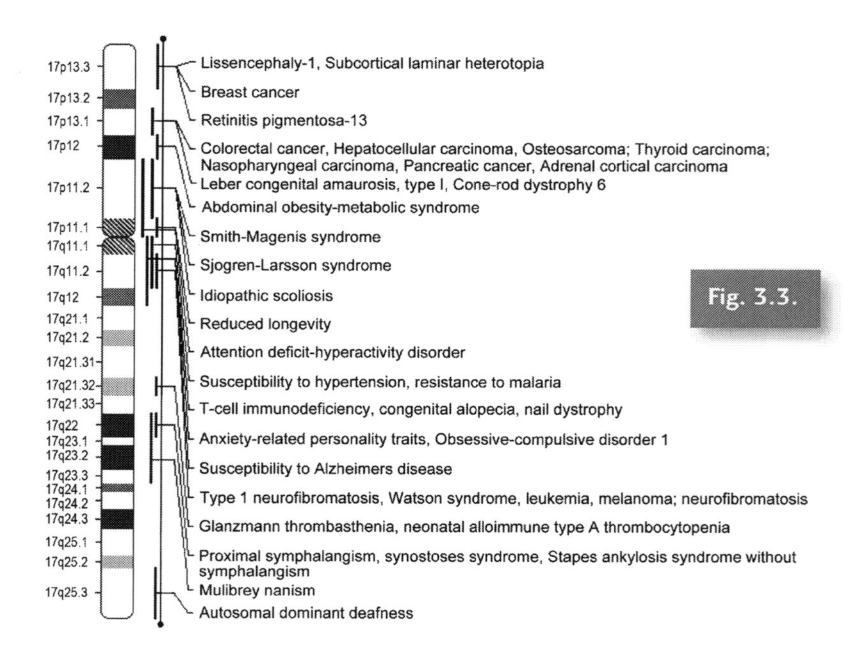

Fig. 3.3.

an advantage, individuals who bear that allele will leave more offspring. That trait and that allele will occur in a greater proportion of individuals in the next generation, and the next generation as a whole will be better adapted to the environment. Sometimes, especially in small populations, chance events can also alter the mix of alleles in the next generation, a phenomenon called *genetic drift*. These incremental shifts in the frequency of alleles in the population—changes in the population's "gene pool"—constitute evolution. Specifically, they constitute a small-scale process termed *microevolution*. Larger-scale evolutionary change, such as the origin of new species or the founding of orders and classes, is called *macroevolution*.

MACHIAVELLIAN GENES

A complex array of varying genes underlies the many different outward manifestations—phenotypes—of many different personality disorders. A person with an unlucky shake of the genetic dice can actually end up with full-blown versions of those disorders right out of the chute—these unfortunates often show their dysfunctional characteristics in early childhood, despite a loving and stress-free environment.

However, a person with a lighter dose of the genetics for a personality disorder is not necessarily predestined to descend into a full-blown, clinical version. There are two routes such a person can follow. With a relatively stress-free environment, the person may simply grow into someone who is "normal" but who can sometimes be difficult to deal with emotionally. The other route involves succumbing to all-out traits of a personality disorder.

How might this happen?

The key, it appears, is often stress. When a person experiences bodily stress, for example, physical exercise, it can turn certain genes on or off—perhaps through the regulatory function of the junk DNA. In the case of exercise, this stress can turn on genes that cause muscle growth—you see the result in the form of bulging biceps. But a body

Intermediate Phenotype

Intermediate phenotype is a concept used by researchers who are wrestling with the relationship between genes and phenotype. To understand "intermediate phenotype," it's helpful to remember that there is often an intermediate case between a full-blown manifestation of a disease and a less harmful variant. In personality disorders, intermediate phenotypes, sometimes called endophenotypes, are used to describe people with subclinical symptoms of diseases like schizophrenia or borderline personality disorder. The stipulation for an intermediate phenotype is that it be found in mildly ill but not "certifiable" siblings and other relatives, and that it even be found in some psychiatrically well relatives. This establishes that the phenotypes are related to risk for an illness and are not the illness itself.

So far, the concept of intermediate phenotypes has been most powerfully developed and used by Dr. Michael Egan and his colleagues at the National Institute of Mental Health for their research on schizophrenia. Schizophrenia, like borderline personality disorder, is a complex disease that results from many causes, including a multitude of genes and environmental factors such as drug abuse, head injury, infections, and even a person's conscious thinking processes—all of which can influence each other.

can experience stress in other ways—for example, by being beaten by a parent, working for a bullying boss, or drinking too much alcohol. All of these activities, amazingly, can switch different sections of one's genetic code from quiescence to an all-too-active state—or vice versa.[7] The resultant proteins, which have different properties from the nonstressed versions, can, in turn, affect our personalities. Depending on the stress and our genetic predisposition, we can be pushed toward depression, eating disorders, drug abuse, or cancer.[8] If a person already has a mild form of a personality disorder, he or she can be pushed into a full-blown version.

Egan's work clarified the relationship between a particular allele related to cognitive function that had previously been weakly and inconsistently associated with schizophrenia. When Egan's group applied the concepts of intermediate phenotype by studying brain function and comparing genotypes in a wide variety of people—including patients with schizophrenia, their healthy siblings, and controls—the suspect allele suddenly popped out as a strong predictor of abnormal prefrontal brain function. This happened in every person sampled, whether or not the person had schizophrenia. Egan's study was one of the first times that a correlation of an intermediate phenotype with a gene was shown to clarify how a gene related to a complex clinical diagnosis.[9]

"What's so surprising," marvels Daniel Weinberger, Egan's colleague at the National Institute of Mental Health, "is that it works."[10] It appears that the next step beyond imaging genetics may relate to the synergistic use of genotyping, neuroimaging, and intermediate phenotypes. It will be exciting to see what future studies reveal when these techniques are applied to antisocial personality disorder, other related syndromes, and their subclinical "intermediate phenotypes."

Faced with the overwhelming variety of phenotypes that can arise from this mixture of genes and environment, it's hard to know where to even begin looking at a person's genome to determine which alleles might be key in motivating behavioral traits. But a fascinating new discipline, *imaging genetics*, has recently arisen that provides precisely the necessary tool. Imaging genetics uses medical imaging techniques to figure out a person's phenotype—the word *phenotype* meaning, in this case, the size and shape of organs such as the amygdala and cingulate cortex—and then evaluating the same person's genes to see how they compare. The value of using medical imaging for the comparison with genetics, instead of old-fashioned questionnaires and interviews, is that genes act much more directly on neural components like the amygdala than they do on a person's ultimate behavior—and they don't lie. You might think of the old research as being the equivalent of trying to figure out how a racing car works by comparing its blueprints (genome) with its performance statistics (behavior)—a dry and thankless exercise at best. Today's imaging genetics allow you to open the car's hood and look around with sophisticated measurement tools on hand. This, in turn, allows you to make comparisons between blueprints and performance even while the engine is running, so you can figure out what's really going on.

But, you might ask, are we seeing the cause of certain thinking patterns? Or the effect? Clearly for some organic brain diseases, such as schizophrenia, we are seeing the effect. But for other conditions, it's often not clear. After all, use of antidepressants can alleviate depression by changing brain chemistry—these changes can be clearly seen with imaging techniques.[11] But the *same* changes in neural chemistry can be seen after a patient has used cognitive therapy techniques to change her thinking patterns![12]

If you ever want to know whether your tax dollars are being used for a good purpose, go take a look at the extraordinary work that the National Institute of Mental Health and other National Institutes of Health are doing in digging out the genetic bases of psychiatric illnesses. Dr. Weinberger, quoted above, also happens to be the director of the National Institute of Mental Health's Genes, Cognition, and Psychosis program. He is one of the leading researchers in this area, as indicated by the number of key papers related to the genetics of personality disorders that bear his name. When I spoke with him about this book, he reiterated his feelings that genes are about risk—not fate—and that no single gene by itself can predict personality.[13]

Most especially, we know that there is no single gene known to create a psychopath, or to cause someone to suffer from antisocial personality disorder, or to generate more sinisterly successful variants of either one of these disorders. But there *are* a number of genes and gene complexes that have been found to affect brain function—most importantly, for our purposes, regarding traits such as impulsivity, mood, and anxiety. Through the use of such sophisticated new techniques and concepts as imaging genetics and intermediate phenotypes, researchers are discovering how alleles of particular genes can help underpin the dysfunctional behavior that can lead to a problematic personality or full-blown clinical pathology. In a sense, you might call these *evil genes.** Let's focus on a few of them.

Serotonin Receptors and Behavior

A variety of different studies have converged on serotonin as being a key communication molecule—"neurotransmitter"—behind the generation and control of emotions. Neurotransmitters are like little flares that carry information across the gaps—synapses—between sending and receiving neurons (see the picture on the next page). It turns out that the serotonin flares can interact with about fifteen different types of receiving cell landing sites, called "receptors." Once serotonin lands on a particular type of receptor, it sets in motion a whole Rube Goldberg–style chain of events. If serotonin hits one receptor, for example, it's metabolically akin to flipping a switch to start a ball bearing rolling down a ramp to bash against your toaster handle and start your morning toast. On the other hand, if it hits another kind of receptor, it's a sort of physiological equivalent of pushing a button that launches the ball bearing out a bedroom window, allowing it to bounce against a plate on a tree and back in the kitchen window below, thus tapping a coffeepot's ON switch and starting your morning cup of cellular java.

Why are there so many different kinds of serotonin receptors? Researchers speculate that serotonin has apparently been used as a

---

*Okay, if you really wanted to be picky here, you could say "problematic alleles," or even "quantitative trait loci that have been affiliated with specific personality disorders," with the caveat that environment can often play an important accompanying role in those with a genome that has set them at risk. But it doesn't have quite the same ring, does it? In any case, it's important to remember that depending on the constellation of other genes in a body, supposedly evil genes might have a neutral, or even positive effect. Or such genes could have mixed positive and negative effects, such as increased intelligence coupled with increased neuroticism.

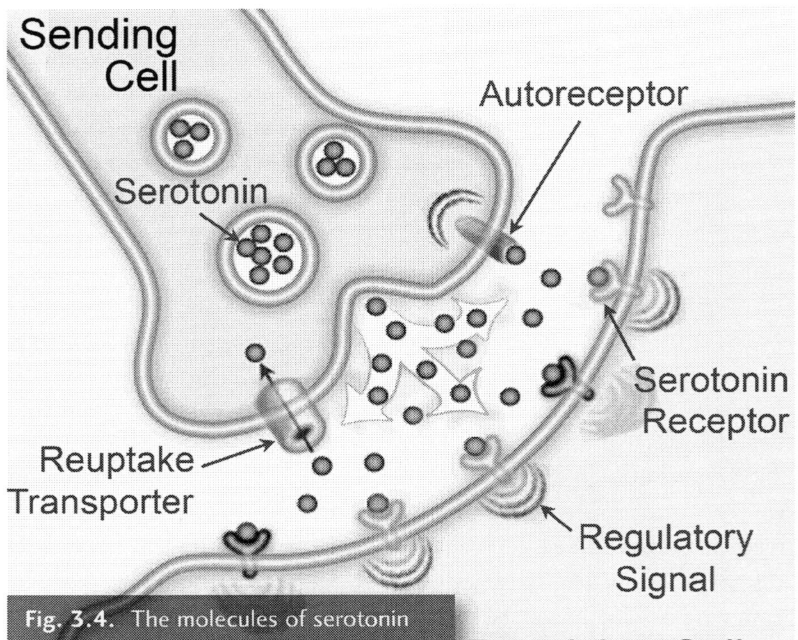

Fig. 3.4. The molecules of serotonin shown in this illustration are given off by the cell that is sending the message. The serotonin binds to "docks" (receptors) in the receiving cell and instructs that cell to either fire or stop firing, among other processes. The amount of serotonin in the gap known as the synapse, as well as the types of receptors (there are at least fifteen types), influences the cell's response. Two different types of sending cell molecules can reduce serotonin levels in synapses. Autoreceptors direct the cells to slow down serotonin production, while reuptake transporters absorb the neurotransmitter back into the sending cell to prepare for the next firing.

communication molecule dating back even to very primitive organisms. This common molecule is found, after all, in pretty much anything with a backbone, as well as in spineless creatures such as flatworms, nematodes, and leeches. For all of these creatures, serotonin assists with sensing, motion-related, as well as cardiovascular functions. Basically, serotonin is a handy molecule for many different purposes.[14] Different receptors on different receiving cells each respond to the little serotonin molecules by kicking their own pathways into gear. Serotonin receptors are each created according to templates set out by specific sets of genes, with each gene often having two, three, or even more different versions—that is, alleles. Obviously, all these different possibilities can make for a dizzying variety of potential genes related to serotonin receptors alone.

Different serotonin-related alleles have been found to be strongly associated with various aspects of personality and temperament, as well as mood disorders. But because these alleles interact and overlap, it's difficult to state definitively that any given allele is responsible for a given personality disorder. Sometimes an allele might help produce a disorder—but if that same allele is found with a constellation of other mitigating alleles, it might not produce the disorder at all. Indeed, the idea that *groups* of genes underlie personality types is an important one and has been given the name "QTL (Quantitative Trait Loci) Model" for behavioral traits.[15]

Our knowledge of how the different serotonin receptors relate to emotions is currently rather limited. We *do* know that one of those receptors, with the cryptic name 5-HT_{1B}, plays a selective role in controlling offensive aggression.[16] Other serotonin receptors have also been implicated in problematic behaviors. Certain alleles of the 5-HT_{2A} receptor, for example, have been found to be associated with self-mutilation, anorexia, and a history of suicide attempts.[17] The HT_{3A} receptor, on the other hand, appears to have a critical influence on the amygdala (the "fight-or-flight" decision-making area of the brain), especially when a person is reacting to another's facial expressions. The HT_{3A} receptor also affects how fast certain areas of the brain process information. Some versions of alleles related to this receptor appear to cause the extremes of neural activity seen in bipolar disorder.[18] Overall, research about serotonin receptors and their associated alleles is just at the "tip of the iceberg" stage—enough to hint that there may be something going on related to precisely the sorts of emotions and behavior that are seen in various subclinical and clinical personality disorders.

On the other hand, if we switch our attention to serotonin *transporters*, as opposed to receptors, we will find that research is far more advanced.

The Long and the Short of It—Serotonin Transporters

Reflecting on the illustration a few pages back showing synapses and serotonin, we are reminded that serotonin receptors are equivalent to docking points that help trigger reactions in the next neuron. Once the reaction is triggered—the switch is flipped—the serotonin is then free to go back and float around in the space between the neurons. But if serotonin is already filling the space, how can a "sending" cell release new serotonin to trigger a new reaction? Somehow, the serotonin already in that space has to be pulled back into the original trigger neuron so that it can be used to trigger the next signal. One of the key molecules that helps do this is a special transporter molecule called SERT (for "serotonin transporter"). You can think of SERT as a cleverly designed protein conveyer belt that helps scoop excess serotonin out of the cleft between neurons and carry it back into the trigger neuron. Two different alleles have been found that help produce SERT—a short allele with fourteen tandem repeats, and a longer version with sixteen tandem repeats. ("Tandem repeats" is short for "repeated tandem base pairs," such as GCGCGCGCGCGC. Sometimes the tandem repeats confuse the cell's DNA copying apparatus, so it's easier for these alleles to be miscopied and made shorter or longer.) The difference between the short and long SERT alleles doesn't lie in the information that codes for the transporter molecule itself but rather in the part of the gene that controls how often the transporter molecule gets produced. The short version of the allele doesn't allow for production of as many transporter molecules. You might think of it as a copy machine that puts out only half the copies you request. The resulting lack of transporter molecules allows serotonin to linger longer and appears to predispose people toward anxiety, impulsivity, suicidal thoughts, affective instability, bulimia, and binge drinking.[19] People with two shorts (one from the mother and another from the father) have fewest transporters and seem to feel these effects most strongly.

The ultimate effect of the short SERT allele is that the part of the brain that is supposed to dampen down your fear responses doesn't seem to be able to do its job very well. Remember that serotonin provides a connection between two neurons. If something happens to that

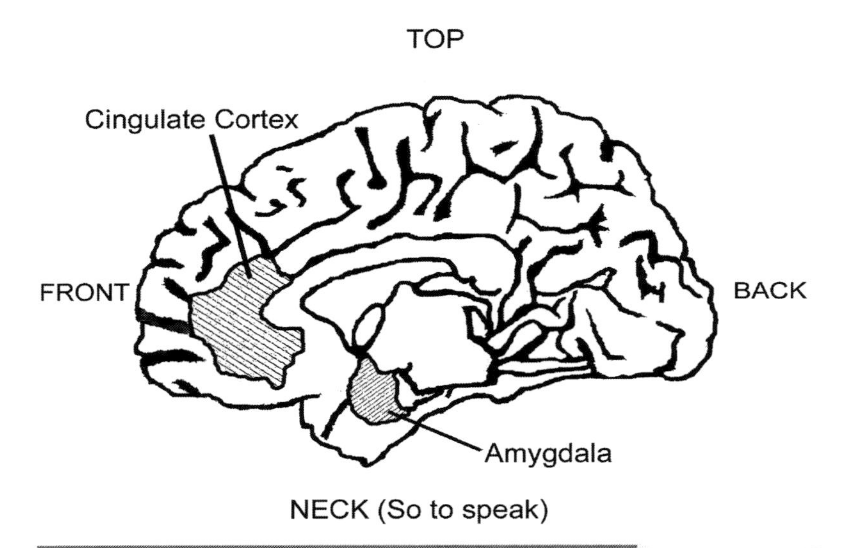

Fig. 3.5. This figure shows two vital, emotion-related organs—the amygdala and the cingulate cortex—that are affected by reduced serotonin transport.

connection, problems are bound to occur—neurons aren't able to communicate clearly. It's like trying to speak (send out serotonin) in a room filled with people who are already talking loudly (serotonin is already in the space between the neurons). It's hard for important messages to get through. The short allele may produce depression simply because natural anxiety and fearfulness aren't restrained.

Researchers are beginning to home in on how all this happens. Using functional magnetic resonance imaging, researchers have found that in normal and depressed people with one or two "shorts," not only were the cingulate cortex and amygdala reduced in size, but the circuits that connected the two organs also appeared to be weakened.[20] (The cingulate cortex is that part of the brain that helps us to focus our attention and "tune in" to thoughts, while the amygdala, again, is the "fight-or-flight" coordinator of the body's emotions.) You can see the weakened "speech" between the amygdala and the cingulate in the illustrations on the next page. The picture on the left shows a normal control loop for fear—the kind of loop that would be seen in a person with long/long SERT genes. The amygdala sends a signal to the lower part of the cingulate, and then on to the upper, and finally, a "calm down" signal is sent back to the amygdala. In real life, this process might relate to something like the fear you feel—due to the amygdala's response—as your plane suddenly begins jouncing up and down. This fear would be communicated by the amygdala to the cingulate cortex, which would turn back around and control the amygdala's response by communicating something like: "*Calm down, it's okay—it's just turbulence as we're flying over the Rockies*."

The illustration on the right, however, shows the circuit response in people with at least one short SERT allele. Although the amygdala is activated, it can't send a strong signal out to the cingulate cortex because of all the other chatter going on. (Remember those serotonin transporters? There are fewer of them, so they don't transport very well—like stagehands who haven't bothered to clear the stage for the next act in the play.) Consequently, the controlling "whoa—take it easy" signal from the cingulate cortex back down to the amygdala is

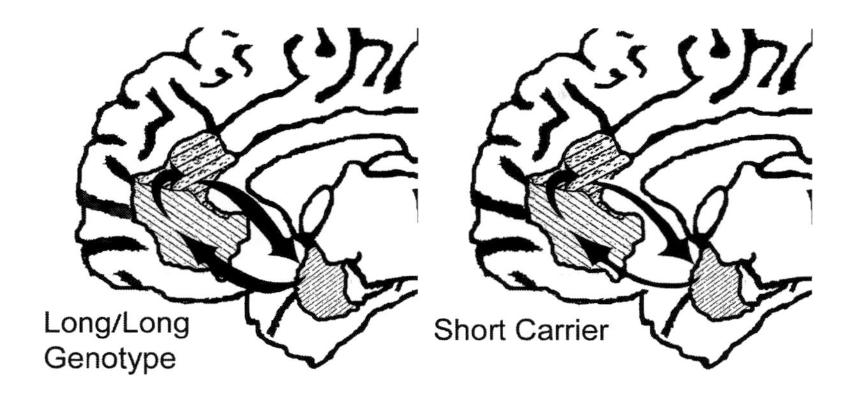

Fig. 3.6. Those with the long/long genotype, as shown on the left, have a full feedback control system that damps down the aroused amygdala. This allows a person to relax after first being startled. Notice how much thinner some of the signaling arrows are in the image on the right. Those with one or two short carriers aren't able to take advantage of feedback—their amygdalae continue to be revved up even after the person has consciously realized there is nothing to feel threatened about.

also weaker. The resulting thought pattern might go like this: "*Calm down—it's just turbulence. I mean, I think it's just turbulence. But . . . the people on Flight 587 thought it was just turbulence, too. My God, that plane shook itself apart midair. Every one of the 260 people on board was killed! I'm going to die!!!*" As you might imagine, this kind of negative thinking can lead to all sorts of problems—depression not the least of them.

Pleiotropy—The Naughty-Nice Aspects of "Evil Genes"

Given the many different negative aspects of the short version of the SERT allele on a personality, why hasn't the allele for the short version of the transporter molecule just died out? Surprisingly, it may be because the long version of the allele can also create problems—not necessarily with emotions, as the short allele does, but in other areas of the body. For example, primary pulmonary hypertension—a serious disorder that causes the heart to essentially overpump—appears to become a problem if a person has received long versions of the allele from both the mother and the father. This double mother-father dose allows for excessive serotonin that spurs the growth of pulmonary artery smooth muscle cells, which eventually blocks the blood's pipeline to the lungs.[21]

Remember, it is a single gene, the gene that produces the serotonin transporter molecule, that has all these varied effects on the body's neural, cardiac, and even the immune systems. The concept that one gene can affect many different areas of the body is so crucial that it even has a name: *pleiotropy*, from the Greek *pleio*, meaning "many," and *tropo*, meaning "turning toward."

An example of pleiotropy can be found in the APOE4 allele (short for *apolipoprotein E 4*). This allele, which is situated on chromosome 19, may have predisposed my father to Alzheimer's disease after his slip from the peaked roof of the covered bridge. Inflammation from his resulting concussion could have caused the allele's activation.[22] (If my father had had a double set of APOE4 alleles, one from each of his parents, he would have been even more likely to have wound up with Alzheimer's, although a number of other genes undoubtedly also play a role.) If all of this wasn't bad enough, the APOE4 allele has another nasty effect. It seems to be associated with high cholesterol—from which my father also suffered.

But it seems that there are several good aspects to the APOE4 allele. One is that, if this allele is switched on by nutritional stress, it may help children survive severe malnutrition early in life.[23] (The trade-off, of course, comes at the other end of the life span.) Another nice aspect of APOE4 is that, although you may lose your memory when you get old, you may actually have a *sharper* memory when you're young.[24]

The flip side of pleiotropy is *polygeny*. Polygeny simply means that a single trait can be influenced by many genes. For example, even though Alzheimer's is associated with the APOE4 allele, other genes may ameliorate the errant allele's effect. This may be why many APOE4 carriers never succumb to Alzheimer's. Polygeny appears to underlie personality disorders that are related to some types of sinister behavior—behavior much like my sister Carolyn's.

Brain-Derived Neurotrophic Factor

Another gene related to mood and anxiety is the gene that produces BDNF—Brain-Derived Neurotrophic Factor. This factor helps support the survival of existing neurons and encourages the growth of new neurons and synapses. There are two common alleles for this gene, dubbed *val* (short for *valine*—an amino acid in the protein coded by the gene), and *met* (short for *methionine*—a different amino acid that is substituted for valine at the same spot in the protein). It turns out that people who have two versions of valine have exceptionally good memories—the double *val* alleles seem to have a stronger effect on memory than any other factor ever studied.[25] Unfortunately, these people are also more neurotic—that is, they have more negative emotionality in regard to anxiety, low mood, and hostility.[26] This instance of multiple effects of a single gene is another example of pleiotropy.

Daniel Weinberger believes the *met* allele may have evolved because a double dose (one from the father and another from the mother) of the *met* BDNF allele just can't "hear" serotonin very well—which could be a real advantage in ignoring the higher anxiety signal that results from short serotonin transporter genes. A double *val* dose, on the other hand, appears as if it may magnify the effect of the short serotonin transporter. Psychiatrist Jim Phelps relays an analogy from Weinberger:

> Imagine that [*val/val*] BDNF alleles, with their memory-improving capacity, make your brain function like a 200 mile-per-hour race car. If you've got a hot rig like that, you'd better be a good driver who's

> capable of handling a fast, but temperamental car. In this analogy, that's the long/long allele pair for the SERT gene: the driver won't get over-anxious and allow the car to get out of control. That's important, because if you smash your car into the wall very often, your car won't run very well. In real life, if you take too many stress-hits, you end up depressed.
>
> By comparison, if you inherit the short/short pair for SERT, and thus are less able to handle anxiety-producing situations such as conflict, trauma, and loss—you are a more cautious and potentially distractible, frightenable driver. In this case, you might be better off with a slower but more crash-resistant car, one that you can smash up against the wall quite a few times without changing how it performs very much. In this analogy, that's the *met/met* allele pair for the BDNF gene.
>
> By this analogy, perhaps the "slower" *met* allele was selected for (evolution-speak) in humans to help people with "two shorts" get through life better. If two shorts make you more cautious, and two *met*'s make you less likely to worry about things, for some people that could make a durable, reliable combination that in dangerous times might be better than the high-performance but "higher-maintenance" *val/val* and long/long combination. Of course at this point that's almost entirely a guess, but it gives us a beginning of a model which might help us understand these genetic variations in humans.[27]

Certain alleles for BDNF receptors have been found to be strongly associated with bulimia and anorexia, which are in turn associated with such personality traits as anticipatory worry and pessimism.[28] Some BDNF alleles are also associated with depression, bipolar disorder, and neuroticism.

Warrior or Worrier? The COMT Gene

Trade-offs—there are always trade-offs. And nowhere is that more clear than with the COMT gene (short for the ungainly catechol-O-methyltransferase), which is a key gene underlying our general intelligence. This gene works by serving as the blueprint for an enzyme that breaks down dopamine and other neurotransmitters. It turns out that the more slowly you metabolize dopamine, the smarter you are, so if you have versions of the COMT gene that don't metabolize dopamine well, chances are you have a higher IQ (other genes and the environment also play a role here, of course). Like the BDNF gene, COMT also has *val* and *met* versions, with the *met* being a slow metabolizer, and the *val* fast.[29] People with *val/val* versions of the COMT gene can be a bit less intelligent—they may also have a slightly increased risk for schizophrenia. *Val/val*s are also at increased risk for antisocial behavior and hyperactivity. None of these detrimental, fast-metabolizing *val* effects are particularly surprising—after all, amphetamines and cocaine, which increase the transmission of dopamine, cause the psychotic, aggressive behavior that is so familiar to emergency room physicians. Compared to *val/val*s, *met/met*s can be smarter, and have a markedly better memory.[30] People with mixed *val/met* versions of the alleles seem to be halfway in between.

Given the advantages of *met*, it would seem that *val* would have died out. Instead, it is common in many human populations, with increases of *met* being balanced by decreases in *val*, and vice versa, in a sort of yin-yang relationship.[31] Why? *Val* versus *met* has been aptly described as "warrior" versus "worrier."[32] It seems that although *val*s may not be as smart on average and they have the mixed blessing of increased aggressivity, they can handle stress better than *mets*. Additionally, *val* cognition, although perhaps not as quick or deep, is more flexible—*val*s can more easily adapt when the rules of the game suddenly change.[33] Conversely, the *met* allele is instead associated with more anxiety or, in research-speak: "lower emotional resilience against negative mood states."[34] It is more frequently seen in individuals with obsessive compulsive disorder, which is characterized by distressing intrusive thoughts and the compulsive performance of rituals. *Met* is also associated with feeling pain more strongly and reacting more negatively to prolonged pain—people with *met* alleles simply don't get the same natural soothing opiates that *val*s get.[35] Although *met* COMT enhances intelligence and memory, it can also add to the effects of a short SERT allele, making a person even more anxious and neurotic.[36] You might think of this as the Woody Allen of genes—producing brilliance coupled with deep neuroticism.

Monoamine Oxidase A

Monoamine oxidase A—MAO-A for short—is a term for an enzyme that helps break down neurotransmitters like serotonin and dopamine so they don't continuously build up inside neurons. As with the SERT gene, it seems that low-functioning versions of the MAO-A gene have been linked to problematic personality traits. These traits involve both impulsive and aggressive behavior, as well as depression, substance abuse, criminal behavior, attention deficit disorder, and social phobias.[37] One recent study has linked the low-functioning versions of the MAO-A gene to those with the dramatic, emotional, and erratic personalities known as "Cluster B" disorders (which include antisocial and narcissistic personality disorders).[38] Differences in neural behavior in those with different versions of the gene have been spotted even in those with mild intermediate phenotypes—that is, in normal or relatively normal people who would fly under the radar of clinical significance for diagnosis of a personality disorder.[39]

One study analyzed one hundred normal volunteer men and women to see whether they carried the high- or low-efficiency versions of the MAO-A gene.[40] These volunteers were then imaged. The upshot was that those with the low-efficiency MAO-A alleles had smaller limbic organs, such as the amygdala and cingulate gyrus. The amygdala reacted strongly when these subjects were given a very mild scare, but the increased amygdala reaction was accompanied by an unexpected *decreased* reaction in the orbitofrontal and cingulate cortices (generally the amygdala would activate these two cortices, which would in turn send signals back to the amygdala to calm it down). These are the types of neurological reactions that are associated with impulsive violence. These reactions display the same sort of tamped down neural control circuitry between the amygdala and the cingulate cortex that we saw earlier in the individuals with short serotonin transporter alleles. It's just that in this case, not only is there a damped connection between the cingulate cortex and the amygdala, there's also a damped connection between the orbitofrontal cortex and the amygdala. Weakening this latter circuit means someone might have trouble with stimulus-reinforcement learning. A typical example of this type of behavior might be the knuckleheaded kid who continues to saunter in late for classes even though he knows he'll get detention.

The low-efficiency MAO-A allele appears to be particularly associated with impulsive violence, as opposed to violence as a purposeful means toward an end. The effects of MAO-A genes were, in fact, first discovered in a Dutch family in which certain males had inherited an unusual mutation that did not allow their MAO-A to metabolize serotonin or other neurotransmitters. Generations of family members had shown bizarre aggressive behavior, such as an attempt to run an employer over with a car; stabbing a warden in the chest with a pitchfork; or entering sisters' rooms at night, armed with a knife, and forcing them to undress.[41]

The MAO-A system is interesting, too, because it was the first neurotransmitter system to reveal how the same environment might have a different effect on people with different genetics. In 2002, Avshalom Caspi and his colleagues gave evidence that indicated why some children who are maltreated grow up to develop antisocial behavior, whereas others do not.[42] The key to the differences, it turned out in this study, lay in the children's genotype. Children who grew up in positive environments generally had no developmental difficulties, whatever their genotype. But those children who grew up being maltreated showed significant differences depending on whether they had high- or low-efficiency MAO-A alleles. Maltreated kids with efficient MAO-A activity weathered the storms of their youth relatively well. However, those with inefficient MAO-A activity developed significant antisocial problems—85 percent of those with a low-activity MAO-A genotype who were severely maltreated developed some form of antisocial behavior. That result was twice as high as the high-activity group under severely maltreated conditions. It was thought that deficient MAO-A activity disposes the kids toward neural hyperreactivity to threat.

How, precisely, might the genes operate differently under different environmental conditions? As mentioned earlier, stress might cause an increase in certain chemicals that in turn cause the DNA copiers to jump their tracks and begin copying from different parts of the DNA strand. This makes slightly different proteins, which in turn cause the properties of the synapses to subtly shift.[43] This type of effect, where a particular allele is problem-free unless the environment (or another gene) kicks it off track, might happen with many different personality-related genes.[44]

Other Moody Genes

A number of other genes also affect mood, although our understanding of how is limited at present. Tryptophan hydroxylase (TPH), for example, is an enzyme that helps in synthesizing serotonin. Various types of TPH and their associated genes appear to be associated with a number of different psychiatric and behavioral disorders—including those, as we shall see, which relate to Machiavellian behavior.[45] Another gene—this one with the cryptic handle of D4DR—has been linked to novelty seeking. D4DR has a variable number of repeats in its nucleotide building blocks that can affect how quickly dopamine is metabolized by the body. The higher the number of repeats, the more novelty-seeking behavior a person seems to exhibit—and the more extroverted a person often is. Shorter forms may be associated with crankier personalities.[46]

And another recent discovery, the DARPP-32 gene, has been found to be associated with both optimized thinking circuitry and, sadly, increased risk of schizophrenia. Daniel Weinberger explains: "Our results raise the question of whether a gene variant favored by evolution, that would normally confer advantage, may translate into a disadvantage if the prefrontal cortex is impaired. . . . Normally, enhanced cortex connectivity with the striatum would provide increased flexibility, working memory capacity and executive control. But if other genes and environmental events conspire to render the cortex incapable of handling such information, it could backfire—resulting in the neural equivalent of a superhighway to a dead-end."[47] It turns out that DARPP-32 is associated with depression and substance abuse as well as with schizophrenia.

Still more genes relate to the hormones vasopressin and oxytocin and help produce the feelings of love we may feel for others.*[48] Perhaps, when set on "high," these genes help produce the kind of person who continually, gullibly forgives all manner of purposeful emotional and physical abuse.

In the end, *all* of the genes mentioned above, and many others as yet unknown, could prove important in any number of personality traits. It will be interesting to watch developments as the field of behavioral genetics unfolds.

EMERGENIC PHENOMENA

A particularly significant concept that we must not ignore here is something called *emergenesis*. This term refers to genetic traits that, surprisingly, *don't* commonly run in families. Examples of this might include leadership, many different types of genius, and psychopathological syndromes like psychopathy and borderline personality disorder.[49] The way this occurs isn't all that difficult to understand. Let's say, for example, that a brilliant executive, known for his extraordinary leadership skills and visionary sense of business, along with his petite wife, who had won a gold medal in the Olympics as a teenage gymnast, has a large family of ten children. What might we expect?

The children would each receive half their genes from their

---

*Indeed, since the first edition of *Evil Genes*, a study by researchers at the Hebrew University in Jerusalem has demonstrated a tentative link between ruthless behavior and variations in a gene that produces vasopressin receptors. For this study, roughly two hundred students had their DNA sampled and were then asked to play a game that (unbeknownst to the students) was called the Dictator Game. Students with shorter versions of the vasopressin receptor gene (AVPR1a) were more likely to behave selfishly.

mother, and half from their father, although there's no telling beforehand what scrambled mixture they might receive. Let's say that part of the father's business acumen and leadership skills relate to an outstanding memory provided by a *val/val* BDNF set of alleles, coupled with a calming long/long serotonin transporter (SERT) allele set. (Of course, there are many other genes that play a role here, but we're simplifying this just to make the point.) The executive's fortunate confluence of BDNF and SERT alleles are part of why he thinks fast and can remain unflustered no matter what might arise. The gymnast wife, on the other hand, may have a short/short serotonin transporter molecule, making her routinely more anxious, but her *met/met* BDNF alleles mean she's still fairly easygoing, although they also ensure her memory is nowhere near as sharp as that of her husband.

With this setup, *none* of the ten children would have any real chance at all of having precisely the same four genes as either of their parents. Just the slight difference in these genes might be enough to take the edge off of each child's ability to replicate either of the parents' successes (or, if the scenario were changed—their failures). And that's for those four genes alone. Multiply this by the as yet unknown number of total genes that have substantive effects on any particular personality trait—two hundred? Two thousand? And remember, there's also that, as yet barely understood, control information hidden in the junk DNA. You couldn't possibly have a chance at replicating precisely the same personality-related genome that the parent has into a child—although you could definitely inherit a confluence of genes that would allow you to show some key factors, including impulsivity, anxiety, intelligence, or extraversion. Put differently, if your father were a Leonardo da Vinci–caliber genius, chances are you'd be pretty smart but no genius yourself.

Inheritance of personality is particularly touchy because other, seemingly nonpersonality-related genes also factor into the equation. Even if a child is given nearly the identical personality underpinnings as the parent, her entire life might be different, for example, if she also happened to inherit a difficult-to-battle tendency toward obesity. Or perhaps she might have inherited genes that gave her a particular delight in music or skill at dancing, or predisposed her toward preternatural enjoyment of the buzz of alcohol, or gave her a low ability to maintain focus even on things that interest her.[50] And of course environment can play a role too. Even within the same family, she might have been treated very differently than her older sister was. Or maybe her mother played with a cat and was infected with *toxoplasma gondii* before she even knew she was pregnant, setting up the conditions for her daughter to become schizophrenic.[51] Or perhaps she was kidnapped and brutally raped as a nine-year-old and dumped bloody and bruised by the side of the road.

All of these reasons, both genetic and environmental, are why virtually all parents with more than one child are amazed at how their children could be so very different. ("Night and day" is the common refrain.) Only identical twins could be expected to have nearly precisely the same genotype—and even that genotype, of course, will be quite different from that of the parents.

Why We Can't Simply Eliminate "Evil" Genes

Some studies have shown that if one of your first-degree relatives has the emotionally unstable condition known as borderline personality disorder, you've got an 11 percent possibility of having the disorder yourself—a substantial increase over the 1 to 2 percent chance for the disorder in the general population, but far from absolute certainty.*[52] The concept of emergenesis explains this spotty evidence for heritability. Having full-blown borderline personality disorder probably requires inheriting just the right confluence of genes related to very different traits involving cognitive dysfunction, mood disorder, and impulsivity, and often needs a final spark from a stressful environment

*Twin studies show that if one identical twin has full-blown borderline personality disorder, there is a 35% chance the other has it, while fraternal twins have only an 8% chance of sharing the disorder. Subclinical borderline personality disorder, on the other hand, showed a concordance of 38% for identical and 11% for fraternal twins.

An Emergenic Prodigy Speaks of Jewish "Smarts"

Norbert Wiener was a child prodigy who received his doctorate from Harvard at age eighteen; he would go on to discover important mathematical properties related to communications, robotics, computer control, and automation. Wiener's father claimed that it was his training methods alone that had made his son so brilliant—otherwise Norbert would have been a perfectly ordinary child. Given the fact that the three other children in the family were not especially gifted, Norbert, with his emergenic genius, found his father's statements extraordinarily galling.[53] In tandem with his intellectual brilliance, Norbert was hypersensitive to criticism and subject to fits of depression.

Due to fear of anti-Semitism, Wiener's Jewish heritage was kept a secret from him as a child—afterward he became interested in Jewish culture and heritage. (Actually, Wiener was fascinated by virtually everything.) Anticipating research findings that would come a half century later, Wiener wrote:

> Let me insert here a word or two about the Jewish family structure which is not irrelevant to the Jewish tradition of learning. At all times, the young learned man, and especially the rabbi, whether or not he had an ounce of practical judgment [shades of the COMT intellect-emotion trade-off] and was able to make a good career for himself in life, was always a match for the daughter of the rich merchant. Biologically this led to a situation in sharp contrast to that of the Christians of earlier times. The

to set things afire. Indeed, studies have shown that the borderline traits of affective instability and impulsivity run in families.[54] But since each of the unlucky borderline traits involving impulsivity, mood disturbances, and cognitive dysfunction is, for the most part, separately heritable, it's unlikely that you'd inherit every one of them, even if one of your parents were to be borderline. We have reason to believe that the spotty heritability situation is similar with antisocial personality

> Western Christian learned man was absorbed in the church, and whether he had children or not, he was certainly not supposed to have them, and actually tended to be less fertile than the community around him. On the other hand, the Jewish scholar was very often in a position to have a large family. Thus the biological habits of the Christians tended to breed out of the race whatever hereditary qualities make for learning, whereas the biological habits of the Jew tended to breed these qualities in. To what extent this genetic difference supplemented the cultural trend for learning among the Jews is difficult to say. But there is no reason to believe that the genetic factor was negligible.[55]

Wiener had no way of coming up with a more concrete, testable model for his speculations. But in 2005, scientist Gregory Cochran, working independently of any institution, along with Jason Hardy and Henry Harpending of the University of Utah, finally proposed just such a model. Cochran and his colleagues suggested that a group of European Jews known as the Ashkenazi commonly carried several genetic mutations that explained their naturally high intelligence. (Ashkenazim score about twelve to fifteen points above the average of one hundred on IQ tests—the highest of any group of humans.) A single dose of the novel alleles can increase the growth and number of connections between nerve cells. As with many seemingly beneficial genes, there are tradeoffs. Those with a double dose of the allelles can wind up with neurological disorders such as Tay-Sachs disease, as well as with cancer.[56] Of course, the above findings are highly controversial.

disorder (which can actually arise from many different causes, some of which are solely environmental), and even psychopathy.

But just what is the difference between, say, the mind of a psychopath and that of a normal person? We are at the cusp of cutting-edge technology that allows us to see. And seeing the differences between psychopathic and normal neurological processes is a crucial step in allowing us to begin to understand the underpinnings of some types of Machiavellian behavior. As we shall discover, understanding the "why" of Machiavellian behavior will also help us to understand why there are Machiavellians at all—and why some of them are so successful.

Chapter 4

Using Medical Imaging to Understand Psychopaths

"In the beginning, there was nothing. And God said, 'Let there be Light.' And there was still nothing. But you could see it."

—Anonymous

Looking Inside the Brain of a Psychopath

"Parents of violent kids think, 'What did I do wrong?'" says Adrian Raine, a "Penn Integrates Knowledge" professor who holds joint appointments in criminology and psychiatry at the University of Pennsylvania. "When the kids come from a good home, the answer may be absolutely nothing. A biological deficit may be to blame."[1]

And now we are beginning to be able to "see" the deficit.

One of the most powerful methods of seeing inside the brain involves magnetic resonance imaging—often shortened to MRI. In this technique, water molecules are tickled to cause them to burp out electromagnetic waves that are only slightly different in frequency from the electromagnetic waves our eyes see as light. Since different types of tissue have different amounts of water, we can use clever technical legerdemain to "see" the different tissue types on photographs that look very similar to x-ray images.

A complete slice of the brain can be seen—"imaged"—in the same way that you are scanning your eyes across this page. Since an image can be produced very quickly (only twenty milliseconds—about one-fifth of the time it takes to blink your eye), it is possible to produce a number of MRI pictures one right after another as if filming a movie. In this fashion, as cells and molecules carry out the work of the brain, the differences between images can clearly be seen. This video version of MRI is called *functional* MRI, or fMRI. MRI compares to fMRI in the same way a picture of you posed with a cheesy smile in your sixth-grade class picture compares to an old home movie of you throwing a snit. You can see what you looked like from the still picture, but you can tell a lot more about how you functioned from the home movie.[2]

Emotion and Language

The following illustrations, from an fMRI study by Yale psychologist and neuroscientist Kent Kiehl and his group, reveal that many regions of the brain in emotionally excited psychopaths function differently than in normal people. The upper two figures of the "generic brain" show areas, highlighted in white, where criminal psychopaths had much less activity than normal subjects when repeating emotionally charged words like *blood*, *sewer*, *hell*, and *rape*.[3] These neural areas relate to limbic and paralimbic levels—very old parts of the brain in evolutionary terms that are also found in fishes and reptiles. The bottom two figures, on the other hand, show areas of the generic brain where criminal psychopaths had *greater* activity (also highlighted in white) related to emotion than normal subjects. This increased activity

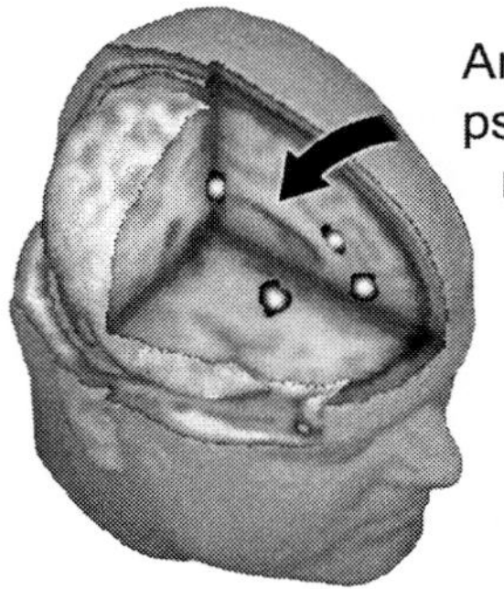

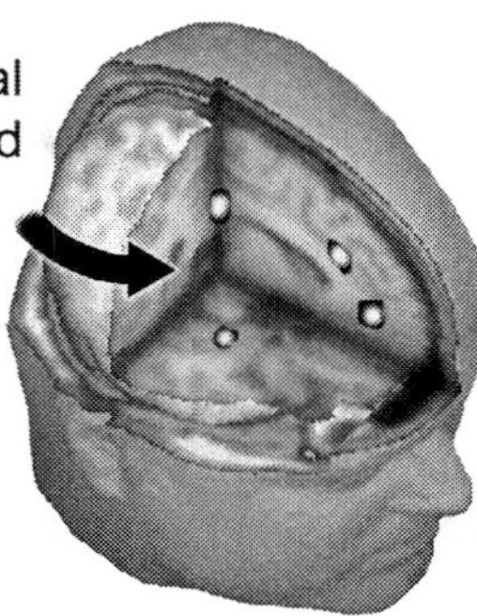

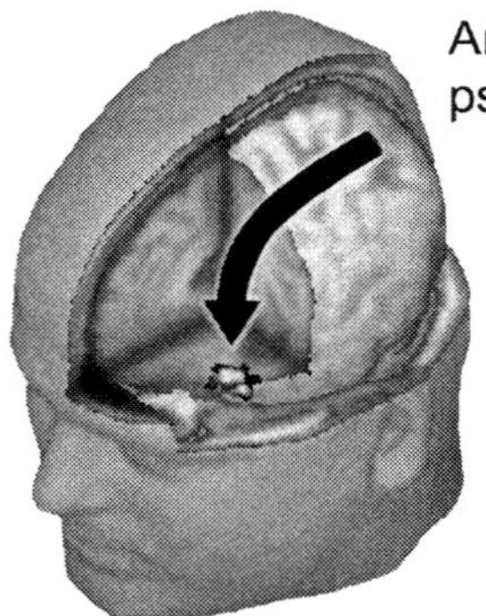

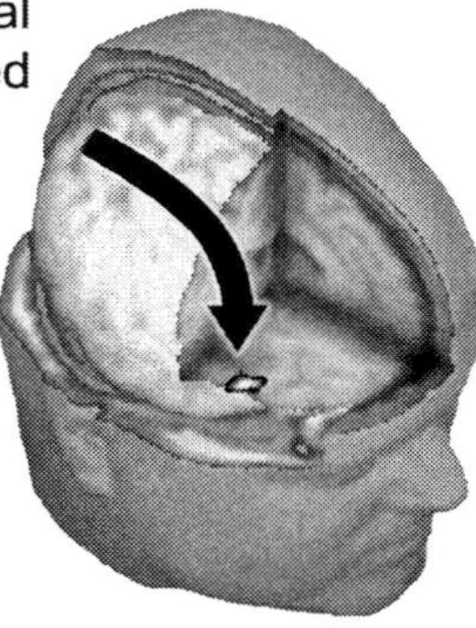

Fig. 4.1. Four different views of a psychopath's brain. The white spots are the areas of interest.

may be caused by the weak input from the limbic system, which forces psychopaths to use alternative neural structures to grapple with emotional information.[4]

Finding neural differences between psychopaths and normal people looks as simple as snapping a picture with a fancy camera—and in some sense, despite the acrobatics involved in processing the data, it *is* simple. But occasionally it's nice to step back and admire the intricate logistics behind these types of experiments. For the images we just saw, simply being able to get the psychopaths to where they could be imaged was no easy matter. The eight psychopaths involved in the study were hauled under heavy guard fifty miles from the maximum-security prison in Abbotsford, British Columbia, to the University of British Columbia Hospital's magnetic resonance imaging unit. As Kiehl wryly understates: "The work is never boring."[5]

Low Response to Threatening Stimuli and Empathy Impairment

Another study has shown that a psychopath is likely to have a weird corpus callosum—the nerve superhighway orchestrating the flow of information between the brain's two hemispheres.[6] The white matter volume of the corpus callosum of psychopaths is nearly a quarter larger than normal, but, like pulled taffy, the highway's length is increased while its thickness is decreased. These differences in neural structure seem to go hand in hand with the curious gaps in emotion and creepy interpersonal skills of psychopaths, along with their low involuntary reactions to stress. Why might the corpus callosum abnormalities cause these personality differences? It might be because a poorly structured corpus callosum inhibits communication between the two halves of the brain—in particular, it doesn't do its normal job of allowing the left brain to inhibit and control the negative moods that can be generated by the right hemisphere. This may cause the expression of the aggressive, unregulated behavior that psychopaths can display.[7]

Researchers feel that corpus callosum abnormalities may be caused by a problem in the development of a psychopath's brain. Pruning—a natural, internal process of tidying up the brain by shearing off unused neurons—may not take place properly. Or the neurons may be too heavily encased in sheaths of myelin—insulating trestles of glial white matter.

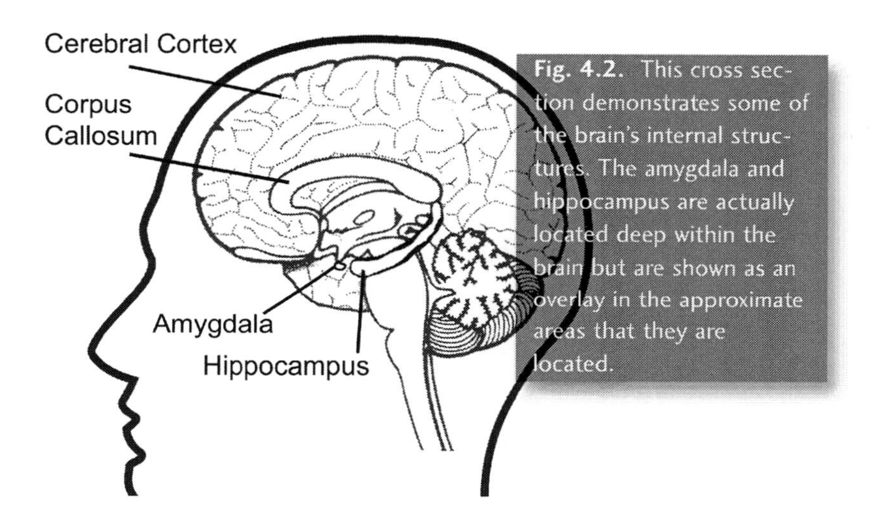

Fig. 4.2. This cross section demonstrates some of the brain's internal structures. The amygdala and hippocampus are actually located deep within the brain but are shown as an overlay in the approximate areas that they are located.

Lack of Anxiety

The amygdala is actually an almond-sized pair of organs deep in the brain that cause a knee-jerk "fight-or-flight" reaction when people are startled. In some circumstances, psychopaths have lazier amygdala than normal people, as was discovered when a group of psychopaths was exposed to the smell of rotten yeast (stinky smells as a rule cause the amygdala to jerk to attention). Researchers think that routinely blasé amygdalae and other related structures in psychopaths may leave them feeling restless, spurring them to raise hell just for the resulting excitement. The psychopaths' nonchalant neural reactions may also be related to their strange lack of fear—which also appears to have a genetic basis.[8] Lack of fear may in turn throttle the development of a psychopath's conscience. Some researchers feel that amygdala dysfunction lies at the very core of psychopathy.[9]

Executive Dysfunction

Peculiarities resulting from injuries or congenital differences in the area of the brain behind the forehead, especially in the orbitofrontal and ventromedial cortices near the eyes, seem to be the cause of the "executive dysfunction" often seen in antisocial children and adults.[10] This means that antisocials often have trouble organizing their behavior. For example, a typical waitress job, which might involve fielding requests from a dozen different tables while interacting with the hostess, cooks, and busboys, would be difficult for someone with noticeable executive dysfunction. Orbitofrontal dysfunction in particular seems to release the normal brakes on aggressive and hostile impulses, while dorsolateral dysfunction contributes to an inability to learn from punishment.

Adrian Raine has conducted many studies that have done much to

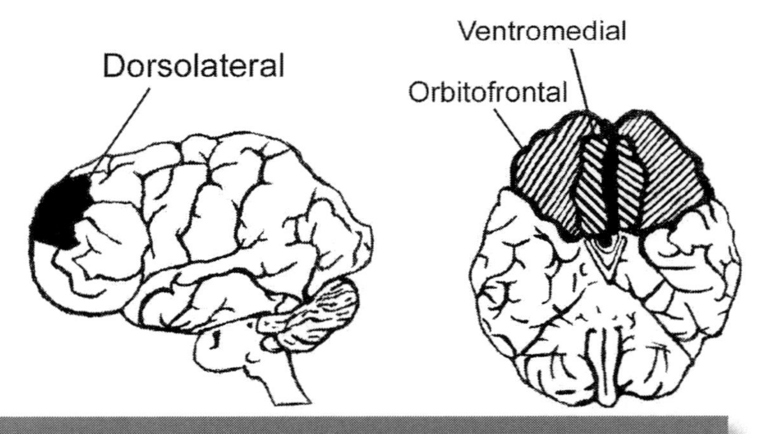

Fig. 4.3. Damage or dysfunction in these three areas of the prefrontal cortex—the dorsolateral, orbitofrontal, and ventromedial cortices—appears to be strongly associated with antisocial behavior. All of the areas indicated are toward the front of the brain, near the eyes. The left picture is a side view, while the picture on the right shows a view looking up toward the brain from its bottom.

revolutionize our understanding of the neurological foundations of psychopathy. Raine has the seemingly mild-mannered look of an accountant—indeed, he *was* an accountant. His first two years in the workforce were as an airline accountant for British Airways. However, after receiving a doctorate in psychology, Raine gained a seasoned understanding of psychopathy by spending four years working as a prison psychologist in two top-security prisons in England. A key to the depth of Raine's research breakthroughs is not only his hardened "real-world" background, but also the way he combines knowledge from this background with an astonishing breadth of hard and soft science academic interests, which include brain imaging, psychophysiology, neurochemistry, antisocial behavior, schizotypal personality, and alcoholism.

"Why doesn't everyone assault others or act violently?" asks Raine.

> One reason is that most of us are good at fear conditioning and we've been punished in childhood for doing minor things like stealing or hitting friends. So we've learned the association between antisocial behavior and punishment and therefore feel fear when we even contemplate an antisocial act. But not everyone is able to form these conditioned responses with equal facility. While some people have biological systems that make it easy, others have biological systems that make it hard. If you are an individual whose right orbital cortex is not functioning well, you're biologically disadvantaged in developing a conscience.[11]

Oddly enough, one study has shown that murderers who have a normal family upbringing have even lower function in their right orbitofrontal cortical areas than murderers who were abused during childhood. Perhaps murderers "without a psychosocial 'push' toward violence require a greater neurobiological 'push.'"[12] In other words, children with less severe neurological problems may be helped by having a normal upbringing—but children with more severe neurological difficulties may not be.

Kind and conscientious people who suffer brain damage in ventromedial areas can suddenly have a complete change in personality and begin to act like psychopaths. Despite seemingly normal intelligence, these "pseudopsychopaths" are often found to have subtly impaired reasoning skills. For example, they may continue to make bad choices in rigged experimental card games even after they understand intellectually how the game is rigged and how they could easily play to win.[13]

Emotional Control— Affective and Predatory Murderers

Imaging is not only allowing us to see into the brains of psychopaths, it is also providing clues about the neural circuitry that underlies their motives. For example, murderers have long been divided into two types: affective and predatory. Affective murderers murder under the influence of emotion. They show little planning to their passionate acts, which often take place in domestic settings. A typical example of an affective murderer is actor Marlon Brando's son Christian, who, while in a drunken rage, murdered the abusive boyfriend of his pregnant, mentally disturbed sister. Predatory murderers, on the other hand, are coldly unemotional, far more controlled, and are more likely to attack a stranger, often using a carefully planned setup. Richard Kuklinski, who "whacked" people for the Gambino crime family, is a good example of a predatory murderer. He perfunctorily explained his motives for killing nearly a hundred people on the HBO film *The Iceman*: "It was due to business." (The first might be seen as committing a crime of passion, while the second is premeditated.)

The two very different sets of neural circuitry involved in affective versus predatory murderers are thought to be similar to the two different sets of circuitries seen in animals such as cats. The affective circuit, for example, is activated if Fluffy the cat is trapped by a barking dog—Fluffy's hackles are raised, her back is arched, and she hisses and spits loudly, pinning her ears back and displaying her teeth. If you are unwise enough to try to pick Fluffy up at this point, she is likely to mindlessly attack you. A completely different set of circuits is involved in predatory behavior, as when Fluffy is stalking a bird. She is snake-silent during this time, with ears and eyes focused; she moves slowly and close to the ground. You can safely pick up Fluffy when she's in this predatory mode, although her paws might pinwheel as she still tries to move toward her target.[14]

As expected, human imaging of affective and predatory murderers has revealed profound differences in how their brains operate—similar to the neural circuit differences known to operate in Fluffy the cat. The affective murderer shows lethargic activity in the prefrontal cortex, the area that normally reins in impulses.[15] The predatory murderer, however, shows good functioning in the prefrontal cortex, which is, as Raine slyly notes, "consistent with the role of an intact prefrontal cortex in allowing him to regulate his behavior for nefarious ends."[16] In short, the predatory murderer is well aware of what he plans to do. Emotion-related areas of the brain (specifically, the midbrain, amygdala, hippocampus, and thalamus) in both sets of murderers were found to be metabolically turbocharged compared to normal people's.* It's thought that the revved-up, party neural atmosphere predisposes both groups of murderers to an aggressive temperament. But only the predatory murderers are able to channel their aggressive impulses into slow-motion torture through bullying, manipulation, deceit—or carefully planned murder.

Problems with Abstract Reasoning

The latest imaging results are showing that psychopaths don't just have dysfunction in neural areas related to emotionally based traits such as aggression and impulsivity. They have trouble processing

---

*Some readers may have noticed that researcher Kent Kiehl's psychopathic subjects showed a *decreased* limbic metabolism while Adrian Raine's subjects showed an *increased* limbic metabolism. This is because Raine's subjects were murderers—not psychopaths. Additionally, Raine's study included those who may have been psychotic (as with schizophrenia), or had other disorders, while Kiehl's study participants had no history of psychosis. This means the two studies cannot be directly compared.

abstract concepts altogether. The next illustration shows the area of the brain where the hitch seems to occur. (For those who need a name, it's the *right anterior superior temporal gyrus*, which means the surface fold on the right, in the front, toward the upper part of the temples of the brain.) Here, neurons in normal individuals are more active than neurons in psychopaths when hearing abstract words such as *justice*. However, when concrete words such as *table* are heard, both normal and psychopathic brains function the same way.[17] This agrees with evidence from other studies proposing that psychopathy is related to abnormalities in the right side of the brain—the artsy, abstract side that, among other tasks, synthesizes the big picture and understands nonverbal cues related to emotion.

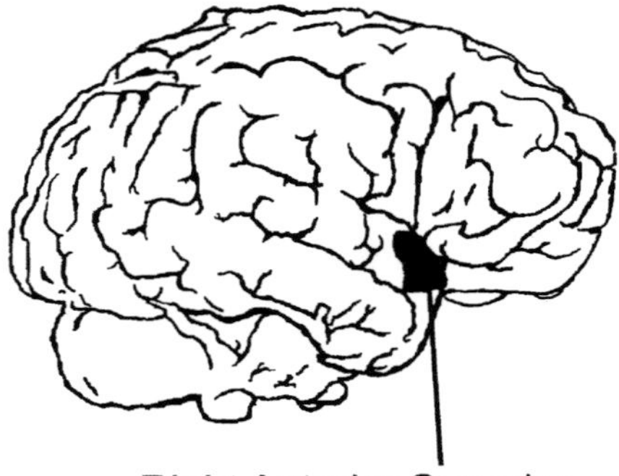

Fig. 4.4. In the area shown, normal individuals showed much more activity than psychopaths in relation to processing abstract ideas.

The Disease of Too Much Trust—Williams Syndrome

Williams syndrome—what might be termed "antipsychopathy" —is perhaps the most endearing of all diseases. Those afflicted are very polite and sociable, show great empathy, and are completely unafraid of strangers. A trained geneticist can instantly pick out the upturned nose, wide mouth, full lips, and the long distance between the nose and the upper lip of a William's syndrome

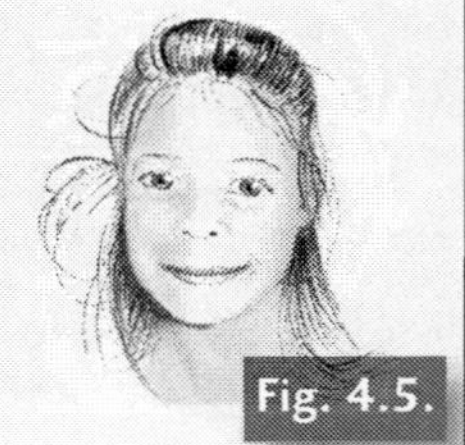

Fig. 4.5.

child. Such children often have heart or blood vessel problems, as well as dental and kidney abnormalities.

The disease is thought to be related to unusual functioning of a genetically controlled neural circuit that handles human social behavior. It is caused by the loss of a tiny snippet of roughly 21 genes on chromosome 7. In Williams syndrome patients, the amygdala, which sparks "fight-or-flight" responses, shows an unusually easygoing response to threatening faces. On the other hand, threatening scenes with no people, such as burning buildings or a plane crash, provoke overly powerful responses, meaning that Williams syndrome sufferers are uniquely "people persons" even as they are afflicted with unending phobias and worries about everything from spiders to heights. Another area, the medial prefrontal cortex, is perennially activated in Williams syndrome patients. This area, which is right behind the center of the forehead, has been associated with empathy and knowledge of how to interact socially, which may explain the heightened capabilities of Williams syndrome patients in these areas.[18]

It appears that the missing snippet is composed of "patterning genes" that tell the brain how to grow. Normally, as a baby's brain develops, there is a push-pull between the dorsal and ventral areas. The dorsal areas relate to mathematics, space, and recognizing the intentions of others; while the ventral areas relate to language, emotion, and social drive. Even in normal people, one area usually grows larger than the other—which is why people are often better at either math, or language, but not both. In Williams patients, however, the ventral area goes into overdrive. The excessive growth in this area directs these patients towards hyper-sociability and rich ability to process emotion, even as they are left struggling with concepts of number and space.[19] It is possible that research on Williams syndrome may provide a better understanding of the overly trusting behavior of some individuals in the presence of Machiavellian and duplicitous behavior in others.

How might these and related abnormalities lead to psychopathic behavior? It may be that the difficulty psychopaths have with processing abstract concepts also means they have trouble processing complex and abstract social emotions such as love, empathy, guilt, and remorse. Therefore, it may be difficult for psychopaths to understand or control behavior involving these areas.

SEEING THE HUMAN CONSCIENCE

Modern medical imaging is showing that the human conscience, a sense of morality, and ethics—all the things psychopaths seem to be missing—aren't just the playthings of philosophers anymore. In fact, neurological regions related to morality itself have been imaged. In 2002, Brazilian researcher Jorge Moll and his colleagues published the results of a functional magnetic resonance study where people were shown pictures that normally evoke a sense of morality, such as abandoned children, physical assaults, and victims of war.*[20] Scanned images of their brains after viewing these scenes were compared to images produced after viewing benign scenes and scenes that triggered disgust rather than morality. Areas activated by moral rather than everyday conditions included critical regions for social behavior and perception. Morality, in other words, involves a tangible neurological process.

Another study, by neuroscientist Scott Huettel of Duke University Medical Center, showed that more naturally altruistic people have ramped-up activity in the *posterior superior temporal cortex*.[21] This is the area that is related to perceiving others' intentions and actions. Altruism, it seems, may be founded on our understanding that others have motivations and actions that are similar to our own. Huettel explains: "Perhaps altruism did not grow out of a warm-glow feeling of doing good for others, but out of the simple recognition that that thing over there is a person that has intentions and goals. And there-

---

*Several years before, a journal reviewer rejected one of Moll's first neuroscience papers on the topic, saying that "morality could never be a topic for neuroscience—it was for philosophy."

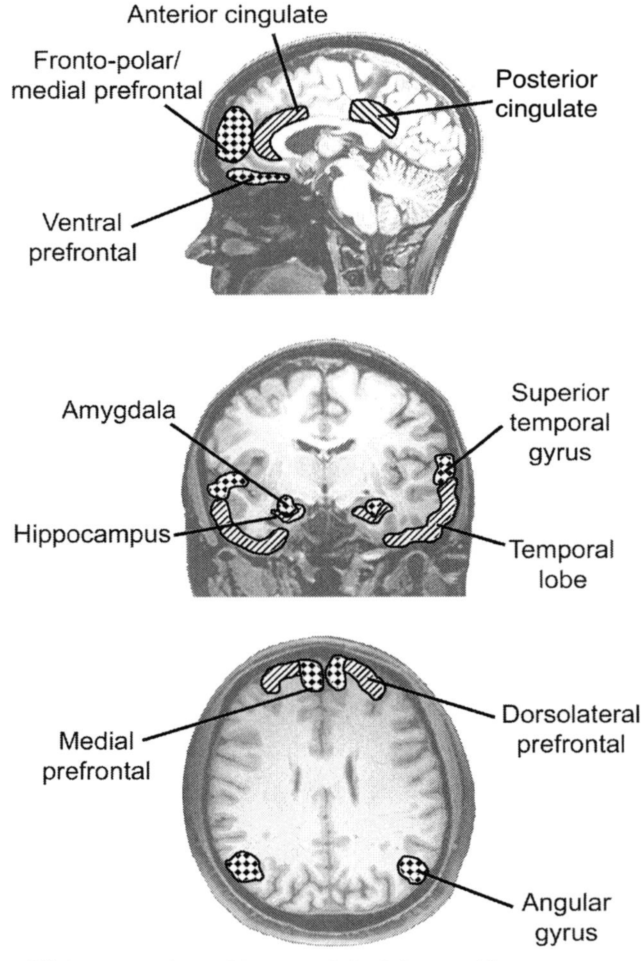

Fig. 4.6. This image, by neuroscientist Adrian Raine and his colleague Yaling Yang, highlights the overlapping areas of moral reasoning and psychopathic dysfunction.

fore, I might want to treat them like I might want them to treat myself."[22] Yet another study showed that the emotions of guilt and shame, like feelings of altruism, activate specific areas of the brain. Shame differs from guilt by a dash of stronger activation in a few additional areas.[23] Even the processes of resentment and forgiveness appear to be related to specific sets of neural circuitry—circuitry that can go awry as psychopathology increases.[24]

It seems that many of the areas that are activated when a person is involved in moral reasoning overlap those that are dysfunctional in psychopaths, as shown in the preceding figure. Impairments to these areas produce impairments in the emotions that comprise the *feeling* of what is moral. Indeed, psychopaths know intellectually what is immoral—they just don't have a feeling of immorality about it.[25]

Why Does Psychopathy Develop?

There are dazzling numbers of theories about how and why psychopathy develops, but most of them involve the idea that children with psychopathic traits have neurological glitches that reduce their moral reasoning and empathic concern for others—the emotional building blocks of conscience. Children with the callous, unemotional traits of psychopathy, along with youth who show the chronic misbehavior of conduct disorder (which has been linked to the development of antisocial personality disorder as the child grows into adulthood), often show additional traits of impulsiveness and narcissism. Such children aren't bothered by the hurtful and even shocking effects of their behavior on others and are less able to recognize expressions of sadness on the faces and in the voices of other children.[26] Forensic psychologist J. Reid Meloy relays the heartbreaking story of a mother of a psychopath that captures the early onset of the disorder's essence:

> "At 18 months [said the mother], it was as if a switch had gone off in him. He started showing tremendous rage, complete lack of

> remorse and an almost complete lack of empathy. His first reaction, when he would see an animal, would be to kill it. He became extremely hateful and vicious." This mother went on to describe his affective and predatory behavior toward her, including starting fires in the house, threatening her with a knife, and sticking straight pins out of the carpeting in front of her dresser, out of her pillow, and into her clothes so they would scratch her when she put them on. . . . As her child has grown older, sadistic behavior is more evident. Once he hanged a cat in the backyard and waited for his mother to come home to watch her reaction. She remembers seeing his pleasure at her horror, and then imitating her horror back to her.[27]

One particularly important theory about the cause of psychopathy has been developed by Joseph Newman, a University of Wisconsin psychologist who has spent the last twenty-five years investigating the wheres and whys of psychopathy. Where many researchers have focused on the lack of fear and other emotions present in psychopaths—hypothesizing that this leads to a psychopath's indifference to the feelings of others—Newman has pursued a very different idea. He believes that "psychopathy is essentially a type of learning disability or 'informational processing deficit' that makes individuals oblivious to the implications of their actions when focused on tasks that promise instant reward."[28] Newman's work is also important in that it is helping us to better understand the different types of psychopathy. One fascinating implication is that psychopathy may not necessarily be related to violence. Forensic psychologists working with violent psychopaths in high-security prisons might be incensed by these ideas, but Newman points out that if you are looking at psychopaths in high-security environments, *of course* those psychopaths show violent characteristics. (It's a bit like the old story of looking for the keys under the streetlight because that's where the light is.) But there are also psychopaths in minimum-security prisons—these individuals show less violent characteristics. "The essence of psychopathy," Newman notes, "is not the violence, or the forgery, or the sex crime, or the many miscellaneous minor crimes, it is the psychobiological process that diminishes regulation and behavior change."[29]

Overall then, rather like intelligence itself, it appears that moral reasoning is controlled by various neurological features of the brain—features that develop differently, if at all, in children with a genotype that predisposes them toward psychopathy. Just as a child needs the neurological structure of the eye to process information from the electromagnetic fields that shimmer through the air around him, a child also needs the structure of the orbitofrontal cortex and related neurological features to have a feeling of compassion. Psychopaths, it appears, may be born preprogrammed with a tendency to grow up "morally blind."

Empathy and Mirror Neurons

Empathy—identifying with and understanding another person—appears to be related to a distributed complex of neural units that are primed in part by mirror neurons. These neurons, believed by many to be the greatest neurological discovery of the 1990s, are triggered not only when humans perform an action but also when a person witnesses another person performing the same activity. Thus far, mirror neurons have been found in the premotor cortex and the inferior parietal cortex, and it is thought they may be located in additional areas of the human brain. It may be that these types of neurons have evolved to allow animals to understand what another animal is doing or to recognize another's action. It appears, however, that human mirror neurons are far more flexible and highly evolved than neurons found in any other animal. Dr. Marco Iacoboni, a neuroscientist who specializes in researching mirror neurons, notes:

> When you see me perform an action—such as picking up a baseball—you automatically simulate the action in your own brain. . . . Circuits in your brain, which we do not yet entirely understand, inhibit you from moving while you simulate. But you understand my action because you have in your brain a template for that action based on your own movements. . . . And if you see me choke up, in emo-

tional distress from striking out at home plate, mirror neurons in your brain simulate my distress. You automatically have empathy for me. You know how I feel because you literally feel what I am feeling.[30]

On the other hand, people with autism often display no firing in their mirror neurons in response to the activities of others. Researchers believe that the "broken" mirror neuron system of autistics lies at the heart of their difficulties with social interaction and lack of empathy, as well as their inability to imitate.

The role of mirror neurons in the development of psychopathy is as yet unclear, since psychopaths seem to have no difficulty *comprehending* the mental state of others. Psychopaths instead seem to have trouble *empathizing* with the emotional signals of others, particularly signals related to fearfulness and sadness.[31] Researchers Linda Mealey and Stuart Kinner have pointed out that psychopaths may simply have trouble projecting onto others feelings they don't experience themselves. One psychopathic rapist commented on his victims: "They are frightened, right? But, you see, I don't really understand it. I've been frightened myself, and it wasn't unpleasant."[32]

Successful Psychopaths

What is perhaps even more disquieting than ordinary criminal psychopaths are "successful" psychopaths, as portrayed in Adrian Raine's findings. By recruiting and questioning ninety-one men from a Los Angeles temporary employment pool, Raine was able to gather a small group of subjects who could be divided into two subgroups: successful psychopaths—those who admitted committing crimes but had never been caught, and "unsuccessful" psychopaths—those who had been caught. The hippocampus, which is important in memory and spatial navigation, played a key role in the study, because this walnut-sized organ is also crucial in the ability to learn both what to do and what *not* to do. Raine theorized that psychopaths with hippocampal impairments would have difficulty learning the obvious (at least to everyone else) cues that would keep them out of trouble.

Sure enough, imaging results revealed that unsuccessful psychopaths had abnormal hippocampi, while successful psychopaths had normal hippocampi. A later study revealed another difference: unsuccessful psychopaths, but not successful psychopaths, had only three-quarters the usual volume of prefrontal gray matter. This area involves the part of our brain that is associated with the ability to determine good from bad; it also suppresses urges that could lead to socially unacceptable or frankly illegal outcomes.[33]

Another of Raine's studies, one that typifies his interwoven knowledge of neurology and the social sciences, sheds light on the structural pathology associated with habitual lying. This study also drew on recruits from temporary employment pools—the recruits were interviewed to find a group of twelve with a history of conning and manipulative behavior, as well as telling lies—for example, to obtain sickness benefits. This group was contrasted with twenty-one normal individuals and a third group of sixteen with antisocial personality disorder but no history of pathological lying. The study showed that prefrontal white matter in pathological liars was a quarter larger in volume than in normal people. Raine and his group had uncovered the startling fact that white matter in the prefrontal cortex is centrally involved in the desire to lie.

How could increased white matter contribute to a deceitful personality? Raine's doctoral student, Yaling Yang, who actually conducted the study, says: "It may just be easier for them to tell lies because the excessive white matter creates an abundance of connections among otherwise contradictory compartmentalized data."[34] Interestingly, children with autism have decreased white matter accompanied by increased gray matter; autistics are also known for being less given to lying.

Robert Hare, who has studied psychopaths for more than a quarter of a century, has recently expanded his studies of criminal psychopaths to include what he calls corporate psychopaths, who are similar in concept to Raine's "successful psychopaths." Hare and his collaborator, Paul Babiak, have developed a questionnaire to help identify these *Snakes in Suits* (the title of their book about their studies). Babiak points out: "The psychopath is the kind of individual that can give you the right impression, has a charming facade, can look and sound like the ideal leader, but behind this mask has a dark side. It's this dark side of the personality that lies, is deceitful, is manipulative, that bullies other people, that promotes fraud in the organization and steals the company's money."[35] Hare points out that knowing your boss is a psychopath can help you to survive: "The most important thing is to be aware," Hare says. "Once you take that position you are in a better position to deal with them."[36]

The upshot of all this, of course, is that there might be a number of unincarcerated psychopaths free among us—unfeeling monsters but with the intact intelligence to avoid detection. Might this help explain why some Machiavellian individuals, like Hitler, Father Geoghan, or Saddam Hussein, were able to pass unsuspected in human social structures?

Perhaps.

My sister Carolyn's last diary entry provides food for thought: *"Back to the real world after panic attack. Must ease Jack out. . . . He is still a good friend to have."*

Panic attack. Good friend.

These are the heartfelt words of a deeply Machiavellian woman—private thoughts, shared with no one, reflecting her internal world. Did her use of the word "friend" carry a different meaning from ordinary usage? That didn't appear to be the case, judging from her final, nearly decade-long, on-off supportive relationship with Jack. Also, the worst psychopaths are typically coldly unemotional—so how could Carolyn be suffering from a panic attack? Maybe she was a lesser, "secondary" psychopath? But some of the details didn't quite jibe. Was McHoskey not entirely correct about Machiavellians being equivalent to psychopaths?

Carolyn was definitely Machiavellian. But if she wasn't a psychopath, well, what *was* she? And where does that leave the understanding of a myriad of other sinister individuals—from the lowly accountant who embezzles his firm's funds, to the nurse who fakes credentials, to those who rise far higher in the socio-politico-economic food chain? Some of the successfully sinister definitely carry the traits of psychopathy, just as Robert Hare describes with his "corporate psychopaths." But others, like Carolyn, show a more puzzling mix of both psychopathic and empathetic characteristics—her easy ability to lie, for example, was coupled with absolute adoration for her cats.

Do neurological disorders beyond psychopathy play a role, or an additional role, in their actions?

Perhaps it's time to peer into the big box my brother sent of Carolyn's belongings. I suspect it might lay the groundwork for better insight into the mind of a deeply Machiavellian—and tortured—soul.

Chapter 5

Insights from My Sister's Love Letters

"I'm really easy to get along with once you people learn to worship me."
—Anonymous

It is a muggy midsummer; a fluke thunderstorm has just turned the roof of the house into a raucous resonator. But I'm enjoying relative peace while the kids are gone, and have commandeered the kitchen for the weekend. Moving to the heavy thrumming of the raindrops, I've hauled the massive cardboard box in from the garage shelf where it's grown a fine sprinkling of dust over the past months. The odds and ends in it had caught my brother's eye while he was in Sequim, trying to find an estate handler to clear out Carolyn's apartment after her untimely, mysterious death. He had mailed the carton to me several months later, mentioning offhandedly that it had some old pictures and "stuff you might be interested in." Bless his heart.

The staccato rain overhead sharpens. I glance up at the skylight—the water sluicing thickly above me makes me feel like I'm standing in a noisy submarine.

Up periscope.

I wonder what Carolyn would think of my looking through her things.

Through the dark gray comes a trembling flash of white, followed instantly by a rifle crack of thunder. The cupboard doors bounce; the fluorescents quiver. I'm left momentarily in darkness.

Hi Carolyn.

The kitchen lights flicker again, then steady; the gray outside begins to recede. The rain scales itself back.

Carolyn's carton is heavy—about the size and weight of a midsize microwave. The dust is gritty. I cut through the duct tape over the top and pry open the flaps. The room fills with a musty, stale smell—I can't help but sneeze as I peer down curiously. Smaller boxes lie inside. Packets. Manila folders. Albums. More diaries. Everything has been neatly organized—the compartments of Carolyn's life.

But I discipline myself now and instead pick up Carolyn's last diary, which had come earlier, separately. After glancing at the final entry, I'd set the diary on the stack of books by my bedside. There it had sat, for months, always demoted to the position of next. I open a page randomly: Saturday, June 26th, 2004: *"Created bitchin' dinner—fillet mignon avec green peppercorn hollandaise, white truffle oiled potato, zabaglione with cherries."*

Carolyn did love to cook. I could never figure out if she actually ate, though. Fifty-eight when she died, she was thin as a skewer, her skin white as bleached bone. The bouffant blonde wig she wore made her look top-heavy—pencil-thin legs lagging as she arched her body forward, suspended over the crutches. It wasn't always like that. There were times, when she was in her teens and early twenties, when she even went to dances, her lopsided frug seeming almost natural.

The Early Years

My sister was born in 1945, more than a decade before me, after my father stopped flying World War II B-17s and began using the GI Bill to slingshot himself through veterinary school in Pullman, Washington. Pullman was where Carolyn was born. And that's where, at three, she took off one lazy summer day and grubbed in the dirt at the far end of the alley with neighbor kids. My mother swore that was how Carolyn caught polio, from "dirty kids," since polio spreads feces-to-mouth from people who generally don't even know they are infected.

But polio is both a disease of poor sanitation and good sanitation. In the dirty old days of open sewers and untreated drinking water, infants were exposed to poliovirus while they were still protected by their mother's own poliovirus antibodies. Thus, being infected with the poliovirus as an infant has a perversely protective effect. Maybe the virtual autoclave of tidy cleanliness my mother kept the house in also played a role—my then three-year-old sister had had no opportunity to catch the virus as an infant, when it would have done little harm.

Once Carolyn had unwittingly swallowed the virus, and it had established itself in her intestines, it seeped into her bloodstream. From there, it crept into her central nervous system, where it killed critical motor neurons, leaving her with flaccid limbs. Two of my maternal cousins also experienced the ravages of polio with subsequent paralysis and withered limbs, although neither shared Carolyn's strangely sinister personality. Polio, as it turns out, runs in families. Identical twins, for example, with their matched sets of genes, are much more likely to both catch polio than fraternal twins.[1] As Dr. Richard Bruno, the world's foremost expert on post-polio fatigue, writes: "You can't get polio if your genes don't allow it. Polio within families, going back generations, is very common."[2]

But by the time my brother and I came along, more than a decade later, Salk's polio vaccine was finally broadly available. It no longer mattered at all if we had a genetic predisposition for polio.

Saturday, April 3rd: "*Got to turn this around and will. The big clock debacle. Am so lucky to have these critters in my world. Leftover sandwich.*"

Each entry in this diary—her last—is short; mostly only a sentence or two. Carolyn wrote with a fine pencil—careful, sweeping strokes that caressed each word. I had admired her artistic skills as far back as when I was seven, in Lubbock, Texas. I sat beside her, watching as cartoon chipmunks flowed from her pencil and gamboled on the paper, as gleefully as if she'd released them from a cage. I would think: *I can hardly wait until I grow up, so I can draw like that!* No such luck.

Big clock. Debacle. Lubbock. My father's Air Force transfers are like my own big clock, making it easy for me to keep track of time. California was when I went to kindergarten. Oregon—first grade. Texas—second grade. Massachusetts—third grade. Pennsylvania—fourth grade. Tick tock. Lubbock is the only place I remember Carolyn living with our family. I remember how she'd hide when one boyfriend would come by, so she could go out with another. She liked to whisper secrets to others, ostentatiously cupping her hand so my little brother and I would notice.

Carolyn told me later that she used to babysit for us in Lubbock all the time. That she practically *was* our mother while she lived with us. I don't remember that at all. I do remember trying to catch horny toads. They had two big head-spines jutting from a forest of smaller spines, like horns on a little demon. Supposedly they could squirt blood at you from their eyes.

I haven't mentioned Carolyn's looks. As a young woman, she was an echo of Audrey Hepburn: pearly lips, delicate chin, aquamarine eyes, wide and trusting as a baby's. Her sense of style set off a sylph-like figure. Even as a youngster, her voice was mellifluous, throaty, sultry. She spoke in low tones with pauses that somehow commanded attention—one had to stop everything else and pay attention to hear anything at all. Boys were nuts about her. And she was crazy about them. I know, because my first diary was a hand-me-down five-year diary first written in by Carolyn.

I can hear her husky whisper as I read her first entry: Sunday, October 21st, 1962. She was seventeen. *"Granny P. came to visit—has my room so I'm a bit disorganized. She gave me this diary which I wonder how faithfully will be kept."* Not very. After two weeks of sporadic penciled musings about boys, the entries stopped, as if she had simply gotten up and wandered away in the middle of a conversation.

My rough ten-year-old scrawl takes over three years later. April 7th, 1966: "*MOM WENT TO THE HOSPITAL. We called the doctor and he called the ambulance. Dad was in New York and wouldn't be home till tomorrow. Suddenly she just started to fall and began to spit up.*" I remember that: squatting over my mother's open, staring eyes, trying to tell her jokes, in case she was still aware, my little brother nudged into the living room to watch *The Munsters*. The next day: "*Today we found out that a main artery broke so they have to operate in Mom's skull.*" Kid-speak for aneurysm. Carolyn wasn't home when the aneurysm happened. This was because several years earlier, at age nineteen, she had disappeared.

A month after the aneurysm, Mom came home. But things changed. For one thing, Mom didn't know who we were. It must have been strange for her: she wakes up in a strange place and after a few days, the doctor introduces a nice man who claims to have been her husband for over twenty years. Then these slobbering, smelly children come up and hug her like she's their, well, *Mom* or something.

For us kids, having Mom back home turned out to be unexpectedly fun. Damage from the aneurysm had changed her personality markedly—she was nicer. She actually focused on us, even if only because she was lost and often, probably, desperately confused. Each night, she'd set out the best plates, saying, "Your father is bringing someone special home to dinner tonight." My father would show up with no guest in sight, and we'd enjoy a gourmet dinner. That summer is the only time I ever remember my father laying a hand on me. Cleverly, I'd asked my newly agreeable mother if I could stay late at a friend's. When I came home that night, a trifle guiltily, my father took me out to the back and paddled me. "How could you ask your mother something like that? You know she isn't right in the head." The words stumbled out hard—he never spoke ill of anyone.

My mother's memory loss and my father's reticence are probably why I never heard details of what Carolyn was like as a child. When I was much older and knew more of the truth about Carolyn, my father confided—while setting up Carolyn's trust fund—that he was deeply puzzled, and ashamed, that he could have fathered such alien spawn. Once, a year before he died, my father mentioned absently, while pouring coffee in his cereal, that he thought the polio had done something to Carolyn's brain. (By that time, the Alzheimer's had done a *lot* to my father's brain.) Of course, being me, I took a look at the research literature.

Apparently, poliovirus not only invades motor neurons of the spine, but also, as if drawn as if by a magnet, always invades another area—the *midbrain*, an ancient neural area that humans share with reptiles.[3] The midbrain includes the *reticular activating system*, the oldest system, in evolutionary terms, above the spine. The reticular activating system is responsible for keeping you awake and focusing attention. Overwhelming damage to this area can result in coma or death. This system ties in with a number of neurotransmitter systems, including those using serotonin, norepinephrine, and, especially, dopamine. It is little wonder, then, that damage to this area can have a broad effect on the brain.

Reticular activating system lesions caused by poliovirus infection can leave adult polio survivors with a perpetual feeling of fatigue, and children who had polio with difficulty staying awake, paying attention, and concentrating. Psychologist Edith Meyer described these effects quite clearly in her 1947 study of fifty-two polio survivors aged eighteen months to fourteen years old. As related in Dr. Bruno's *The Polio Paradox*:

> For three years, Meyer followed these children's performance in school and measured their mental abilities. She discovered that "a high percentage of children clinically recovered from poliomyelitis insofar as motor disability is concerned, had qualitative difficulties in mental functioning which, as a rule, do not appear in the conventional type of intelligence test." Through special psychological tests, or merely by observing their performance in school, Meyer found that the children had "fatigability and fleeting attention" for months after the polio attack. When tested, she discovered that the children

had short attention spans, difficulty concentrating, and poor memory for visual designs. These problems were "present in cases in which the medical history notes drowsiness, severe headache, and, in some cases, only nausea during the polio attack." Meyer found that even children who had "non-paralytic" polio, who had no paralysis or even weakness, had symptoms of poliovirus damage to the brain activating system.

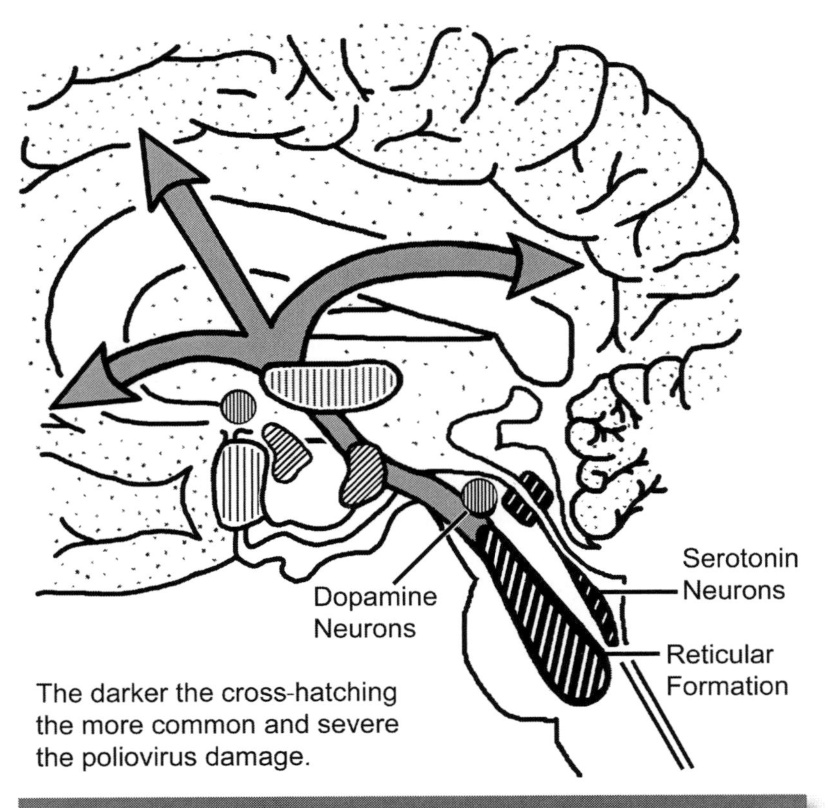

Fig. 5.1. Poliovirus attacks neurons in *very* specific locations. It particularly likes to attack the reticular formation—the pivotal area of the reticular activating system that is responsible for focusing attention, arousal, and vigilance. Neurotransmitters released by the reticular activating system (represented by the gray arrows and crosshatched areas) "turn on" the brain.

Bruno further notes, "Children who'd had polio had the equivalent of what today would be diagnosed as attention deficit disorder."[4] He suspects that poliovirus-damaged neurons recovered to some extent and eventually were able to send out new sprouts, which compensated for the widespread damage in the reticular activating system.

Strangely enough, although the poliovirus invades the reticular activating system along with the cortical region motor neurons, it never invades the *non*motor "thinking" neurons in the cortex. Thus, polio leaves higher-order cognitive processes intact, which is why polio survivors often attain "social, educational, and professional achievement as high or higher than those of the general population."[5]

As far as I could determine then, polio *could* influence personality, if only in its effect on alertness and one's ability to pay attention. But there was an occasional hint there might be something more going on. Edith Meyer's study, for example, revealed that nine of fourteen mothers of three- to five-year-old polio survivors were certain that their child had become more irritable and disobedient since the illness. "Unsatisfactory emotional adjustment was observed," noted researcher Edith Meyer, "which improved gradually after the child's discharge to the home, but in most cases left some marks upon the patient's personality development."[6]

Overall, however, I could find little evidence that polio produced the kind of dysfunctional, Machiavellian traits Carolyn showed. In fact, most research shows just the opposite effect: "Polio survivors from around the world have transcended mere normalcy to become the world's best and brightest," writes Richard Bruno, going on to cite polio survivors who are chief executives of international corporations, artists, sport heroes, members of the British and Canadian parliaments, the US Supreme Court, as well as an American president.[7]

But my father had raised an interesting point. In the years after his passing, I would continue to scour emerging research. Aside from polio's understandable link to disorders such as depression and anxiety, however, few new studies emerged related to polio and psychiatric dysfunction.[8]

Perhaps this was because there was little real relationship to be found.

THE MISSING DECADE

The dust from the carton sets me to coughing. I place Carolyn's last diary down among the packets and randomly grab a black-edged photo album with the words "Polaroid Land Camera" printed in gold on the cover. The album appears to be from Carolyn's late twenties—she's with people, partying. There she is on a seedy orange couch, wine glass in hand, cozily holding hands with a dark-haired man who is leaning toward her—more as if he *wants* to know her than as if he actually does.

There she is in a virginal white dress, bouffant blonde wig perfectly coiffed, dwarfed beside a gigantic bouquet of red roses. There she is smiling alluringly again from a bed—a companion picture shows a look-alike for Steve McQueen grinning from under the same set of sheets.

There's a gray cat with a little bow around its neck. The cat walking along the floor. Cat. Cat. Cat.

Next she's in a bikini by the pool, lounging with a drink in one hand, cigarette in another, obviously relaxed

Fig. 5.2. Carolyn in her midtwenties, dressed to impress in virginal white.

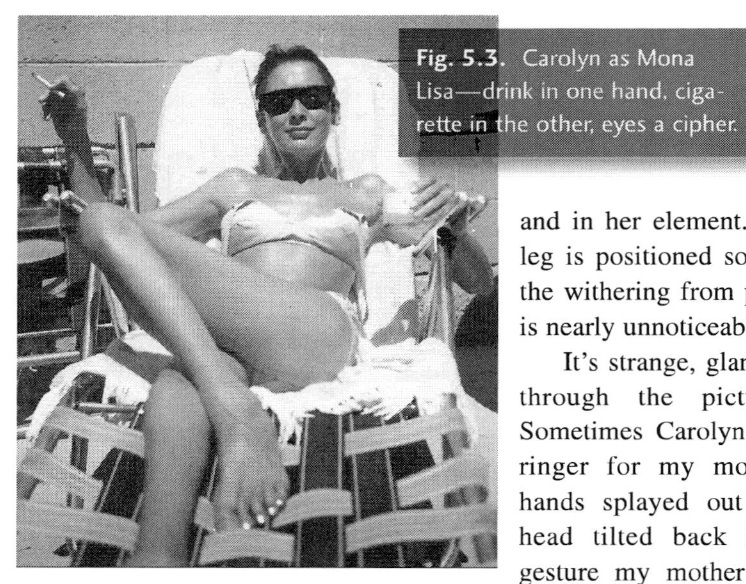

Fig. 5.3. Carolyn as Mona Lisa—drink in one hand, cigarette in the other, eyes a cipher.

and in her element. Her leg is positioned so that the withering from polio is nearly unnoticeable.

It's strange, glancing through the pictures. Sometimes Carolyn is a ringer for my mother, hands splayed out and head tilted back in a gesture my mother frequently made. Carolyn's features reveal the expressive lips, deep-set eyes, porcelain skin, and high cheekbones of our mother's Welsh blood. Then, with a slight change of camera angle and sunlight, Carolyn is suddenly a feminine version of our father, with the dark hair, smooth olive skin, and slanted elfin eyes of what is known in Scandinavia as the "black Norwegian"—my father's father's people. It's like one of those novelty lenticular photographs—move the image slightly, and you see something completely different.

I reach for a second photo album in the same dark, peeling style. The party theme continues. There's Carolyn in a rainbow-colored shirt, seated on a couch, body turned away from a bearded man who is hugging her. She has a satisfied, feral smile as she glances over her shoulder toward him. Her hands are clasped over his, as if to help his hands hug her more tightly.

I flip through quickly. There are many pictures of this bearded man.

Incongruously, there are several pictures of children—a boy of five and a girl of perhaps fourteen. Obviously brother and sister. The boy has the wide-eyed, frenzied glee of overstimulation; the girl has the long, straight hairstyle of the '70s. She gazes forlornly toward the camera and Carolyn—a lost look that makes me want to reach through the years to hug her. But I can't see what the girl sees on Carolyn's face—my sister's back is to the camera.

Here's Carolyn by a pool with four scruffy-looking men in their mid-thirties smiling uncertainly into the sun. There's Carolyn braless in a revealing white dress, legs straddled around a man as she sits on his lap facing him. In the background sits a beautiful blonde—clearly a friend—with a cigarette, drink, and flaming red pantsuit with black choke collar. Some of the photos are dated: November 1971.

The obvious finally occurs to me: These are pictures from the lost years.

My sister had vanished for a decade immediately after my father had accidentally discovered that she hadn't been earning the straight A's she'd been reporting from her studies at Berkeley. Instead, Carolyn had dropped out of college early on with nary a word to anyone, and had spent the next two years living an unfettered life on the money Dad was sending. Despite the subterfuge that preceded Carolyn's disappearance, my parents had agonized for the ten years of her absence, not knowing whether she was alive or dead. My aunt eventually tracked her down to a seedy "escort service" in Las Vegas.

Memories of Carolyn

I was twenty when I first remember having a conversation with Carolyn. After my aunt had found her, Carolyn had called my father and asked for the money to come and visit. After sending money twice, which she spent on other things, he sent her an actual plane ticket, which did the trick.

Carolyn planned to visit for a week, to get to know the family again. Apparently she regretted her decade-long disappearance from our lives. I remember greeting her as she limped past my father to give a hug, smiling that broad, artificial smile, enveloping me in the smell of cigarettes and perfume. Her false eyelashes were so large her eyelids looked like roaches. She spoke slowly, enunciating clearly in that melodious voice of hers as we stumbled through our greetings.

Later that day, when the others had left, she confided with a sultry, intimate whisper: "God, I feel so lucky you are my sister. You are the real reason I came all the way here—just to spend this week with you. You're the special one. This will be so fun spending time together."

Actually, *I* didn't think it was much fun, sitting beside this feeble brown-nosing stranger inside the chinked log walls of the cabin my father had built from scratch. But it *was* morbidly interesting. If nothing else, I figured, over the coming week I could perhaps learn about Carolyn's perspective on her early years, and what life was like in my family before I was born. And there was something else, something I could barely admit to myself. What if she really did want to rejoin the family? Maybe something had changed because of the hard years away, and the big sister I had admired was finally, really and truly, coming home.

After a few minutes of chitchat, my sister excused herself to go downtown to pick up a few things. Actually, it must have turned into quite a few things, because I didn't see Carolyn again for another five years. Later, we found she'd spent the rest of the week living with a man she'd bumped into at the grocery store.

The Letters

The rain resumes outside, drops slapping gently against the skylights. I close the albums and my memories and move on within the carton, randomly sampling to get a feel for how everything is organized. Each of the smaller boxes, each large envelope, each packet, seems to revolve around one man: whoever was obsessed with her at that period of her life. Added to each packet are assorted extra mementos from other admirers, like condiments with a main course.

I take a deep breath of the musty air and begin to carefully pick through each piece.

At the top of a thick packet of poems accompanied by dried flowers, I find a sheaf of thick parchmentlike double-sized paper:

> March 18th, 1969
>
> My Darling,
>
> I am a lucky guy! I have searched for you all these years, slowly dying little by little and more and more. Now finally to find you and to know you exist. The distillation of my yearnings, the well of never-failing interest and awareness.
>
> You command my attention at all times. Though we may be physically apart, I think of you, your vibrancy and zest, your cool calm. Your moods, gay, somber, in all their degrees of intensity, are a part of me, a part of *you* that is always with me. I am so very lucky to have found you . . .
>
> Do I pursue you overmuch? Could I having been loved by a Goddess, feel love again with a mortal woman? . . . I have only my love to offer you at this time, but that is total and eternal for I shall never stop loving you darling.
>
> Forever yours,
> Ross

Several days later, Ross had apparently thought of some amendments. On the next page—neatly labeled "Part II," he continues:

> We have an intellectual and philosophical compatibility that I have never experienced before and almost despaired of finding. How precious and wonderful it is, Miss Wonderful! You are inscrutable to most people and perplexing to the rest. Yet we are so much in rapport. It is "natural" and reasonable that you should hesitate to live with me and share your life. . . . Also it is "natural" that you should be hesitant from the standpoint of possible feelings of responsibility for placing my children elsewhere.

Interesting. Ross appears not only to be head over heels for Carolyn—he's ditching his kids to clear the path. I check the dates. *His children must be the kids in the photographs.* The bearded man must be Ross.

But, as I search and shuffle through the materials, aside from the small stack of sweet love poems, there is nothing more from Ross. No notes, no letters, no memos. No more photographs of the bearded man or his children.

Another letter, in an envelope inscribed "Ki." That was Carolyn's self-chosen nickname. Inside the card is a velvet heart with an arrow through it. Preprinted inside are the words: *For My Wife.* A handwritten poem followed.

I'd forgotten she'd once had a husband.

There is one other brief card professing love from him. That's all. No wedding photo album. No marriage license. No mention of him in her will. After all these years, only these two tantalizing bits of paper give witness to the marriage, like bits of flotsam after a ship has foundered.

I go back to the albums, paging. There's my father, holding the infant Carolyn. He looks almost surly, but I know what that look really means: he's sneaking into an easy grin. And my mother stands shyly on a porch, toddler Carolyn smiling in the foreground. Two parents who obviously loved my sister deeply, who were at that point in their lives, and for many years to come, a loving, stable couple.

For an outsider, it may be difficult to believe that my sister was able to bilk my parents of college money and live an unbroken life of deception and subterfuge without my parents having done something—anything—to intervene. Yet the way Carolyn interacted with my own parents was probably little different from the way other similarly afflicted children have interacted with their parents.

Was Carolyn's skewed temperament congenital? Certainly our family has had its share of troubled characters: My mother had her problems with alcohol—she would eventually escape suicide through a hairbreadth rescue. One great-uncle, proudly dry now for nearly three decades, let slip that mom's father—my grandfather—was probably drunk when he steered into a rocky embankment and died. (All

Fig. 5.4. Dad holding Carolyn on his lap.

four of Grandpa's brothers were apparently alcoholics.) Mom's oldest brother, who I never met, hip-hopped from place to place as an off-beat, charismatic flimflam man.

On my father's side, there was the pair of great-uncles who, legend has it, stalked each other for three days with pitchforks before being banished from the household and disappearing forever. And the great-grandfather who only spoke one sentence in his entire adult life (the absurdly inconsequential "Bring me my pipe."). Both my father's brother and his father—my paternal grandfather—were alcoholics. Dad's sister was a gorgeous, eccentric slip of a woman who refused all of her many suitors. Instead, she walked the equivalent of a marathon each day in her work as a postwoman, then came home to continue her strange, unending exercise routine—jogging, playing tennis, bicycling—and to consume the gallons of spiked coffee that left her slurring and twitching.

Fig. 5.5. Mom stands demurely on the porch while Carolyn frolics in the foreground.

I spent the day before her death with her in the unventilated basement room she'd chosen for her final days. The air was woozy with smoke as she lit cigarette after cigarette—she paused our reminiscences to reassure me that her decades of chain-smoking had nothing to do with her terminal lung cancer.

This aunt came by her odd behavior naturally. Her mother—my grandmother—was a whale-sized Valkyrie with nearly supernatural shrewish powers. Grandma spoiled us grandkids with sugar-laced tea, knäckebröd, and mounds of Swedish meatballs. I loved her very much and, since I was only five years old when I last saw her, was oblivious to the insidious Ragnarök she enjoyed creating in my parents' marriage. (Grandma refused, for example, to even speak to my mother for the first years after my parents had wed, despite the fact that she was living in my parents' house with them at the time.)

My family is large—there are twenty-one maternal cousins in my generation alone. Thus, although the eccentrics tend to create legends, both sides of the family are also filled with loving, talented, larger-than-life inventors, artists, writers, doctors, and businessmen, who run the gamut from flamboyant public figures to reclusive hermits. In my family, at least, the seeds of the eccentric, not to mention those of the successfully sinister, seem closely related to the seeds of success.

But what about Carolyn herself? What was life like for her—the once bright-eyed toddler who spent helpless weeks poised on the edge of death in an iron lung? Carolyn's earliest years—literally *years*—

were spent living in hospitals. For months after the time of the initial infection, she, like many other young polio victims, was undoubtedly terrorized—stringently isolated from my parents and others, and totally dependent on the care of strangers.[9] At the time, hospitals were frequently overwhelmed with polio patients—little thought was given to including children on treatment decisions, which were often simply imposed, like torture. Questions or complaints could often bring on punishment. Some polio patients felt that, if they did not develop a non-childlike servility in every aspect of their behavior, their lives could be placed in jeopardy by a sometimes arbitrarily cruel staff. Even after Carolyn arrived home from the hospital, there would have been little improvement in her quality of life—she was shuttled back to the hospital for operation after painful operation, and frightening,

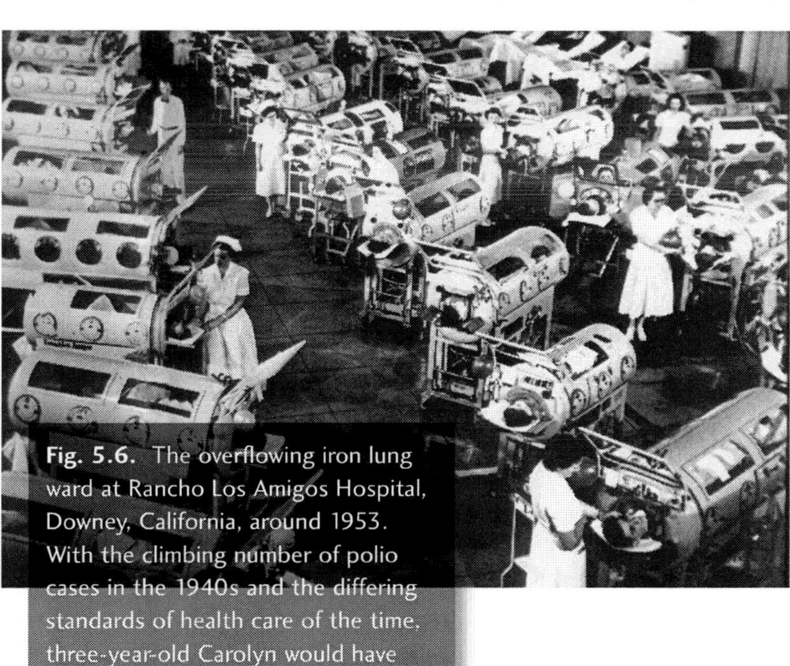

Fig. 5.6. The overflowing iron lung ward at Rancho Los Amigos Hospital, Downey, California, around 1953. With the climbing number of polio cases in the 1940s and the differing standards of health care of the time, three-year-old Carolyn would have received scant attention during one of the most terrifying periods of her life.

almost medieval therapies. And when she was able to at last play with other children, she would have suddenly found herself with a new identity—the little crippled girl who other children no doubt bullied and teased. Many polio survivors became pariahs—others feared that they somehow remained infectious. All of this would have further stressed Carolyn's already tormented psyche.[10]

As the intermittent hospital stays continued, year after year, my mother could only visit occasionally, as we younger children began tugging at her time and attention. In any event, hospitals at the time allowed families to visit for only several hours a week.[11] No doubt feeling helpless himself at the many hospital separations, my loving, outdoorsy father lost himself in his work—spending eighteen hours a day for weeks on end—roping problematic cattle for vaccination (he was a wizard with a lasso), or being bitten by surly dogs as he tended their broken legs, or pulling calves at two in the morning. The grinding work must have taken his mind off the athletic games he would never play with his oldest daughter. These were the very games he would later play so happily with us younger children—fortunately for us, our good-natured father had escaped the alcoholism and personality disorders that were so rampant in the other members of his family.

No, despite my parents' unquestioned love, if anyone could be thought to have endured childhood stress, it would have been Carolyn.

Who *was* Carolyn? Who was this person who could appear so normal and yet be so disturbed?

On a little red heart dated December 25, 2000—four years before her death—I find:

> Santa—I know I have been naughty and I don't expect any presents but since you have opposable digits, could you please unscrew this #%$# jar of Parmesan?

There is a little scrapbook Carolyn has put together: snapshots, newspaper clippings, and pictures of friends from eighth grade. Sandwiched among the labored signatures of friends, I find:

Good luck in High School, but with your mind you won't need it.

—Marilyn. ('60)

Get to the top even if you have to climb over somebody's back.

—John.

In the late 1980s there are a flurry of letters from Ron and a rival. Ron writes:

MY KILO

My Kilo with an articulate soft musical voice, a wonderful choice of words, laughter that sounds like the soft tinkling of silver bells, the graceful mannerisms of a princess, the poise of a queen. When she enters a room it brightens and seems to glow with her personality. If you enter a room expecting to find Kilo and she is not there, there are no words to express the dismal loneliness of that room. She is affectionate, warm, tender and considerate of others. If crossed she can be cold, disdainful and cruel. Her face has a wonderful animation when she is happy. She is feminine from the top of her pretty head to the tips of her toes. She needs no jewelry to show off her loveliness. Her beautiful ivory white skin and beautiful dark hair accentuates her loveliness. A very lovely woman is my Kilo.

When I have forgotten everything else, I will remember my Kilo and she will live in my heart and pay no rent.

No rent. Sounds like a bargain. Meanwhile, Ron's rival, Ed, writes counterpoint: "*There are lots of fish in the water and a lot of them would love you. Including me.*" A smiling fish with hearts fluttering around its nose swims past the signature.

Yes, indeed, things were heating up with Ron. I find a memo dated May 7, 1990, signed by Ron and Carolyn: "*Ki and Ron have decided to meet on the mount in the year 2000.*"

But the packet of letters instead fast-forwards to Ron's neatly clipped obituary. He died in Sequim on June 30, 1996, at the age of eighty-three. Carolyn was fifty-one. The obituary makes note of two sons, a daughter, a brother, two sisters, six grandchildren, one great-granddaughter, numerous nieces and nephews, and a "special friend."

Not Carolyn.

L'affaire avec Ed didn't end so well, either: *"I don't think we would ever get along so well, but I'd like to stay friends . . ."*

***

How can I make sense of all this? Besides the letters, there are literally dozens of birthday, Valentine, Christmas, and what-not cards from a bounteous assortment of men. There is one from my mother, signed with the spindly writing of her last years, wishing happy birthday to *A Wonderful Daughter*—probably one of the last communications before my mother couldn't bring herself to speak to Carolyn again. There are recipes, report cards, childhood awards, more love notes with flattened, perfumeless flowers, mash notes. They tell me everything and nothing about Carolyn—mirrors reflecting mist.

But the letters and notes I've found *do* tell me that Carolyn was deeply loved, not just by my parents, but by many different people. That there was something captivating about her—something that allowed men, for a while at least, to think that she was the answer to their dreams.

I think back to the question that has bothered me for so long—how could a disturbed individual attract so many people? The letters tell me. Beauty. Intelligence. Charisma.

I nudge the last of the packets back into place in the carton and close the flaps. Although I've set aside the diaries for careful reading later, the box feels somehow heavier as I manhandle it around, duct taping the spirits of the past. My weekend of investigation has left me feeling as if I've been shoveling water. But one thing has come clear. Carolyn showed "a pattern of unstable and intense interpersonal relationships." Not only with us—her family—but with practically everyone she ever became close to.

I also know that, in 2001, Professor John McHoskey, who had previously done such interesting work relating Machiavellianism to psychopathy, found something very interesting about those unstable and intense relationships. In one of his final, crucial studies, McHoskey found something that might help explain not only the writings from Carolyn's box, but Machiavellian behavior in general.

CHAPTER 6

THE CONNECTION BETWEEN MACHIAVELLIANISM AND PERSONALITY DISORDERS

"Welcome to the Psychiatric Hot Line:

—If you are obsessive-compulsive, press 1 repeatedly.
—If you are schizophrenic, listen closely and a little voice will tell you which number to press.
—If you have borderline personality disorder, hang up. You have already pushed everybody's buttons."

—Anonymous

The year 2000 loomed with warnings of a Y2K disaster and biblical prophecies of doom. But nestled in his relatively remote Eastern Michigan University surroundings, Professor John McHoskey continued to gnaw at his Machiavellian research. To the unpracticed eye, it might seem the research bones had been picked clean. After all, McHoskey's previous work had shown that psychopathic traits corresponded tongue-in-groove with Machiavellian traits. The pathological lying of a psychopath, for example, matched the Machiavellian trait of duplicity. A grandiose sense of self-worth, found in both psychopaths and narcissists, corresponded to Machiavellian feelings of entitlement, superiority, and arrogance. Psychopathic "shallow affect" was echoed in the numerous studies showing a cool and aloof posture by those who received a high score on the Mach-IV test. And just as psychopaths displayed glibness and superficial charm, those with high Mach scores were found to be more persuasive and more likeable than their low-scoring counterparts. Even guilt was similar—neither psychopaths nor Machiavellians seemed to be troubled by it, but both personality types were perfectly happy to induce guilt in others to manipulate them more easily.

Despite the neat dovetailing, however, McHoskey was aware of one nagging detail. His seminal study had begun by simply assuming that the Machiavellians whom Christie had described were largely equivalent to psychopaths—the tests McHoskey had conducted as part of the study had merely confirmed that initial assumption. But a potential problem was that other personality disorders hadn't been checked. It seemed unlikely, but what if there was another syndrome that might even more closely match the characteristics of Machiavellianism?

There was one relatively straightforward way to find out. McHoskey could give a large group of people Christie's Machiavellian test. He could also administer a second test—one that showed predispositions for common psychological disorders. The results from both tests could then be compared to see which psychological disorder correlated most strongly with Machiavellianism.

The Dimensional Approach to Understanding Personality Disorders

As it happened, the perfect tool had been developed for easily assessing possible clinical diagnoses: the Personality Diagnostic Questionnaire, or PDQ-4+. By looking at the answers to ninety-nine true-false questions, a reasonably intelligent guess could be made as to whether the test-taker reported any clusters of behaviors consistent with DSM-IV Axis II criteria for personality disorders.

The PDQ-4+ is based on a *dimensional* view of personality dysfunction. That is, it contends that dimensions, or symptoms, related to personality disorders can be measured within samples of normal people. For example, some people might have a mild version of the impulsivity often associated with antisocial behavior, while others might have much stronger versions, or different varieties altogether. Each disorder could be described by a number of symptoms, and each symptom could come in mild, medium, hot, or extra hot versions. People with the hot or extra hot version of enough symptoms of a particular personality disorder could be said to have that disorder. Others, with slightly milder symptoms, or with only a few extreme symptoms out of the many that might characterize a disorder, could be considered to have a tendency for the disorder. Such individuals might not qualify for a clinical diagnosis, but an inherent predisposition was definitely there.

For this new study, McHoskey decided to examine what is probably the most extensively studied group of humans on the planet: undergraduate college students. He administered the twenty-question "Mach-IV," as Christie's Machiavellian test was known, as well as the ninety-nine-question PDQ-4+, to nearly three hundred students, sweetening the pot a bit by giving them extra credit for taking the tests.

McHoskey's Findings

To understand the importance of what John McHoskey found in this, perhaps his most important research contribution, it's helpful to have a little background in how psychology analyzes people with problems. The DSM-IV—that ever-handy diagnostic manual—segregates people's various dysfunctions into different categories, or axes. Axis I includes all mental health conditions *except* the personality disorders, while Axis II describes the personality disorders. (Another axis has recently been proposed to incorporate personality-related genetic information.)[1]

There is some controversy about why a disorder might be classified as Axis I instead of Axis II. Fundamentally, the two axes are thought to be the same, but a distinction was drawn between those disorders that were thought to be due to biological factors, which were put into Axis I—and other conditions which were thought to be due to psychosocial stresses, which were put into Axis II. Thus diseases such as schizophrenia and bipolar disorder were classified as Axis I, while seemingly purely psychological conditions such as dependent and narcissistic personality disorders were classified as Axis II. Now, of course, it is recognized that many of both the Axis I and Axis II disorders are rooted in biology and genetics.[2]

The decision to use two separate axes was intended to draw attention to the more subtle disorders. But instead, the two axes had the unpleasant side effect of providing insurance companies an excuse to pay only for diagnoses involving the more florid—and thus seemingly more severe—Axis I disorders. This in turn meant that a person suffering from an Axis II disorder would often instead be diagnosed with an Axis I disorder so as to ensure insurance coverage. Thus, rather than drawing attention to personality disorders, the Axis I/Axis II distinction resulted in *less* interest in, and research on, personality disorders.[3]

Axis II is further broken down into three "clusters" of personality disorders that seem to share certain similarities. People with Cluster A disorders, for example, are often seen as odd or eccentric—they have difficulty relating to others. Cluster B sufferers tend to act in a dramatic, emotional, and erratic fashion. They frequently have difficulty with impulsivity and often violate social norms, as with antisocial personality disorder. Those with Cluster C personality disorders are often seen as anxious and fearful; they are frequently afraid of social relations. The specific personality disorders fall into the three different clusters as shown below:

CLUSTER A (ODD, ECCENTRIC) PERSONALITY DISORDERS:

- *Paranoid Personality Disorder:* Pervasive suspiciousness, distrust, and resentfulness of others; vindictive, rigid, and good at avoiding blame.

- *Schizoid personality disorder:* Indifferent to social relationships; attempts to avoid interpersonal interactions; lacks empathy. Also has difficulty with emotional expression.
- *Schizotypal personality disorder:* Suffers from perceptual dysfunction, depersonalization, interpersonal aloofness, and suspiciousness; shows "magical thinking" with a belief in special powers.

CLUSTER B (DRAMATIC, EMOTIONAL) PERSONALITY DISORDERS:

- *Antisocial Personality Disorder:* Problematic sense of right and wrong; deceitful and manipulative; easy willingness to lie; not bound by laws and social norms; irresponsible and impulsive; superficially slick and polished; potential for violence; enjoys humiliating others.
- *Narcissistic Personality Disorder:* Possessed of grandiosity and exhibitionism; lacks empathy, hypersensitive to criticism, possesses a constant need for approval and admiration.
- *Histrionic Personality Disorder:* Overly dramatic and theatrical; throws frequent tantrums; always wants to be the center of attention; manipulative and demanding; vain; sexually provocative.
- *Borderline Personality Disorder:* Rapid mood swings; emotionally unstable with very troubled relationships that include intense fears of abandonment; inconsistent attitudes and behaviors; no clear goals or direction; frequently considers self-harm.

CLUSTER C (ANXIOUS, FEARFUL) PERSONALITY DISORDERS:

- *Obsessive-Compulsive Personality Disorder:* Possessed of perfectionism that makes task completion difficult; rigid and inflexible; has a need for control; preoccupied with details and rules; unwilling to compromise; extreme "workaholic."
- *Avoidant Personality Disorder:* Inhibited; introverted; intense feelings of inadequacy; hypersensitive to rejection; socially awkward.

- *Dependent Personality Disorder:* Submissive and clingy; passive; extreme lack of self-confidence; has difficulty making decisions and an intense desire to be taken care of.

Using these DSM-IV disorders as a foundation, McHoskey was looking for a correlation with Machiavellianism.[4] As expected, those who fell into the dramatic, emotional, and erratic cluster B seemed to correlate most strongly with Machiavellian types of personalities. But oddly enough, antisocial personality wasn't the only disorder that correlated with Machiavellianism. Another disorder did as well—*borderline* personality disorder.

Borderline Personality Disorder

Preeminent psychiatrist Robert Friedel was to enter into his lifelong study of borderline personality disorder because of his sister, Denise. In his book *Borderline Personality Disorder Demystified*, he describes a youthful experience:

> I was cleaning my golf clubs in preparation for the spring, but one was missing. I looked everywhere in the house—no club. Denise walked by, so I asked her if she had seen it. She calmly said yes, she had broken it in two and thrown the pieces into the snow behind the house. It seemed we had argued over something a few weeks earlier, and she had done it then. At first, I thought she was taunting me. She knew how hard I had worked and saved to buy those clubs. Surely no one would do such a thing, not even Denise. . . . Later, when the snow melted and I found the broken club, I realized that something was truly different about Denise, and that it was probably best not to provoke her in any way, for any reason.[5]

Indeed, borderline personality disorder, or BPD, is a disorder that can be so profoundly confusing that even today, despite thousands of studies, it remains little known and poorly understood. This is despite the fact that one to two percent of the population—some three to six million people in the United States alone—are thought to suffer from its most profound effects.[6] A still higher percentage show symptoms of the disorder but are able to maintain conscious control of their impulses when necessary and are thus able to avoid diagnosis.[7] Ten percent of psychiatric outpatients and twenty percent of psychiatric inpatients—hefty and expensive percentages!—have the disorder.[8]

Although women are three times more likely to have the disorder than men, preeminent borderline researcher Joel Paris has noted, "If it were not inconsistent with clinical tradition, we could have described a single gender-neutral disorder that covers the present ground traversed by the criteria for antisocial personality disorder and borderline personality disorder."[9] In other words, antisocial and borderline personality disorders shade into each other.

Borderlines often show a variety of symptom complexes and coping characteristics, as shown in the sidebar.[10] These characteristics can be perceived as being extremely manipulative and are often, in fact, malevolently Machiavellian in their effects. Underneath all of these behaviors is a tendency to make situations more explosive and emotional than the facts would warrant.

Even the term *bor-*

Borderline Personality Disorder

Coping Characteristics

- splitting behavior
- projection
- blame shifting
- control issues
- interpersonal sensitivity
- situational competence
- narcissistic demands
- "gaslighting"
- chameleon behavior

Symptom Complexes

- impulsive and self-destructive behavior
- rapid mood swings with anxiety and depression
- feelings of boredom and isolation
- intense and unstable personal relationships

Borderline Manipulation—and Very Real Remorse

Some experts in borderline personality disorder object to the word *manipulation* as applied to their patients. Well-known BPD expert Marsha Linehan, for example, particularly dislikes the word "manipulative" when referring to borderline patients. She feels that "this implies that they are skilled at managing other people when it is precisely the opposite that is true."[11]

In fact, psychologists and psychiatrists commonly retrain themselves as well as their patients to view "manipulative" behavior from a different perspective to aid in therapy. As one treatment program states: "Rather than viewing themselves [individuals with borderline personality disorder] as someone who is attempting to manipulate, is attention-seeking, or is sabotaging treatment, the trainees [individuals with borderline personality disorder] learn to view themselves as driven by the disorder to seek relief from a painful illness through desperate behaviors which are reinforced by negative and distorted thinking."

These are enormously effective therapeutic strategies. However, my purpose is not to aid in therapy but rather to analyze the ultimate

derline itself is a result of the confusion that midcentury psychoanalysts felt when treating such patients—they theorized that the borderline syndrome was a form of pathology lying on the border between psychosis and neurosis. Although analysts no longer believe that patients with borderline personality disorder have an underlying psychosis, the name *borderline* has stuck. A much more descriptive label would be *emotionally unstable*. The central feature of the borderline condition is instability, affecting sufferers—and those around them—in many sectors of their lives. Additionally, borderlines have significant issues with boundaries. As psychiatrist Joel Paris writes: "[Borderlines] become quickly involved with people, and quickly disappointed with them. They make great demands on other people, and easily become frightened of being abandoned by them. Their emotional life is a kind of rollercoaster."[12]

effects of borderline-like and related behavior. Consequently, if the behavior and effects described here amount to manipulation, even if the person is not consciously aware of his or her intentions, it is called manipulative. In fact, it is clear that part of the reason that Machiavellian researchers did not recognize the affiliation between borderline personality disorder and Machiavellianism earlier is the use of understandably empathetic terms by borderline specialists that obscured the nature of borderline symptoms.

I would like to emphasize here, however, that although borderline traits can underpin some Machiavellian activities, many deeply dysfunctional, clinically diagnosed borderlines can be cognizant of and deeply remorseful regarding their problematic behavior. It's just that these borderlines often don't have the emotional toolkit they need to stop that behavior. The remorse these borderlines can feel is quite different from the remorselessness of psychopaths. But when borderline and psychopathic traits combine and are lit by the additional fuse of narcissism, as with some of the deeply sinister individuals who are featured players in this book, the resulting conflagration of a personality can be devastating.

For years, the borderline diagnosis was denigrated as a wastebasket diagnosis—"a label to be used when the doctor simply did not understand the patient or could not 'fit' the patient's symptoms into any other, more acceptable disorder."[13] However, the many recent breakthroughs that have emerged from neuroimaging and other data have provided strong evidence that borderline personality disorder is indeed real—and devastating to both the disorder's sufferers and the people around them.*[14]

Paul Mason and Randi Kreger, authors of the classic book about borderline personality disorder *Stop Walking on Eggshells*, write: "It's

*In this book we shall follow the example of borderline expert Jerold Kreisman and others who, for the sake of clarity and efficiency, refer to individuals by their diagnosis. Thus, "borderline" is shorthand for someone who exhibits symptoms consistent with the DSM-IV diagnosis of borderline personality disorder.

no secret that [nonborderlines] often feel manipulated and lied to by their borderline loved ones. In other words, they feel controlled or taken advantage of through means such as threats, no-win situations, the 'silent treatment,' rages, and other methods they view as unfair."[15] Psychiatrist Larry J. Siever describes this behavior differently: "Although [people with borderline personality disorder] can be apparently manipulative, they don't think about the behavior as such. They're trying to meet their needs in the only way they know how. . . . They are trying to elicit a response to soothe them, to help them feel better."[16]

Borderlines have rapid mood swings with a tendency to experience anxiety and depression; they also often show impulsive and self-damaging behaviors and suffer from chronic feelings of boredom and social isolation. Borderlines also have intense and unstable personal relationships, which they devalue and manipulate frequently. Coupled with these symptoms is a propensity to "self-medicate" with drugs or alcohol, or to have difficulties with food; they sometimes become bulimic or anorexic.

Looking back at my sister, I can see that she had problems with all four of these typical borderline symptom complexes, as well as the accompanying addictive and anorexic behavior. Her diaries lay out the symptoms in snippets:

> Terribly depressed the past couple of months. Today seemed worse than most . . .
>
> I believe I ate *aubergine et jambon et fromage*. Must have become fairly inebriated and crawled to bed—shame.
>
> Am determined to do cleaning of premises in a first rate fashion, as I plan to handle the remainder of my life. Tomorrow is another day—we rather hope. It will be better; but better it won't be unless self-foibles are examined!
>
> Finished cleaning chore, no mean task! It's what I call bumbuster. Vacuuming on the scooch gives one an appreciation of housework!

> Unfortunately, Gary fed me info on Jack guaranteed to touch my heartstrings—ain't going there anymore!
>
> $84 check for Jack from Sprint. Will sit on it for a while. Dinked for several hours with dinner prep, but too late to eat.
>
> . . . a phone call from "Bob" of the American Legion speaking for Jack looking for refund check from Sprint. I'm sure am being set up as ogre and Jack, the poor victim.
>
> . . . on a more hurtful note, apparently Janine [a former caretaker] is alive and well with her complaints of my dissolute behavior.
>
> I must have said something to Peg yesterday on the phone that really rankled her. She called this morning and lit into me.
>
> . . . sad letter from Mary—apparently the accident her daughter fomented is causing a ¼ mill suit . . . I would love to be able to supply her with some legal resources.
>
> December 25th: Stomach revolt . . . Nougs [the cat] did not care for toys Emma and Sam gave her. She was agitated by the entire Christmas experience.
>
> I hate having to stabilize with alcohol first thing in the morning . . .

What comes clear from the diaries is not only the range of Carolyn's symptoms, both physical and mental, but the heartbreaking depth of her suffering as she attempted to cope with those symptoms. I can only imagine her scuttling crablike around her apartment as she attempted to clean—her polio-withered leg dragging limply behind. Her many manipulations over the decades had ultimately left her abandoned by her family (*my* family—*me*), and her few friends were either at an arm's length or seriously troubled themselves. Carolyn's last Christmas found her completely alone, her cat—not Carolyn—"agitated by the entire Christmas experience."

The "*Dinked for several hours with dinner prep, but too late to eat*" comment was not atypical. According to her caretaker, Carolyn had an ironclad personal rule—she ate nothing at all, all day, until after 8:00 PM. Carolyn was five foot five in her stocking feet. A minimum healthy weight for a woman her size is around one hundred and twenty pounds. She weighed eighty-one pounds.*[17] (The forensic autopsy listed "malnourishment.") And, of course, she was an alcoholic.

Even as she spoke of examining her "self-foibles," she carried a curious cognitive blindness about herself—the very heart of the borderline experience. She clearly felt for Jack, for example, but was also purposefully manipulating him—keeping his much-needed money, then playing the victim when he became upset about it. Carolyn cared about Mary's problems and would have loved to help her, but at the same time, she didn't bother to explore what she might have said to her friend Peg that would have made her so angry. I am reminded of the words of borderline expert Jerold Kreisman: "The borderline is capable of great sympathy and comforting but often may lack true empathy, the ability to put himself in the other person's shoes, in appreciating how others are impacted by his behavior. Additionally, when they are hurt, their rage at those who have hurt them may be intense and cruel and devoid of concern or understanding for the other party."[18]

In sum, Carolyn presented the classic enigma of the borderline—intelligence mixed with a surreal, well-defined pattern of dysfunction.

Was it the polio that had somehow brought about the dysfunction—perhaps through the trauma it brought into her childhood? Early

---

*To give a little perspective, Kelly Ripa, star of two hit television shows (*Live with Regis and Kelly* and the sitcom *Hope and Faith*), made headlines for being "Pin Thin!" at 105 pounds. Kelly is an inch shorter than my sister was. Interestingly, one study of neurotransmitter activity has found that ritual religious fasting enhances the transmission of serotonin between neurons, thus perhaps helping to induce a spiritual state. Perhaps my sister's anorexia was an attempt to self-medicate. Indeed, anorexia is in part a cognitive disorder—those who have it are often in denial, and see their bodies differently from reality.

On a side note here, some may wonder whether Carolyn suffered from symptoms related to bipolar disorder or hypomania. I don't believe so—neither syndrome has appeared in any of my relatives, and while Carolyn often showed symptoms of depression, she never displayed any symptoms of being hyperdriven, excessively chatty, or having an-idea-a-minute. Be that as it may, there is evidence for considerable overlap between borderline and bipolar diagnoses—both disorders show unstable self, impulsivity, affective instability, anxiety, irritability, and episodes of intense anger.

trauma, it seems, can bring about borderline personality disorder. But if it was truly polio alone that was the cause of Carolyn's disorder, why didn't many other polio survivors suffer the same psychiatric symptoms as Carolyn?

I continued, month by month, year by year, to check the research literature. But I couldn't see a pattern.

Borderline Coping Behaviors

Fig. 6.1. Carolyn ready for Halloween in this undated photo. She was ever the enigma.

The diverse coping behaviors of borderlines can be devastating for those around them—just as my sister's behavior caused decades of desolation in my parents' lives. Splitting, for example, is a coping process where the borderline swings between idealizing and devaluing people in relationships. A person is seen to be either all good or all bad, with the borderline unable to reconcile that there can be both good and bad within a person. On a larger scale, splitting behavior is shown when the borderline pits people against one another, making one group the "white hats" and the other the "black hats"—although who is considered as good or bad can shift from day to day, or even hour to hour. Authors Steven Leichter and Elizabeth Dreelin recount the following example of dealing with diabetic patients who also have BPD:

> [A] 26-year-old phlebotomist enters the patient's room to obtain the venous blood specimen for [a] test. After one futile attempt at venipuncture (made futile because the patient wrenched her arm away from the technician during the process), the patient demanded that the venipuncturist leave her room and requested that the head nurse for that floor see her. When the head nurse arrived, the patient complained about how inadequate her care was because an inexperienced venipuncturist was sent to draw her blood. The head nurse then called the senior phlebotomist, who came to see the patient.
>
> The patient induced the senior phlebotomist to admit that her younger associate was inexperienced. The patient separately got the head nurse for the floor to admit that it was unprofessional for the senior phlebotomist to comment negatively about her junior associate. In the end, the patient never allowed the blood sample to be obtained. When asked by the attending physician why the blood was not sampled, the patient responded that there seemed to be discord among the staff, and they failed to remember to draw her blood.[19]

Leichter and Dreelin add that it doesn't matter what specific issue actually caused the borderline patient to act that way. From the perspective of the borderline, what *is* important is that a problem is identified around which the borderline can set various health professionals in opposition to each other.

Splitting has been related to the poorly regulated emotions so often seen in borderlines. One theory holds that the phenomenon may be tied to how we store memories; memories made when we are in one mood are easier to bring back to mind when we are in the same mood. Mood instability would lead to fragmented memories of other people (and of the borderline herself), that would lead to the changing perceptions involved in splitting.[20] Splitting may also involve overactivation of emotional processing (ventromedial prefrontal cortex) as opposed to rational processing (dorsolateral prefrontal cortex).

Borderlines also often have significant problems with their personal identity. One common manifestation of this symptom is a chameleon-like personality shift. A sufferer of this symptom writes: "I have a *chameleon-like* ability to take on the coloring of the individual I am with. But the act is done more to fool me than to fool them. When I 'become' a persona it's not worn over the real me as a cloak. For the time being, I have become who I'd like to be . . . I am not some kind of a *Machiavellian manipulator* with nothing better to do than ruin lives. The process isn't even really conscious" (italics added).[21] Note how the writer feels obligated to point out that she is not a Machiavellian manipulator whose intent is to ruin lives—the implication, of course, is that most people perceive her to be doing so. (And that may prove her all the more Machiavellian in her desire to dispel that perception.)

Another borderline trait, projection, happens when a characteristic possessed by the borderline is projected by the borderline onto someone or something else because the characteristic is just too painful for the borderline to accept. For example, Carolyn's cat, not Carolyn, was "agitated by the entire Christmas experience." Those who deal with a significant other with borderline personality disorder often know instinctively not to try to tag their partner with the label of "borderline." A true borderline's knee-jerk response would be: *I don't have borderline personality disorder. YOU have borderline personality disorder.* A common related defense mechanism is blame shifting: whatever has happened, no matter how culpable the borderline, it is always someone else's fault.

It's thought that borderlines may "attempt to establish control of their own emotional states by manipulating or controlling the behavior of others."[22] Typical borderline control strategies involve "putting others in no-win situations, creating chaos that no one else can figure out, or accusing others of trying to control them."[23] Another common borderline trait is "an amazing ability to read people and uncover their triggers and vulnerabilities. One clinician jokingly called people with BPD psychic . . . [Borderlines use their] social antennae to uncover triggers and vulnerabilities in others that they can use to their advantage in various situations."[24]

One of the most difficult aspects of the borderline condition is that people with the disorder can be competent and in control in some situations, yet out of control in others. As Mason and Kreger put it: "[M]any [borderlines] perform very well at work and are high achievers. Many are very intelligent, creative, and artistic. This can be very confusing for family members who don't understand why the person can act so assuredly in one situation and fall apart in another. This ability to have competence in difficult situations while being incompetent in seemingly equal or easier tasks is known as situational competence."[25] What can make things all the more difficult is that others may not be exposed to the borderline's out-of-control or chameleon-like behavior, so that the nonborderlines would not be believed if they tried to explain how differently the borderline can act when out of the public eye.

Borderlines can often be very self-involved and will try, sometimes using outrageous methods, to bring or maintain the focus of attention on themselves. These traits are also characteristic of narcissistic personality disorder—a common borderline personality co-disorder. Many borderlines see their relationships with others as being only about themselves. They may feel threatened if their partner or children have other friendships and may act to sabotage these relationships.

"Gaslighting" involves the denial by a borderline (or other person with a conscious or unconscious desire for manipulation and control) that certain events occurred or certain things were said.* The borderline may deny another's perceptions, memory, or very sanity. Here is an example of gaslighting with a college roommate:

---

*The term "gaslighting" comes from the 1944 film *Gaslight* in which Ingrid Bergman, a newly married heiress and soon-to-be victim, remarks to her ne'er-do-well husband Charles Boyer that the gaslights in their home seem to be dimming. "No, they aren't darling," Boyer lies suavely. "You are imagining things." In reality, the gaslight of their Victorian household *was* dimming—Boyer was sneaking up to the attic and causing the pressure to dim in the room below. Gradually, it unfolds that Boyer is trying to convince his wife she's losing her mind—that she inherited bad genes from her mother, who died insane. Boyer hides Bergman's possessions, for example, to make it appear her memory is fading, and plays other tricks to disorient and confuse her. In the restricted Victorian household of Bergman's character, with no telephone, television, or computers, there were few touchstones to reality. Bergman's self-confidence and emotional equilibrium were destroyed as, over time, what she perceived as reality was not being validated by her seemingly loving husband.

> I asked my roommate a question, she guessed at the answer but didn't know for sure if she was correct. So I emailed my mother the question and she verified my friend was right. So I showed her the email from my mother, the whole point being "Look, you were right!" Well somehow this triggered a 15 minute long screaming, yelling, belittling, venomous verbal tirade that ended with her screaming "And then you show me this letter from your mother to tell me I'm wrong!" THE EXACT OPPOSITE OF WHAT I SAID! I started yelling at her to look at the email to see that I was, as I said, showing her she was right and she refused to look at the paper, actually turned her head away so she wouldn't see the printed words. For whatever reason, she NEEDED to feel I was trying to tell her she was wrong and therefore stupid.[26]

As Mason and Kreger point out, nonborderlines base their feelings on facts, while borderlines change the facts to fit their feelings. Gaslighting denial can also occur because a borderline has been dissociating—the borderline does indeed remember the scene differently—and not accurately. Sometimes, however, it is very difficult not to conclude that the borderline is intentionally lying. As psychiatrist Theodore Dorpat writes in *Gaslighting, the Double Whammy, Interrogation, and Other Methods of Covert Control in Psychotherapy and Analysis*, "Gaslighting is probably the most commonly used and effective type of verbal communication individuals have for manipulating and controlling other persons."[27] One book on how to use gaslighting techniques on one's enemies suggests a succinct synonym: "mind-fucking."[28]

The Impact of Borderline Personality Disorder on Others

Randi Kreger's passion for understanding and communicating information about borderline personality disorder came after seeing a therapist about her own failed relationship. Her intelligent and educated former boyfriend would often tell her she was talented and wonderful. At other times, however, he would careen into a different personality and scream that she was contemptible and the cause of all his prob-

Theta Rhythms

The effects of gaslighting on normal individuals can be extraordinarily unsettling and can contribute to confused behavior and scattered thinking patterns in those who have been subjected to the phenomenon. How might this confusion occur?

Decision making generally requires the coordinated activity of several or more brain structures. Structures that need to work together to accomplish a given task, such as the hippocampus and medial prefrontal cortex, show synchronized neural firing, like the internal clocks in computer chips. These synchronized rhythms, which are strongly affected by neurotransmitters, appear to help the brain to focus on the task at hand by allowing it to ignore structures that are not tuned in to the same frequency. The occasional inability to shut down certain brain wave communication channels is like having to listen to someone talking who won't shut up, says Matthew Wilson of MIT, who has contributed significantly to research in this area.[29] Out-of-synch brain waves may be related to mood disorders and diseases such as schizophrenia.

Could it be that, by sending conflicting signals as with the difference between reality and what the borderline falsely insists is reality, desynchronization might occur in neural structures that normally work together? Such desynchronization might account for the confused short-term reaction and the depressed long-term reaction to gaslighting behavior.[30]

lems. His insidious manipulations had left Kreger feeling as if she was "walking on eggshells"; the end of the relationship had left her with low self-esteem and a marked feeling of distrust. In-depth discussions with her therapist revealed that Kreger's former boyfriend had, in all probability, suffered from borderline personality disorder. This shocked Kreger into learning more about the disorder and, eventually, cowriting what has become a near-gospel self-help handbook for those loving, living, or working with those showing traits of borderline personality disorder.

Health care provider organizations have found that individuals with borderline personality disorder who see a doctor for routine or emergency reasons can take up an extraordinary amount of resources and staff time; efforts to resolve the many subsequent customer service complaints can occupy considerable administrative time and effort. Additionally, because the patient does not comply with requested measures, it is often difficult to achieve clinical improvement with borderline patients. At the same time, the patient denies any self-responsibility for these failures—and is more likely to litigate.[31]

Borderlines can be devastating to deal with in business environments. According to Dean Knudson, a psychiatrist for Behavioral Medical Interventions, a disability-management and workplace-intervention company in Minneapolis:

> People with BPD often have failing relationships and difficulties getting along with others, including co-workers. . . . They are prone to depression and abrupt anger and can engage in vicious personal attacks. . . . They often create chaotic environments, doing so in the workplace by pitting supervisors and workers against each other. Their self-identity and beliefs fluctuate frequently, and they can lack empathy for others or fail to feel remorse for their misdeeds. . . . People who are borderline are simply very, very difficult to live with, to speak with, to work with.[32]

It is the "now you see it, now you don't" borderline propensity for normal, loving behavior, alternating with denial in the face of bewilderingly spiteful and manipulative activities, that can make borderlines so difficult to understand—or diagnose—whether in the workplace, a hospital, or at home.

In fact, it is precisely those bewildering personality traits that led my family and me to an unlikely but intimate introduction to the consequences of Machiavellian—and deeply borderline-like—leadership.

And to two new members of our family.

Chapter 7

Slobodan Milosevic

The Butcher of the Balkans

"The hypothalamus is one of the most important parts of the brain, involved in many kinds of motivation, among other functions. The hypothalamus controls the 'four F's':

1. fighting
2. fleeing
3. feeding; and
4. mating"

—Psychology professor in neuropsychology intro course

Bafti and Irfan, then in their teens, first arrived at our house not long after the 1999 war in Kosovo. A professional we knew had agreed to serve as their sponsor for high school and college, but somehow, when it came time for the promised tuition, room, and board, the sponsorship evaporated. By that time, because the erstwhile sponsor had farmed the boys out to us, we had gotten to know them well. It seemed a horrific shame to send them back to a life of makeshift day labor in Kosovo, especially when they had such bright dreams of a college-educated future.

So, with a deep breath and some soul-searching, we adopted them, and set their goals for college as one with our family. Our two daughters suddenly had two big brothers to squabble with over the use of hair dryers, television, computer time, and dirty dishes. Now all four kids are, well, typical brothers and sisters, off and doing their own thing most of the time but relishing the opportunity to get back together for bouts of merciless teasing. Although Bafti and Irfan are brothers, their physiques reflect the melting pot of the Balkans. Bafti's light coloring echoes that of a blonde, green-eyed maternal aunt, while Irfan's features are like those of their father, descended from a well-known imam. (What with his black hair, dark eyes, and mild, difficult-to-place accent, Irfan is often misidentified as Mexican, which ruffles his Albanian pride.)

It took me several years to disentangle the politics behind the boys' semi-refugee status. Kosovo, a small province immediately next-door to Serbia, was peopled largely by Muslim Albanians, an ethnic and religious group who are quite different from Serbia's Orthodox Christian Slavs. The two mutually antagonistic peoples—who had gleefully worked toward each other's destruction for centuries—had been subsumed in the greater Yugoslavia in the historic equivalent of a howitzer wedding. Much as Hitler had used (and murdered) the Jews as a pretext to help unite the "Aryan" Germans, the situation in Yugoslavia was ripe for using minority Muslims as a pretext to unite the Serbs. Areas such as Kosovo and Bosnia were particularly useful in this regard—Serbs were present, but only in smaller numbers, so they could be portrayed in Serbia proper as being persecuted and thus in need of military intervention.

All that was needed was a Machiavellian to light the fuse.

The Quintessential Machiavellian: Slobodan Milosevic

Slobodan Milosevic, the "Butcher of the Balkans," used his sinister brand of Serbian nationalism to tear Yugoslavia apart in a bloody four-war decade of ethnic cleansing. By the time he was forced to stop, over two hundred and twenty-five thousand people were dead and millions of refugees—our sons Bafti and Irfan among them—were scattered worldwide.

After the unsolved mystery of Milosevic's prison cell death in March 2006, Jeffrey Fleishman neatly encapsulated Milosevic's devious life for the *Los Angeles Times*: "Sipping plum brandy and puffing Dutch cigarillos, the silver-haired Milosevic was defiant and arrogant, relishing his role as the key to stability in the Balkans. The Yugoslav leader frustrated a parade of U.S. and European diplomats by making promises he often broke. His government—circumventing years of international sanctions—took on the aura of a tawdry, gangster-run enterprise. Former US Ambassador Warren Zimmermann once called Milosevic 'the slickest con man in the Balkans.'"[1]

Milosevic maintained a connection with Dessa Trevisan, the *London Times* Balkan correspondent and doyenne of the Balkan press corps.

> The doughty Trevisan confronted the Serb leader. "I said to him: 'Mr. Milosevic, you have so much power, you have the whole nation behind you. You have to make a speech of reconciliation.' He listened to me, and he said, 'You mean a conciliatory speech.' I said, 'No, no, one of reconciliation.' He said it was a good idea. He would always agree with you . . . [H]e would agree, and do nothing. He is like an eel, he would look at you with those piggy eyes, he would flatter you and make it seem like he is listening, that what you say is going in, and then he would do the opposite."[2]

Milosevic had extraordinary people sense, with an uncanny ability to judge how serious his opponents were. Said one senior US official with extensive experience with Milosevic: "He was a real student of human nature. We might say ten times that he had to do X, Y and Z. He knew the one time out of ten when there would be consequences if he did not."[3]

Milosevic biographer Adam LeBor notes that in Dayton, Ohio, negotiating the final peace settlement for Bosnia, Milosevic became just one of the guys—an "ebullient rumbustious Serb," instead of a sinister fanatic. This, LeBor points out, was a clever move to disguise the fact that the war Milosevic had initiated in Bosnia had caused the deaths of two hundred thousand. Milosevic skillfully played his politician negotiators against one another, mocking the Americans when he was with the Europeans, and likewise contemptuously mimicking the Europeans when he was with the Americans. (Throughout his life, in all his dealings with people, Milosevic's excellent memory was a great boon.)[4] The "Butcher of the Balkans" walked out of Dayton feted as a peacemaker.

Arch dissembler that he was, however, Milosevic wasn't always the life of the party. He clicked on his charm only for those who counted—with others, he could be insufferably rude.[5] The ever-perceptive US Ambassador Warren Zimmermann stated: "As with all natural actors, it was impossible to tell how much consciously he deceived others and how much he deceived himself."[6]

Milosevic's biographers Dusko Doder and Louise Branson remarked on his ability to charm and flatter power people, disguising his true intentions and ambitions. "Orthodox Marxists in the party hierarchy regarded him as a staunch Bolshevik. At the same time, Western diplomats saw him as a young and energetic bank president who was pragmatic, reasonable, and pro-Western. He played his roles well: he talked liberal economics to one audience while he emphasized the need to maintain Marxist orthodoxy to another."[7] Later, both the Serbian nationalists and the diametrically opposed Communists would each be convinced that Milosevic was on their side. "The most striking thing about Milosevic was the absence of any ideological motivation at all," note Doder and Branson. "He was a chameleon."[8]

IDENTITY DISTURBANCE

"'Milosevic could switch moods with astonishing speed,' former envoy Richard Holbrooke wrote in *To End a War*, a diplomatic chronicle of the Bosnian conflict and Dayton negotiations. 'He could range from charm to brutality, from emotional outbursts to calm discussion of legal minutiae. When he was angry, his face wrinkled up, but he could regain control of himself instantly.'"[9] Milosevic's chameleon-like ability to shift moods and identities is, as mentioned earlier, associated with borderline personality disorder. This shape-shifting has been given a psychological term: *identity disturbance*, which means "a markedly and persistently unstable self-image or sense of self." In some individuals, such identity disturbance shows itself as a tendency to be obstinately inflexible; in others, it can be revealed as an opposite tendency to be overflexible in values, attitudes, and preferences to please others.

And indeed, both counterposing tendencies were apparent in Milosevic. Perhaps surprisingly, Milosevic's wife, Mira, exerted an extraordinary, almost Svengali-like influence over him. One of Mira's friends related that Mira's ideas influenced Milosevic so much that he would begin to "utter her thoughts and assessments as his own unaware of where she ends and he begins."[10] Mira was deeply complicit in Milosevic's activities—so much so that an entire book has been written about the joint actions of the pair: Slavoljub Djukic's *Milosevic and Marković: A Lust for Power*. In the foreword to Djukic's book, Mihailo Crnobrnja writes: "[U]nderstanding Milosevic without understanding his wife, and the special bonds that held them together, is next to impossible. It might not be an exaggeration to say that his enormous and unwavering love for her, together with her lust for power, was the unfortunate combination that triggered the tragic events in Yugoslavia." Mira suffered from her own deeply Machiavellian personality characteristics. Many feared her.

But identity disturbance is an odd trait. As alluded to previously, in some individuals it can show itself as inordinate flexibility, in others, or even in the same individuals at different times, such a disturbance manifests itself as being very critical, inflexible, and dogmatic about certain beliefs, to the point of offending others.[11] Guy Lesser, who studied Milosevic intensely during The Hague war crimes tribunal, described precisely those traits: "[U]ltimately the most interesting challenge in watching the proceedings is less about trying to sort out the daily testimony as it is presented by the prosecutors and more about trying to size up Milosevic as a man—and to ponder the enduring human capacity for evil . . . [O]ne can glean clues to his personality from the pro forma way in which he usually greets the crime-based witnesses, particularly those whose stories are the saddest. 'I am sorry for what happened to you' he'll say, in a harsh baritone. Then, virtually without a pause, he'll add, '*if* it happened to you.'"*[12]

Tess Wilkinson-Ryan and Drew Westen have found that the many aspects of identity disturbance might most succinctly be described as having several different dimensions, including:[13]

- *Role absorption*—the tendency to define oneself in terms of a single role, label, or reference group. The person's identity seems to revolve around a "cause" or shifting causes, and the person tends to define himself in terms of a label that provides a sense of identity.
- *Painful incoherence*—where a person tends to feel like a "false self" whose social persona does not match his or her inner experience. Psychiatrist Salman Akhtar has described this as *lack of authenticity*, which manifests itself as a tendency to take on the characteristics of others and a chameleon-like tendency to change one's personality in different situations.[14]
- *Inconsistency*—the person shows an inconsistency in his or her

*I myself saw this dogmatic inflexibility firsthand, so to speak, while visiting in Kosovo with Bafti and Irfan's relatives, watching Milosevic testifying on television during The Hague Tribunals. When asked about the massacre of Racak, where my sons' young cousins were butchered along with the rest of the village, Milosevic responded to the effect that the deaths were due to an artillery barrage—a by-product of war. When asked how an artillery barrage could produce marks of mutilation and torture, Milosevic's response was that his translating earphones had cut out.

behavior and attitudes that would make any coherent rendering of who the person really is difficult; for example, an individual who is a strong proponent of conventional sexual values for others while being personally promiscuous.

These facets of identity disturbance show themselves in the more difficult-to-understand personality quirks that Machiavellian characters often display.

THE DSM-IV DESCRIPTION OF BORDERLINE PERSONALITY DISORDER

As mentioned earlier, psychologist John McHoskey found that Machiavellian traits correlate not only with antisocial personality but also with borderline personality disorder. And so it is interesting to learn that identity disturbance is one of the key criteria that the DSM-IV uses to define the disorder. A positive checkmark on any five of the nine criteria is enough to categorize a person as having the disorder.

But *identity disturbance* isn't the only borderline trait that Milosevic seems to have carried. Another symptom was that of *alternating between extremes of idealization and devaluation*, also known as *splitting* or *black-and-white thinking*. As described above, splitting is when the borderline sees another person or group as either all good or all bad—there are no nuanced shades of gray to ponder. Milosevic, for example, was well known for feeling that "anyone . . . was either for him or against him, with no middle ground."[15] Likewise, Milosevic's chameleon-like persona at the Dayton talks made it easier for him to "split" his opponent negotiators, alternately seeing one group, and then the other, as the "bad guys."

In relation to another criterion, *paranoid ideation* (that is, having frequent paranoid thoughts), many borderlines generally expect others to behave badly toward them. Along these lines, Milosevic's most striking characteristic, according to one prominent biographer, was his complete lack of trust. Milosevic even had a saying that "if his hair

***DSM-IV Diagnostic Criteria for Borderline Personality Disorder*[16]**

Mood-Related Criteria

- Inappropriate, intense anger or difficulty controlling anger (e.g., frequent displays of temper, constant anger, recurrent physical fights).
- Chronic feelings of emptiness.
- Affective instability due to a marked reactivity of mood (e.g., intense episodic dysphoria, irritability, or anxiety usually lasting a few hours and only rarely more than a few days). [Dysphoria is the opposite of euphoria; it is a mixture of depression, anxiety, rage, and despair.]

Cognitive Criteria

- Frequent paranoid ideas or severe dissociative symptoms [dissociative symptoms can include hallucinations or a sense of depersonalization or unreality].
- Identity disturbance: markedly and persistently unstable self-image or sense of self.

knew what his intentions were, he'd have to shave it off."[17] He disliked committing anything to writing and was fondest of verbal agreements with no witnesses. As Doder and Branson further note:

> Milosevic liked to compartmentalize his activities, never giving any one subordinate too much control or understanding of the bigger picture. He took extra care to keep a formal distance, to make it seem as if others, outside his control, were responsible. He was the type of politician who leaves no traces. He never wrote an article under his name; his short speeches contained no plans; his interviews were slogans designed for the primitive nationalist ear. His style was conspiratorial. Everything was moved by word of mouth—without a

Behavioral Criteria (Forms of Impulsivity)

- Recurrent suicidal behavior, gestures, or threats, or self-mutilating behavior.
- Impulsivity in at least two areas that are potentially self-damaging (e.g., spending, sex, substance abuse, shoplifting, reckless driving, binge eating). (This does not include suicidal or self-mutilating behavior.)

Interpersonal Criteria

- Frantic efforts to avoid real or imagined abandonment that do not include suicidal or self-mutilating behavior.
- A pattern of unstable and intense interpersonal relationships characterized by alternating between extremes of idealization and devaluation. [This also includes the concept of "splitting," where a person is either "good" or "evil," with nothing in between.]

> paper trail. He never delegated to those around him defined areas of responsibility that could be regulated by a clear-cut statutory code.[18]

In relation to the characteristics of impulsivity and inability to control anger, biographer Slavoljub Djukic noted that Milosevic was "extremely impulsive, never [making] any effort to strike a balance between what he wants and what can actually be accomplished. Many of his decisions have been poorly considered, either premature or devastatingly belated, and lacking in foresight."[19] After the war began and NATO bombing sorties contributed to the ongoing destruction around him, Milosevic began "to crack, staging temper tantrums, screaming at aides, and throwing documents in the air."[20] Even under less extreme conditions, Milosevic's outbursts could be brutal. "I always knew you were a cunt!" Milosevic hissed in livid anger at an Albanian Communist who understandably refused to deliver a large block of votes to Milosevic.[21] (The only way to influence Milosevic, apparently, was to "threaten force, which he respected and feared.")[22]

The criteria related to feelings of abandonment and suicide also have bearing on Milosevic's life. After Milosevic reached adulthood, his father, mother, and a much-admired maternal uncle each died by their own hand. (Recent research has implicated a region on chromosome 2 in a genetic tendency toward suicide—this region is also associated with alcoholism, major depression, and bipolar disorder.)[23] Forensic psychologists "have speculated that Milosevic was a depressive, scarred by a family history of suicide and abandonment."[24] Indeed, Milosevic's parents separated not long after he was born, which could undoubtedly have placed stress on a boy with a certain genetic predisposition. (Interestingly, Milosevic's brother Borislav grew up with a very different personality—he "not only excelled in high school but thoroughly enjoyed life, and effortlessly went on to establish a prominent career.")[25]

Milan Panic, a Serbian-American, Yugoslavian-born businessman whom Milosevic seduced into serving as prime minister in an effort to avert sanctions against Serbia, related an anecdote that was more to the point. Soon after Panic was invited into the government, Milosevic became surprised and increasingly nervous at Panic's burgeoning popularity. One evening, Milosevic arrived unannounced at Panic's house, where they moved to sit out on the veranda overlooking Belgrade and drink wine. Panic laid forthrightly into Milosevic:

> "You are responsible for what is happening in Bosnia. You are responsible for the catastrophic economic conditions in Serbia. Resign. You have to resign!" Milosevic listened in shock, unused to such direct and hyper-critical talk. Then, as Panic relates the story "suddenly he looks at me and says, 'Enough.' I have never seen him so despondent. So he takes a revolver—he always carries a revolver—and hands it to me. 'Shoot me,' he says. 'Get it over with.' And I am stunned, of course. I can't believe my ears."

> "'Are you crazy?' Panic cried. 'You've got to be sick! You have children, you have family. You want me to shoot you? You are sick. Resign!'"[26]

In general, Milosevic's emotions were poorly regulated. In times of stress, he turned to alcohol. When sanctions against Serbia were imposed by the West and domestic opposition was reaching a crescendo, he watched

> huge crowds of protesters on television and asked Dusan Mitevic, his friend and propaganda chief, with genuine incredulity, "Who are those people?" . . . Rumors began to circulate about his state of mind. Some attributed his mood swings to diabetes, type II, from which he had suffered since the early 1980s, but in reality few people saw him in his black moods, when he would hurl streams of profanities at those around him. He was given to seizures of apoplectic fury when crossed or when confronted by the brutal stupidity of his proxies which he thought gave him a bad name. He was exceedingly vindictive, his soul full of one black passion—to get even, to avenge . . .
>
> Several of his former friends and classmates also spoke of pathology. The man, they said, was subject to serious depressions, sometimes staying at home for a couple of days to hide the condition.[27]

Milosevic was "a man who wanted to be in total control of every detail. . . . And by all accounts he took great pleasure in the misery he brought to others and in the sense of power and control this afforded him."[28] This taking pleasure in the misery he brought others was more characteristic of psychopathy, or outright sadism, than borderline personality disorder. In point of fact, Milosevic ignited wars in Slovenia, Croatia, Bosnia, Kosovo—and ultimately would set Serb against Serb.[29]

Milosevic also exhibited the frequent narcissistic and antisocial characteristics of a pattern of grandiosity, need for admiration, and lack of empathy; Serbian psychologist Zarko Trebjesanin described him as a "cold narcissus."[30] Oddly enough, the exaggerated sense of self-importance associated with narcissism—the slippery twin sister of both antisocial and borderline personality disorder—can motivate Machiavellian manipulation and lying. After all, for many people—even normal people with no personality disorder—a laudable end can justify many means. And for a narcissist, nothing is more laudable than the grandiosity of the narcissist himself.

As the *Economist* pointed out prior to Milosevic's death: "Mr Milosevic may be facing 66 separate charges of the gravest crimes imaginable, including genocide, at the UN's war-crimes tribunal in The Hague, but he appears to be enjoying himself nevertheless. Berating or bullying prosecution witnesses with relish, and peppering the judges with objections, he has turned his right to act as his own lawyer into a bravura performance. . . . Richard May, the presiding judge, constantly reminded Milosevic to stick to the issues: 'Attacking the other side is not a defense,' he has repeatedly explained to the defendant."[31]

Was Milosevic a Borderline?

The Machiavellian Milosevic had significant issues related to at least five of the criteria of borderline personality disorder, according to the DSM-IV. He also possessed a number of related borderline-like coping mechanisms. But, perhaps surprisingly, that does *not* make Milosevic a borderline. To be so judged, one's symptoms must each be found "clinically significant"—that is, a trained clinician must determine that five of the nine criteria in question are severe enough to cause significant distress or impairment. This is not quite as straightforward as it might seem.[32]

Say, for example, a person throws a phone occasionally, or curses someone out in traffic. Is this enough to classify him as having difficulty controlling anger? Probably not. But what if he loses his temper frequently at his son for the pettiest of reasons but spoils his wife and daughter no matter what they do? That's perhaps a little more problematic—worthy of further investigation by a therapist. Now, what if this person loses his temper, chews out his boss, subsequently loses his job—and does this three times over a period of a year? Does he now have difficulties controlling his anger? Probably. But it's important to note that these facts might never come to the attention of a clinician unless our man gets a divorce or goes to jail.

The bottom line is that, if one uses a categorical DSM-IV approach to analyzing borderline personality disorder, a person has to be so severely disabled to achieve a definitive diagnosis that he essentially can't function effectively in society. Diagnosis exists primarily to facilitate treatment in a clinical setting, but a number of problematic individuals—even if they do have symptoms that would reasonably qualify as clinically significant—just don't come into a clinic to receive the attention of psychiatric or forensic services.[33]

Another problem with diagnosing borderline personality using the DSM-IV criteria relates to the fact that the criteria are written in clinical, dispassionate fashion that obscures as much as explains borderline symptoms. And there is yet another problem: the DSM-IV assumes that "all nine criteria are equally contributory, and allows for the seeming paradox that someone with the supposedly enduring diagnosis of BPD could suddenly be 'cured' of the illness by overcoming even one defining criterion."[34]

THE DIMENSIONAL APPROACH TO DESCRIBING BORDERLINE PERSONALITY DISORDER

A number of psychologists and psychiatrists have advocated a different method of diagnosing BPD and subclinical BPD, one that involves a dimensional, or symptom-based approach, similar to that of the PDQ-4+ test McHoskey used for his seminal Machiavellian study. A fairly typical dimensional approach to BPD (advocated by BPD specialist Robert Friedel, whom we noted earlier regarding his borderline sister Denise), involves identifying varying degrees of the following four symptoms:[35]

- poorly regulated emotions
- impulsivity
- impaired perception and reasoning
- markedly disturbed relationships

As you'll notice, these four dimensional symptoms are quite similar to the typical borderline symptom complexes we described in the last chapter. In fact, those four dimensions appear to relate to something integral to our basic humanity, because *all* of the ten personality disorders of the DSM-IV show up as deviations from normal in a very similar list of traits.[36]

Milosevic would register "tilt" on all four of Friedel's dimensional symptoms of borderline personality disorder. Three of Milosevic's personality traits—those involving poorly regulated emotions, impulsivity, and markedly disturbed relationships—have already been discussed in relation to the DSM-IV criteria for BPD. But the fourth dimensional symptom—impaired perception and reasoning (a more general form of the frequent paranoid thoughts or severe dissociative symptoms)—deserves special attention.

As it happens, Milosevic showed impaired perception and reasoning on a number of different occasions. For example, while international outrage about the brutality of Serbian ethnic cleansing was turning Yugoslavia into an outlaw state, US Ambassador Warren Zimmermann met with Milosevic one evening to attempt to negotiate a reversal of the cleansing and a withdrawal from Bosnia. Shockingly, Milosevic denied that the Serbs were even in Bosnia. LeBor notes: "It is hard to know what Milosevic was thinking. Did he really believe that the ambassador of the most powerful country in the world, with extensive intelligence services, and spy satellites that could read a numberplate, did not know what was happening in Bosnia and who was responsible?" Zimmermann was left speechless as the gaslighting

Milosevic pleaded, "I'm not so bad, am I? Am I such a black sheep?"[37] Later, at The Hague, Milosevic would smugly testify that "the 1995 massacre of 7,000 unarmed Muslims at Srebrenica was carried out not by Serb militiamen, but by French intelligence."[38]

Milosevic's chief of the general staff, General Perisi, would later describe Milosevic as "not living in reality, rejecting 'competent opinions and proposals,' and resorting to 'fraud and lies . . . to change and shape the people's perception of reality.'" Milosevic had completely convinced himself that the West was bluffing about bombing. When the US ambassador to Macedonia, Christopher Hill, suggested that bombing was a wholly realistic option, Milosevic retorted: "Anyone who does that—bomb—is going to spend the rest of his life on a psychiatric couch."[39]

Milosevic would constantly repeat the statement that "Serbia is not at war," even as the death notices of Serb soldiers killed in Bosnia filled up newspapers and were pinned to trees all over Yugoslavia.[40] The mayor of Belgrade, Zoran Djindjic, recalled telling Milosevic: "'You really have problems; there are one hundred thousand people on the street demonstrating against you.' [Milosevic] looked at me and said, 'You must be watching too much CNN. There aren't.'"[41] Congressman Rod Blagojevich, a Serb American from Chicago's Northwest Side who was helping to negotiate the release of three American servicemen, said that Milosevic reminded him of "a defendant who had prepared himself never to admit anything and who had repeated his version of events consistently so that he essentially came to believe in it."[42] (The idea that Milosevic could have reprogrammed his memory may be possible; a recent study has demonstrated that people can consciously choose to forget certain memories. It turns out that the same neural circuits that are used to consciously suppress movement can also be used to suppress memories.)[43] Another former associate observed that Milosevic "decides first what is expedient for him to believe, and then he believes it."[44] Biographer LeBor describes Milosevic's thinking, which bordered on delusional: "Confronted with the disastrous reality of his policies, Milosevic reverted to denial, outright mendacity and fantastical talk of wonderful economic opportunities."[45] Even on a personal front, Milosevic's thought processes could conflict notably with reality: after his father committed suicide, Milosevic avoided all mention of it, except for one occasion—in which he denied that the suicide had occurred.[46]

The dimensional approach to understanding borderline personality disorder is useful because it relates much more directly to what is going on neurologically. And neurological research related to the disorder has made significant breakthroughs over the past few years. Most researchers agree that a dysfunction of the emotional regulation system is a core component of the disorder and that this emotional dysregulation manifests itself most often, and dramatically, as a pattern of unstable and intense interpersonal relationships.[47] It is thought that a dual-brain pathology in both the prefrontal and limbic circuits might underlie borderline symptoms.[48]

Was Milosevic a Psychopath?

But, you might say, *not so fast.* Doesn't Milosevic look to be an ideal candidate for a diagnosis of psychopathy rather than borderline personality disorder?

Yes and no. It's clear that Milosevic was "compassion-impaired" with regard to anyone who wasn't one of his lockstep supporters. But at the same time, he was "well regarded for his loyalty to his relatively small circle of friends," although that loyalty was contingent on absolute loyalty to him.[49] More than that, Milosevic was so infatuated with his wife that, even when the two were sixteen-year-olds, they were dubbed "Romeo and Juliet."[50] Decades after they first met, while imprisoned at The Hague, Milosevic would spend the lunch hour together with his wife, "holding hands, kissing each other, and stroking each other's faces."[51] Milosevic's brother, Borislav, related how: "[Milosevic] is a man of strong will, he has his own beliefs, his own positions, but on the other hand he is a man devoted to his friends

Fig. 7.1. The "Butcher of the Balkans" returns from Dayton in winter 1995 with a big hug for his wife, Mira.

and family, he is a very good paterfamilias. He is a good father and he is not a cruel person, as he is portrayed."[52] For his children, Milosevic would do anything. British diplomat David Austin noted that Milosevic once ended negotiations at five o'clock sharp because he had to go home for his daughter's birthday party.[53]

It seems, then, that Milosevic might have had some characteristics of a psychopath—but not all. We can get a better handle on this by looking at the dimensional-like traits of psychopathy. They are defined by researchers David Cooke and Christine Mitchie as:[54]

- arrogant and deceitful interpersonal style
- deficient affective (emotional) experience
- impulsive and irresponsible behavioral style

You can see there is quite an overlap in the dimensions of psychopathy and those of borderline personality disorder. Moreover, each of these dimensions is intentionally a bit fuzzy—after all, how can a few words of definition catch the full range of possibilities of neural behavior that might be associated with a certain personality characteristic? In the end, the two disorders can shade into each other. People like Milosevic might have an arrogant and deceitful interpersonal style and an impulsive behavioral style; yet they could also sometimes act quite responsibly, and their emotional experience could be deficient only toward some—not all—people.

Two British psychiatrists, Nicholas Swift and Harpal Nandhra, see so many patients with combined borderline and antisocial traits that they coined a new term for the syndrome: *borderpath*—a fusion of borderline and psychopath.[55] Many others have noted that psychopaths who experience anxiety—so-called secondary psychopaths—have a symptom set that solidly overlaps with borderline personality disorder.[56] (Secondary psychopathy doesn't necessarily mean that the symptoms are any less severe—several studies have shown that secondary psychopaths are more aggressive and disruptive in institutions than are primary psychopaths.)[57]

Information about borderline personality disorder appears to provide special insight into certain unusual symptoms and traits. Yet research from psychopathy and other disorders is also relevant. As the grand maestro of psychopathy research Robert Hare states: "[I]t is not surprising that there is substantial comorbidity [co-occurrence] of psychopathy with antisocial, narcissistic, histrionic, and borderline" personality disorders.[58] We will therefore continue to explore research that relates to both borderline personality disorder and psychopathy. (Those with borderline personality disorder alone, after all, can often feel very real remorse related to the emotional damage they inflict on others.) If you are left occasionally confused as to when one disorder shades into another, don't worry—even experts often face the same dilemma. We're talking about human beings, after all—sloppy business!

Personal Impact: Milosevic and My Family

It's fascinating to learn about different psychological characteristics and how even subtle distortions can profoundly shape personalities. But such analyses can often miss the most important aspect of all—the effect of such personality disturbances on others. I was able, for example, to observe the impact of Milosevic's devastating personality traits through its influence on the early lives of our adopted sons.

For Bafti and Irfan, life had been increasingly grim since Milosevic had revoked Kosovo's autonomy in 1989, putting the region under the direct and draconian rule of Serbia itself. Schools and colleges were closed—all Albanian professors were expelled from the University of Pristina. Denied any education past eighth grade, our sons ended up in an illegal, privately organized high school, seated on the floor of a neighbor's house, crowded with fifty other students in a sixteen-by-sixteen-foot room, the teacher squeezing past to reach the makeshift chalkboard.

Many Albanian language newspapers and television stations were closed, and hundreds of thousands of workers—including Bafti and Irfan's parents—lost their jobs. There was no money for food, clothing, shoes, medicine, or fuel during the frigid winters. Racism against "primitive" Albanians was rampant, and the country, once a model of integration, was reduced to a de facto apartheid. The tiny income the boys made selling cigarettes—fleeing at the merest glimpse of the corrupt and brutal Serbian police—was the only thing that kept the family alive. Random arrests increased, along with beatings, prison sentences, and outright murders by the Serbian paramilitaries.

The leader of the Kosovar Albanians, Ibrahim Rugova, modeled his resistance after the peaceful, passive techniques of Gandhi. But Rugova's advocacy of passive resistance seemed, if anything, to encourage Milosevic and the Serbs in their terrorism and ethnocentrism. As Rugova would eventually discover, Milosevic and the Serbs were not like Lord Mountbatten and the British in India, and Rugova himself was nowhere near as manipulatively adroit at nonviolence as Gandhi.

January 15, 1999, was the beginning of the Muslim holiday of small Bajram, widely known as Eid—a day of celebration that marked the end of the austere fasting of Ramadan. Extended machine gun fire had taken place early that morning in Racak, a village roughly half a mile from Bafti and Irfan's home in Shtime. Arising early to celebrate the holiday, Bafti had heard the gunfire but thought nothing of it. Both Shtime and Racak, where a number of Bafti and Irfan's cousins lived, were routinely exposed to such weapons firing in the increasingly terrorized province.

US Ambassador William Walker's explicit description of the events surrounding what Bafti had heard near Racak that January morning helped galvanize international opinion and would eventually lead to the NATO intervention against the Serbian military infrastructure. Walker—head of the Kosovo Verification Mission, the international monitoring group sent to Kosovo to guarantee the human rights situation in the beleaguered province—recalls the day as follows:

> We entered the village [of Racak]. . . . There were a lot of women around in tears and crying. We came out of the village. . . . After about 500 yards, we came across the first body. . . . I was a little shaken by this thing with the head gone. . . . We saw about 10 bodies while going up the hill. We finally reached a pile of bodies, maybe 17, 18, 19 bodies just helter-skelter in a big pile, all with horrible wounds in the head. All of them were in these clothes that peasants in that part of the world wear when they're out in the fields doing

Fig. 7.2. Counting bodies on the hillside beside the village of Racak after the massacre.

> their jobs. A good number of them had lost control of their bodily functions, and so their clothes were stained, and that sort of thing. This had not been concocted by anyone, even though this was later the claim of the government.[59]

NATO commander General Wesley Clark recounted a "red-faced Milosevic's description of the Racak massacre as a provocation. 'This is not a massacre,' he said. 'It was staged. These people were terrorists.'"[60] Milosevic's words were a prelude to his defense at the International Criminal Court in The Hague. A description by *Newsweek*'s Michael Leverson Meyer catches Milosevic's expertise in twisting the truth. "'Why are you inventing this?' Milosevic asked a witness who told of a villager with his chest hacked open and heart ripped out, a photo of which I also have. 'I saw it,' the man replied, telling how he and others emerged from the forest to find the bodies after the Serbs had left. . . . What's dismaying is Milosevic's insouciant disregard for the truth of what happened in that village that day. It's not that he knows, or doesn't. It's that he considers it irrelevant, a laughing matter."[61]

***

By the end of Milosevic's dictatorship, lines of bedraggled vendors were reduced to hawking bottles of petrol or logs for firewood in downtown Belgrade, the capital of Serbia—all a result of Milosevic's brutally craven bilking of every possible penny from the populace. Milosevic was one of Serbia's top bank managers before he came to power nationally. He could have "used his knowledge of capitalism to introduce free market reforms and privatization. Instead, he ran the Serbian economy the same way he ran the Serbian state, setting up a network of trusted loyalists who either took over or sidestepped the established financial institutions."[62] When Milosevic was directly accused of taking the Serbs back to the Middle Ages, his response was simply a smug "I know."[63]

CHAPTER 8

LENSES, FRAMES, AND HOW BROKEN BRAINS WORK

"Don't believe everything you think."

—Anonymous

"Do you know who Stalin was?" Irena asked. Irena was the officers' waitress—she'd noticed me eyeing the picture of "Uncle Joe" Stalin screwed into the wall beside the porthole. Captain Shevchenko shot her a warning look. It was the early 1980s, and I was a hundred miles off the coast of the Pacific Northwest on a Russian trawler—technically then part of the "Evil Empire" of the Soviet Union. As far as Shevchenko was concerned, Stalin was not a suitable subject for discussion with me—the sole American onboard. I glanced down; my tumbler had been unobtrusively topped off with more Stolichnaya, straight from the freezer.

"Of course I know who Stalin was," I replied. "*Na zdorovye.*" To your health. I cocked the tumbler as the rusty two-hundred-foot trawler lurched starboard. After months at sea, I could no longer smell the belching, acrid smoke of the fish meal plant or the rotting fish on deck. The perennial thrumming of the ship's engines and salt spray that often ruffled my hair just seemed natural now—like home.

I followed Irena's question with one of my own: "Did you know Stalin was responsible for the deaths of at least twenty million people during his purges?"

"Have you ever known anyone who lost somebody during those so-called purges?" Shevchenko scoffed. The captain was a true believer in communism.

"Yes," I said, "Most of my teachers lost at least one member of their family."

"Oh," said the captain. He'd thought he had me. "Well . . . everybody makes mistakes."[1]

Lenses and Frames

It has been nearly twenty-five years since my days as a translator on Soviet trawlers. I still miss the daily ration of palatalized consonants and the meandering, caboose endings I once slung so glibly. As Vladimir Nabokov muses in "An Evening of Russian Poetry":

> Because all hangs together—shape and sound,
> heather and honey, vessel and content.
> Not only rainbows—every line is bent,
> and skulls and seeds and all good worlds are round,
> like Russian verse, like our colossal vowels:
> those painted eggs, those glossy pitcher flowers that swallow
> whole a golden bumblebee,
> those shells that hold a thimble and the sea.[2]

The study of language, it turns out, has often played a key role in understanding neurological processes. Linguist Noam Chomsky used his ideas about the formation and learning of language to help pick apart flaws in Skinner's ideas, which helped begin the long overdue overthrow of Skinnerian behaviorism in psychology. Chomsky gave credence to the idea that the brain was composed of a modular set of units, with specialized, innately unique areas that were responsible for learning different things, such as language, mathematics, or the various motor skills. Another revolutionary investigator, Harvard psychologist Steven Pinker, cut his professional teeth on research related to language before moving on to write *The Blank Slate*. Pinker's brilliant book, along with the Judith Rich Harris's seminal *The Nurture Assumption*, was to help redefine psychology so that nature—genetics—was firmly shown to play an equal or even more crucial role than nurture—that is, the environment. Pinker, along with John Tooby and his wife, Leda Cosmides, and others, also stood on the shoulders of Chomsky's ideas to eventually conclude that the human brain, including its module for learning language, evolved by natural selection, just like other body parts.

The ability to learn a language, any language, is related to innate "wiring" that's built in to nearly all human beings. As a language and surrounding culture is learned, however, it subtly shapes perception, often in ways in which people aren't consciously aware. Those who grow up speaking Chinese, for example, process mathematics in different areas of the brain than those who grow up speaking English as their first language. Both groups use the inferior parietal cortex, but Chinese speakers also use a visual processing area, while English speakers use a language processing area. Richard E. Nisbett, co-director of the University of Michigan's Culture and Cognition Program, notes that studies involving this type of phenomena are important because they tell us "something about the particular pathways in the brain that underlie some of the differences between Asians and Westerners in thought patterns."[3] Other studies have shown that gyri in the frontal, temporal, and parietal lobes develop differently in Chinese speakers than in English speakers—acquisition of a different language appears to cause anatomical differences in the brain.[4] Chinese speakers literally see the world differently than English speakers—eye-tracking studies show that English speakers tend to first focus on individual items in the foreground of the picture, while Chinese speakers tend to first take in the background and the picture as a whole.[5]

In some sense, then, language and culture might be thought of as helping to structure the neurologically based lenses that people use to perceive reality. But of course, language and culture aren't the only influences on our neurological lenses. Family upbringing, religion, political persuasion, educational background, work experience—all help create the different framing lenses people use.[6] Practiced expertise with a musical instrument, for example, can change the structure of the musician's primary motor cortex; London's experienced taxi drivers develop enlarged back ends of their hippocampi as a result of the intricate mental map of the city that they develop and store.[7]

James Surowiecki's *The Wisdom of Crowds* drives home his counterintuitive thesis that multiple viewpoints from individuals with a wide range of backgrounds, rather than the restricted viewpoints of experts or specialists, are crucial in reaching informed decisions on complex topics. Such successful problem solving almost certainly reflects the value of using a wide variety of framing lenses. In some sense, getting input from a broad variety of people is like getting input from a wide variety of devices—microscopes, telescopes, litmus paper, tensile testors, ultrasound devices, and weighing scales. There is indeed a shared physical reality out there; but for complex problems, no single one of us—experts included—has the all-encompassing set of tools or ways of perceiving that are necessary to truly understand its every aspect. In fact, experts receive such similar training that occasionally they can be unaware, as a group, of shared inadequacies in their approach.

Living and working among the Russians was a terrific way for me to, in some sense, broaden my neurological frame. (Later, the study of engineering would prove enticing precisely because I knew it would provide yet another, very different, frame with which to shape my experiences.) I've a good ear for language—mirror neurons in my language module fire quite nicely, thank you. After a few months out on the boats, my Russian took on the soft Ukrainian flavor of

Nakhodka's fishermen, mixed with the salty language of fishermen everywhere. I learned that Russian-speaking brains slice life differently than my English-speaking brain. Reality may be the same in Russian—but it feels different. And reality feels different in another way beyond that of language.

***

"The terrible things your own people say about their country," said Captain Shevchenko one night as we sat up drinking after another late-night trawl was tucked into the hold. "No self-respecting person should ever say things like that about where they live. Not if they have any respect for their history and their culture and their race. Not if they have any patriotism."

"You can't teach patriotism," I began.

But Shevchenko interrupted contemptuously, as if I'd just drooled. "Of course you can teach patriotism. We do it all the time."[8]

The conversation rolled on, but that part of it stuck, bothering me.

I remembered dozens of one-sided tipsy Slavic arguments, which from the Soviet's perspective involved clear-cut dichotomies of good against evil. Excessive Western personal freedom, for example, versus sacred duty to the state. All-pervasive Western drug addiction versus minor Russian drinking habits (not quite!). The wicked American invasion of Vietnam versus the high-minded Russian invasion of Afghanistan, which was solely for the good of the Afghans.

I could see the crumbling decay of the Soviet Empire all around me, from the rotten fish processing plant below decks, to the hollow-eyed fear of the political commissar and the KGB, to the "who cares" attitude toward work, to the rampant alcoholism of the crew. But despite the all-pervasive rot, it was a rare Russian who could see—much less admit to—any problem with the system of government.

I'd tried to lure the Soviets into dispassionately viewing both the Soviet and the American systems by expressing my own genuine admiration for the positives of Soviet society. (After all, how could I not love a culture where even the lowliest deck crew worker was enamored of chess, and where vulgar public displays of wealth were nonexistent?) I related positives about my own country and modeled criticism as well. "Of course the United States has made mistakes," I admitted, proudly showing off my naive open-mindedness. I played

Fig. 8.1.

right into their hands. The Soviets almost literally could not hear me when I said anything good about America—it just didn't mesh with anything they'd ever been taught, and anyway, I was obviously brainwashed. They tuned in only when I criticized my country—criticism of the United States was easy for them to understand because I was only reinforcing for them what they'd already learned so deeply. In psychology, such a phenomenon is called "confirmation bias." It involves behavior where one looks for and notices things that confirm one's beliefs, while ignoring, not looking for, or undervaluing the relevance of what contradicts one's beliefs. Selective thinking. A bit of voluntary blindness. Dangerous stuff.

As Shevchenko had inadvertently pointed out with his crack about patriotism, I was a cultural chump. If you've been raised from childhood to think a certain way about things like the greatness of Mother Russia and the Soviet Union, if you've had a one-sided education about the superiority of communism and the evils of Western decadence, a few conversations with a foreigner to the contrary won't amount to a hill of beans of difference. Despite the incipient decay and fearful, police-state behavior visible everywhere around me, despite the deaths of tens of millions of Russians in horrendous purges that lasted half a century, Shevchenko could mindlessly, in almost cultlike fashion, assert that theirs was the best system in the world.

I'd learned one of the most valuable lessons I would ever learn—that deep-rooted emotional reasoning can often trump logic.

How does that happen?

Neurological Systems and How They Function to Regulate Emotion

The Cerebral Cortex

Emotion involves two very different structures in the brain, the first of which is the cerebral cortex—the newer part of the brain in evolutionary terms. But where, you might ask, *is* the cerebral cortex? Well, if you happened to have a preserved dead body lying around, you could use a handily vibrating Stryker saw to whir your way around the upper portion of the skull. Then you could chisel away at the bone until you were in a position to use real muscle to pry the skullcap off. In an old anatomist's trick, you could tuck the removed skullcap under the remaining part of the skull to serve as a pillow—this lifts the rest of the corpse's head into a more "comfortable," easily viewed position. Peel away the remains of the meninges, the fibrous membranes that cover the brain, and you would see the cerebral cortex lying directly before you—the entire region just below where the corpse's hat used to sit.

If you weren't terribly squeamish after your first cut and didn't mind doing a bit of careful scalpel and forceps work, you could lift the brain out. The three-pound grayish-brown concoction of lobes and ventricles would nestle nicely in your hands, rather like a ropy, congealed pudding. It is hard to believe this compact mass once directed the cuddling of babies or dreamed of starting a business.

Next, you could take a scalpel and slice the brain in half, so you could easily compare the two halves with the areas shown on the next illustration. Poking a bit with a probe, you could pick out the four specialized areas that are particularly important in processing and controlling emotions. These areas have clunky names that can roll surprisingly swiftly off an anatomist's tongue: the anterior cingulate, ventromedial prefrontal, orbital prefrontal, and dorsolateral prefrontal cortices. We noted some of these regions earlier in relation to psychopathy. Most of these areas are in the prefrontal cortex at the very front of the brain, near the eyes and forehead. Problems involving any of these regions or the pathways connecting them may result in strange emotional behavior—impulsivity, moodiness, or the inability to weigh the soundness of a decision.

But what precisely does each of those four areas do?

The first of the four—the orbitofrontal cortex—is designed to inhibit inappropriate actions. This allows us to set aside our urges and

Dorsolateral Prefrontal Cortex

Orbitofrontal Cortex

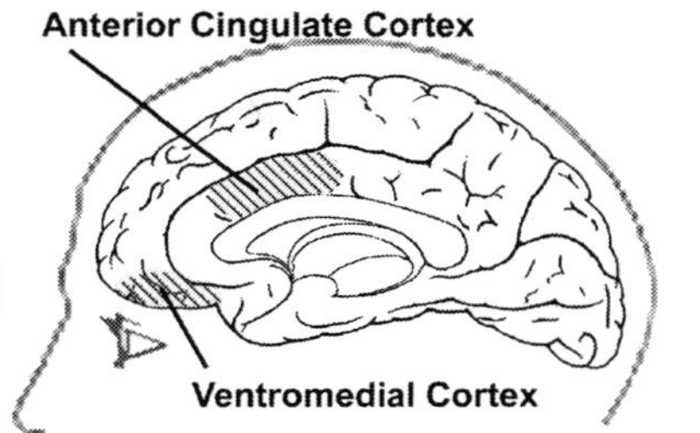

Fig. 8.2. The illustration on the left shows an external view of the brain, while the right shows a cross section. The four areas shown are those that relate most strongly to processing and controlling emotions.

put off immediate reward in favor of long-term advantage. When your stomach is growling but you wave away your favorite type of cookie, it's your orbitofrontal cortex that's helping you say no. (The term *orbito-*, incidentally, refers to the area right above the orbit of the eye.) As you might expect, the orbitofrontal cortex plays a significant role in controlling impulsivity. A dysfunctional orbitofrontal cortex doesn't "play nicely" with the rest of the brain—this may propel a person willy-nilly toward explosively impulsive behavior.

The nearby dorsolateral prefrontal cortex is where plans and concepts are held and manipulated. This is the area that would, for example, process plans related to your family's trip to Disneyland—from purchasing plane tickets, to getting your hotel, to planning what you'll be seeing and doing each day. This is also the area that seems to choose to do one thing rather than another. At the grocery store, this area would help you decide to select Fuji apples rather than Golden Delicious. Subtle differences in the operation of the dorsolateral prefrontal cortex may explain the differing styles, for example, of the habitual grocery store ditherer as opposed to the fast shopper who's in and out of the store in seconds. The dorsolateral prefrontal cortex is also deeply involved in the ability to think logically and rationally about various topics. People with slight problems in their dorsolateral prefrontal cortex appear to act normally; however, they may confidently, even arrogantly, draw bizarre and irrational conclusions. Problems related to this area may help cause the gaslighting and projection seen so frequently in borderline-like behavior.[9]

The ventromedial cortex is located near, and has very dense connections with, an area completely separate from the cerebral cortex known as the limbic system. The limbic system composes the subconscious part of the brain where emotions are born. The ventromedial cortex therefore allows us to consciously experience our emotions, and helps link conscious to unconscious thought. (You might say that the ventromedial prefrontal cortex plays a role in emotional cognition, whereas the dorsolateral prefrontal cortex is involved in rational cognition.) The ventromedial prefrontal cortex also gives meaning to our perceptions. For example, depressed people who find no meaning in anything they do often have inactive ventromedial cortices. On the other hand, bipolar individuals in acute manic phase who find meaning in everything they do have hyperactive ventromedial cortices. As we shall see, emotion relates in important ways to our ability to make wise decisions. Not surprisingly, problems with the ventromedial cortex, much like problems with dorsolateral prefrontal cortex, can lead to subtly irrational behavior.

Finally, the anterior cingulate cortex, nestled on the underside of the cerebral cortex, helps us to focus our attention and "tune in" to thoughts. As such, this area of the brain is related to the ability to focus on boring, difficult, or unpleasant subject matter. Dysfunction here may inhibit the ability of borderlines and subclinical borderlines to focus on something they do not wish to hear. The anterior cingulate cortex also plays a role in helping to make new memories permanent and in producing feelings of empathy.

The Limbic System

If the cerebral cortex, with four of its subunits, is the first of the two major areas involved in emotional processing, what's the second? It is the limbic system—a far older area (in evolutionary terms) nestled deep within the brain. If you haven't happened to have tidied up yet and still have your dissected brain lying around, you could turn one of the bisected halves of the brain on its side and start picking apart the deep interior to see the main components of the limbic system—the amygdala, thalamus, and the hippocampus. Most emotions are born in the limbic system, along with our appetites and urges. Even though this part of the brain is below our level of awareness, its constant feeding of impulses to the conscious cortical areas profoundly affects us.

Neural Connections

Meanwhile, as you might expect, the *connections* between all the different areas we've just discussed are also critically important in handling emotion. Those connections are generally made through neurons

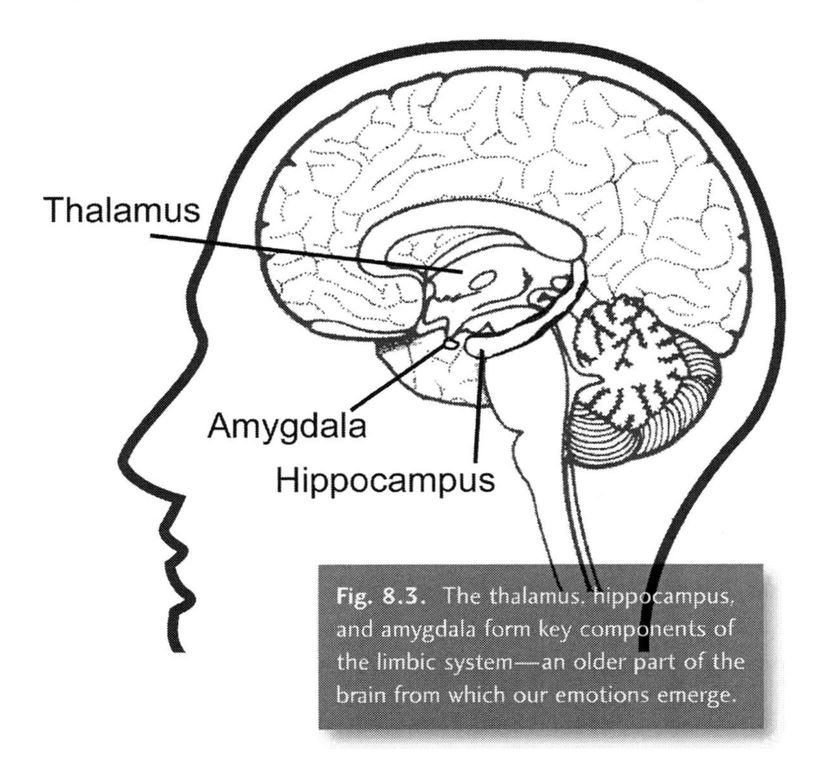

Fig. 8.3. The thalamus, hippocampus, and amygdala form key components of the limbic system—an older part of the brain from which our emotions emerge.

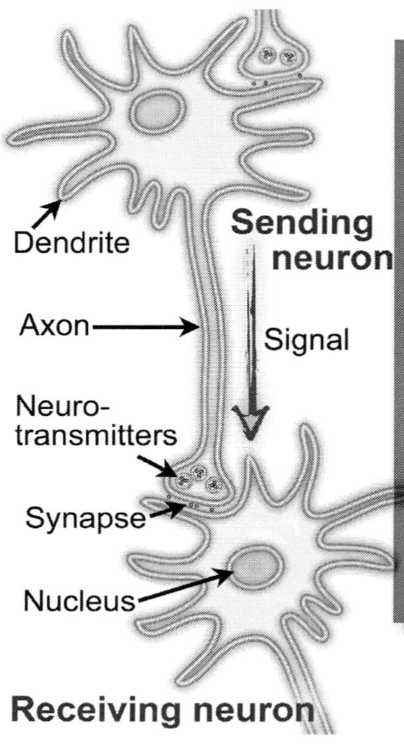

Fig. 8.4. This drawing shows the connections between a neuron that is sending a signal and the receiving neuron. Chemical flares known as neurotransmitters carry the signal across the tiny gap between the neurons. Neural signals progress in this hopscotch fashion to allow different areas of the brain to communicate all sorts of information—from sensory information related to sight, hearing, or touch, to high-level processing related to deliberative thought, to emotional reactions such as fear or aggression.

—spidery cells that act like electronic wires to carry information in a syncopated rhythm. As the information travels down the neural wires, it leaps small gaps—*synapses*—between the neurons in the form of chemical flares known as neurotransmitters. Neurotransmitters come in dozens of different flavors, but, as we've already seen, two of the most well-studied and significant are serotonin and dopamine. Serotonin, as we know, plays a critical role in disorders such as depression, borderline personality disorder, bipolar disorder, and anxiety; it is also thought to be involved in sexuality and appetite. Dopamine is thought to relate to control of the brain's reward mechanisms, as well as the control of movement. Problems related to the dopamine system have been strongly related to psychosis and schizophrenia.

To see how the neurons make connections between the different

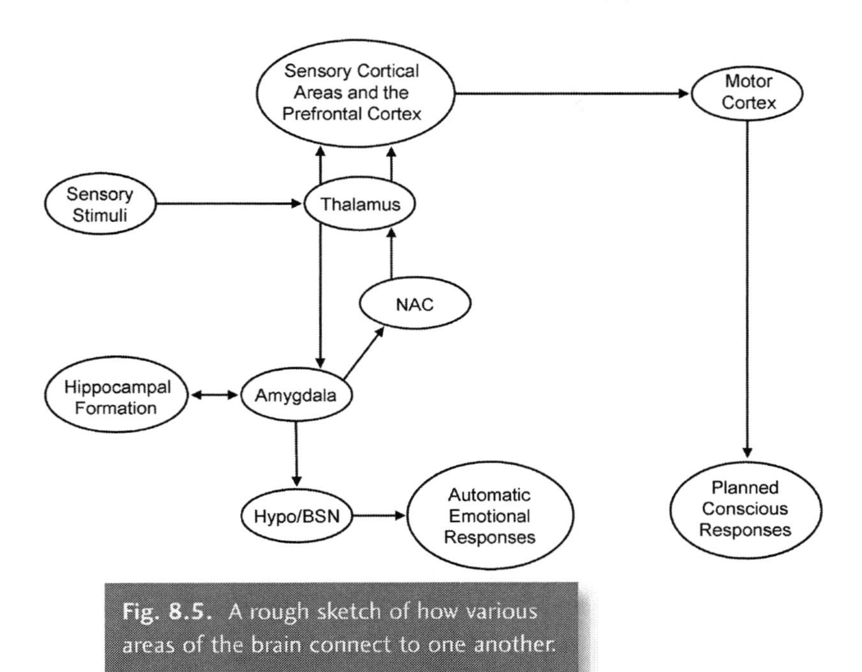

Fig. 8.5. A rough sketch of how various areas of the brain connect to one another.

areas of the brain, we might want to think about a map. For example, when we want to see how cities in the state of Michigan are connected to each other, we often pull out a road map and trace the connections we're interested in to figure out how to get from, say, Kalamazoo, Michigan, to Hell (yes, there is a Hell in Michigan). In just the same way, we can see how the different regions of the brain connect to each other by following a "brain map" like that shown above.

To get a feel for how this diagram relates to physical processes that are occurring in the brain, start at the left, with the oval labeled "Sensory Stimuli" (that's our "Kalamazoo," from which we can reach the different points in our neural Michigan). Information from the sensory stimuli—eyes, ears, touch—is sent undulating along neurons to the limbic structure of the thalamus. The thalamus, an egg-shaped gray mass deep in the center of the brain, is a routing station for these sensory signals, some of which are sent on to the amygdala. At the amygdala, the emotional significance of the information is determined. For example, if you were to have lesions on your amygdala so that they had to be surgically removed, you wouldn't be able to understand the meaning of growls, screams, angry voices, or other negative signs. From the amygdala, signals are sent on to the nucleus accumbens (the NAC), which determines the appropriate levels of motivation and reward—this is the common site of action for drugs such as cocaine that produce euphoria.

All of this processing at the subconscious limbic-system level is not very precise, but is fast—fast enough to get your body revved for reaction to danger. For example, if you hear a nearby gunshot, a signal from the amygdala might be shunted through to the hypothalamus and brain stem nuclei (Hypo/BSN) to produce an automatic emotional response: *"Run!"* Ultimately, all of these deep-brain limbic structures transmit their information up to the prefrontal cortex, where we consciously become aware of the emotional responses we refer to as feelings.

But there is one last piece to the process. Some of the original sensory information that was split at the thalamus is sent directly on to the cerebral cortex (including the four areas—the anterior cingulate, ventromedial prefrontal, orbital prefrontal, and dorsolateral prefrontal cortices—that we mentioned before). Here, slower but more precise conscious evaluation of the directly routed sensory stimuli takes place. But even so, the emotionally preprocessed signals, along with components from memory, eventually arrive at the higher, conscious, prefrontal areas, and can strongly affect, or even overwhelm, our conscious thought processes. These signals are what trigger that surge of affection we feel when we see our beloved dog, or the fear and revulsion we feel when we encounter a rattlesnake in our path. These are also the emotional signals that influence how open we are to receiving and understanding new information. This emotional overlay is intimately related to why my Soviet colleagues and friends on the trawlers often could simply not understand when I hinted that there might be something wrong with their government, or something right about the West.

"Feel Good" Politics: How Machiavellians—and Altruists—Manipulate Emotions

The role of emotion in shaping "rational" thinking is tremendously underrated. Strong evidence shows that human behavior is the product of both the rational deliberation that takes place in the front areas of the cerebral cortex and the "emote control"—emotional reasoning—that originates in the limbic system.[10] These two neural systems operate in radically different fashions and often are in conflict with one another. As Princeton sociologist Douglas Massey writes: "Emotionality clearly preceded rationality in evolutionary sequence, and as rationality developed it did not replace emotionality as a basis for human interaction. Rather, rational abilities were gradually *added* to preexisting and simultaneously developing emotional capacities. Indeed, the neural anatomy essential for full rationality—the prefrontal cortex—is a very recent evolutionary innovation, emerging only in the last 150,000 years of a six-million-year existence, representing only about 2.5 percent of humanity's total time on earth."[11]

In an article on the effect of emotion in foreign policy and international law, law professor Jules Lobel and psychologist George Loewenstein expand on Massey's sentiments:

> Human behavior . . . is not under the sole control of either affect or deliberation but results from the interaction of these two qualitatively different processes—like a computer that has two different types of processors it can draw upon that process information in qualitatively different ways. Emote control is fast but is largely limited to operating according to evolved patterns. Deliberation is far more flexible—it can be applied to almost any type of task or problem one might encounter—but is comparatively slow and laborious. Deliberation involves what psychologists call "controlled processes" that involve step-by-step logic or computations and are often associated with a subjective feeling of effort. . . . *Emote control is the default mode, while deliberation is invoked in special circumstances.*[12] (italics added)

Emote control is not necessarily a bad phenomenon. It can lead to harrowing rescues from burning cars, loyalty to our friends even when the costs far outweigh any benefits, and the impassioned leadership of Winston Churchill in his defense of Britain against the evils of Nazism in World War II. But it can also lead to other, less happy results—especially with regard to Machiavellians.

One such example is the twenty-five-year crusade to prove that a Virginia man, Roger Coleman, was innocent of the rape and murder of his sister-in-law. Coleman was a likable, good-looking man who resolutely insisted on his innocence. Thus, despite a large body of evidence—that is, rational facts—that proved beyond a reasonable doubt that Coleman was guilty, death penalty opponents rallied to his cause. Jim McCloskey was Coleman's principal advocate—he fought for years to save Coleman's life and even founded a group, "Centurion Ministries," to help get the falsely convicted out of jail. McCloskey says, "I promised Roger Coleman the night he was executed [that] I would do all within my power to prove that he was innocent. Those were my last words to a dying man."[13] Eventually, the state was convinced to make use of new DNA technology to reexamine the case. Hopes were high among death penalty opponents that Coleman's name would prove to be an effective rallying cry to help prevent future executions. When Coleman's DNA analysis came back, however, he was shown to have been guilty as charged. (This is not to suggest that all those convicted of murder are in fact guilty. The Innocence Project at the Benjamin N. Cardozo School of Law at Yeshiva University has proven otherwise for dozens of the poor and forgotten.)

But why was McCloskey so certain that Coleman was innocent? The ultimate source of McCloskey's certainty is revealed by his statements after Coleman's "guilty" DNA results came back. McCloskey "felt betrayed by the man whose last words included the statement 'An innocent man is going to be murdered tonight.' 'How can somebody, with such equanimity, such dignity, such quiet confidence, make those his final words even though he is guilty?' McCloskey said."[14] McCloskey had made an "emote control" decision that Coleman could not have been guilty—this decision had been deeply confirmed by Coleman's body language. The intrusion of reality in the form of Coleman's betrayal must have been devastating. Machiavellians such as Coleman often take advantage of an emotionally based—perhaps even genetically predisposed—desire on the part of some honest individuals to believe that others are also honest. This can occur despite sometimes overwhelming evidence to the contrary.

A recent imaging study by psychologist Drew Westen and his colleagues at Emory University provides firm support for the existence of emotional reasoning.[15] Just prior to the 2004 Bush-Kerry presidential elections, two groups of subjects were recruited—fifteen ardent Democrats and fifteen ardent Republicans. Each was presented with conflicting and seemingly damaging statements about their candidate, as well as about more neutral targets such as actor Tom Hanks (who, it appears, is a likeable guy for people of all political persuasions). Unsurprisingly, when the participants were asked to draw a logical conclusion about a candidate from the other—"wrong"—political party, the participants found a way to arrive at a conclusion that made the candidate look bad, even though logic should have mitigated the particular circumstances and allowed them to reach a different conclusion. Here's where it gets interesting.

When this "emote control" began to occur, parts of the brain normally involved in reasoning were not activated. Instead, a constellation of activations occurred in the same areas of the brain where punishment, pain, and negative emotions are experienced (that is, in the left insula, lateral frontal cortex, and ventromedial prefrontal cortex). Once a way was found to ignore information that could not be rationally discounted, the neural punishment areas turned off, and the participant received a blast of activation in the circuits involving rewards—akin to the high an addict receives when getting his fix. In essence, the participants were not about to let facts get in the way of their hot-button decision making and quick buzz of reward. "None of the circuits involved in conscious reasoning were particularly engaged," says Westen. "Essentially, it appears as if partisans twirl the cognitive kaleidoscope until they get the conclusions they want, and then they get massively reinforced for it, with the elimination of negative emotional states and activation of positive ones." Interestingly, a more extreme version of this type of behavior may underlie borderline-like splitting.[16]

A completely different process occurred when a participant had no emotional investment at stake, as with statements concerning the "neutral" Tom Hanks. In this straightforward, rational process, only the dorsolateral prefrontal cortex was activated—both Democrats and Republicans were swayed toward reaching the logical conclusion by the mitigating statement. Dorsolateral activation is, notably, the part of the brain most associated with reasoning as well as conscious efforts to suppress emotion.

Ultimately, Westen and his colleagues believe that "emotionally biased reasoning leads to the 'stamping in' or reinforcement of a defensive belief, associating the participant's 'revisionist' account of the data with positive emotion or relief and elimination of distress. 'The result is that partisan beliefs are calcified, and the person can learn very little from new data,'" Westen says.[17] Westen's remarkable study showed that neural information processing related to what he terms "motivated reasoning"—that is, political bias (in this case, at least)—appears to be qualitatively different from reasoning when a person has no strong emotional stake in the conclusions to be reached.

The study is thus the first to describe the neural processes that underlie political judgment and decision making, as well as to describe processes involving emote control, psychological defense, confirmatory bias, and some forms of cognitive dissonance. The significance of these findings ranges beyond the study of politics: "Everyone from executives and judges to scientists and politicians may reason to emotionally biased judgments when they have a vested interest in how to interpret 'the facts,'" according to Westen.[18]

But is emote control really that common—particularly in such areas as public policy, which cry out for reasoned and rational discourse?

Absolutely.

For example, well-intentioned, emotionally based concerns about the sanctity of human life have led to the United States withdrawing support for birth control programs to third world countries—despite the fact that many of those in favor of withdrawing support have never lived in those countries and have absolutely no idea of the magnitude or devastating effects of overpopulation there. Interestingly, people who are against such birth control programs, based on the sanctity of human life, are often also firmly pro–death penalty. When I point out the inconsistency of being pro–death penalty but anti–birth control to these friends, they suddenly decide that not quite *all* human life is sacred. Then they change the subject.

Similar emotional reasoning has led kindhearted individuals to support "feel-good" programs such as busing, which seemed, on the face of it, to be an outstanding method to integrate school systems. Opponents of this program—whatever their reasons—were seen as racists, which meant that rational concerns about the program were discounted.[19] The result was that cities such as Detroit were devastated as the well-to-do moved to the suburbs, out of range of the mandated busing system. This worsened the segregation the busing had been designed to remedy. Similarly, a laudable desire to eliminate shabby housing, drug use, and crime in poor areas led to "the projects," which were to house even more highly concentrated areas of drug use and crime. Such government-mandated programs as busing and the projects, often generated by emote control related to genuinely altruistic considerations, have wasted billions of taxpayers' dollars and led to a worsening of the very conditions they were meant to solve.

No one can claim to be truly unbiased. We all come at issues through our experiences and values, filtered by the emotional and cognitive processes of our hardwired neurological makeup. But if we socialize only with members of our own particular religious persuasion; if we work in an environment with only one-sided political input; if we read only Web sites or other news sources by writers who echo our views, then we strongly reinforce the emotional, rather than logical, basis for our beliefs. After all—if "everyone" we know believes what we believe, we find an emotional reinforcement that helps close off consideration of other perspectives. (So much for the wisdom of crowds.)

Aren't there times when we as citizens should respond with healthy emotions, to fight for what we believe in, especially when we feel policies are causing people actual harm? Of course. But simply looking at the research results, one must conclude that people's first emotional responses about what's wrong, who is to blame, or how to proceed, particularly in relation to complex issues, must always—*always*—be considered suspect.[20] There is no simple algorithm for teasing rationality from emotion. An ardent Democrat or Republican, a dyed-in-the-wool communist union organizer, a young devotee of Scientology, a Palestinian suicide bomber, or a KKK grand kleagle could each read the above paragraphs and think, *I'm not irrational—it's those other idiots who can't see the obvious.* But we all have pockets of irrationality, some large, some small, no matter if we are mathematicians who make our living doing proofs, wealthy philanthropists, or stay-at-home housewives.

If there is one thing that is important for us to know, it is that emote control allows our best traits—love, caring, loyalty, and trust—to be used as manipulative levers. Me-first Milosevic-like Machiavellians, with their convincing masks of integrity and charm, climb in every social hierarchy, schmooze in every community, saunter through every neighborhood. Whether we care about children, students, families, factory workers, fellow followers of Christ, brothers in Islam, blacks, whites, Mongolians, or Democratic or Republican political planks, the successfully sinister have no compunction about using our best intentions to further their own purposes—and themselves. By believing a heartbreaking speech about how important it is for us to be treated "fairly," or a tale of how we've been victimized, or a plea to put our hearts and minds toward helping others, we may be doing our tiny part to stoke the fires and empower a Machiavellian. It is bitter balm indeed to learn how easily Machiavellians can use our own neurological quirks to fool us into actively working against the very ideals we hold most dear.

Seeing Subtle Defects in the Emote Control System—Borderline Personality Disorder

If an irrational emote control is the default mode on even normal people, what happens when there are subtle defects in the emotional system?

A lot, as it turns out. Some of these defects allow us to be more easily manipulated, as with the Alzheimer's victim who is conned into giving all her money to an "investor," leaving her with nothing. But other defects can lead, it seems, to some of the insidious, duplicitous, sometimes irrationally self-serving thought processes of the successfully sinister.

Probably the best way to begin to understand these latter effects—and certainly one of the ways most relevant to Machiavellianism—is to take a careful look at the dysfunction that occurs in borderline personality disorder. In this condition, three sets of neural circuits appear to be disturbed—all involving the neurological areas we've explored thus far. Here's the list:

- emotional dysregulation (moodiness, depression, anxiety, feelings of emptiness): *limbic* system
- impulsivity: *anterior cingulate* and *orbitomedial prefrontal* systems
- cognitive-perceptual impairment: *dorsolateral*, *ventromedial*, and *orbitofrontal* systems

Superb recent results from imaging and other studies are providing information that can help each of us intuit a sense of what can go wrong with these circuits. We'll examine each in turn.

Emotional Dysregulation: Limbic System

In those with borderline personality disorder, we can actually *see* how some poorly tuned portions of the limbic system seem to cause the characteristic fluctuating moods, depression, anxiety, and feelings of emptiness. The amygdala and the hippocampus of the limbic system play key roles in memory and emotional responses. Magnetic resonance imaging has shown that both organs are noticeably smaller than usual in borderline patients.[21] This decrease in the size of the hippocampus is intriguing in that this organ seems to be associated with a person's ability to "catch" contextual cues. Abnormalities in the hippocampus may explain why borderlines don't seem to be able to pay attention to important but placidly unemotional task-relevant information. Instead, their brains seem to key in on emotionally related cues—especially if these cues are negative.[22] Just because these cues can be detected, however, doesn't mean the borderline reacts to them in the same way as a person without the disorder.

In fact, it appears that the amygdala, as well as other parts of the

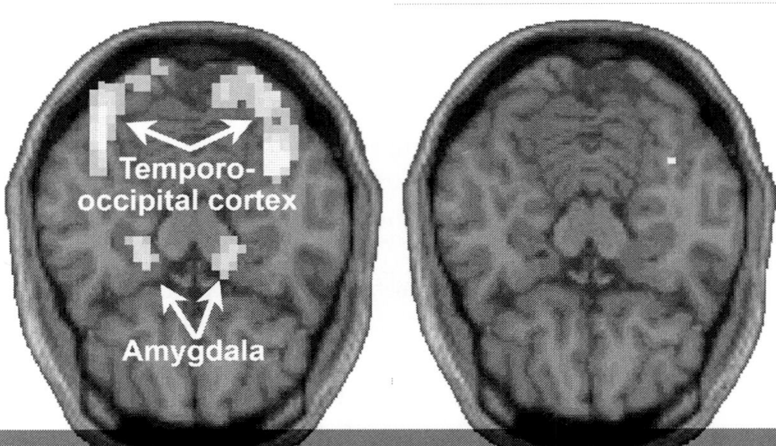

Fig. 8.6. In this view peering down from the top of the head at a cross section of the brain, activated areas are shown in a lighter color. You can see a big difference in the neural responses of patients with borderline personality who viewed repulsive slides (left), compared with healthy volunteers who viewed the slides (right). The borderline patients showed high activity in the amygdalae and the temporo-occipital cortex, while the healthy volunteers showed a normal, subdued response.

brain, actually function differently in borderlines under conditions that provoke emotional responses. The functional magnetic resonance imaging results above show what happens to cerebral metabolism in borderline patients, as opposed to normal controls, after viewing repulsive images on slides (imagine, if you will, something grotesque, like roadkill.) The light spots indicate unusual excess activity found only in the borderline patients. These spots are related to increased metabolism in the amygdala as well as the prefrontal and temporo-occipital cortex.

Researchers believe that hyperactive amygdalae are a cause of the intense and slowly subsiding emotions experienced when borderlines suffer even minor irritation. The increased activities in the prefrontal and temporo-occipital cortices—which indicate increased attention to emotionally relevant input from the environment—may be due to the boosted signal from the amygdala. Translated into practical terms, this would explain why a borderline might overreact to a minor constructive criticism by a spouse or friend, evoking an angry response that leads to a major argument.

Similarly revved-up amygdalae were found after borderline patients were exposed to faces showing various types of emotion. The overly hefty amygdala response is likely to be a key component of borderline emotional vulnerability, especially in the context of disturbed interpersonal relations and the crucial role of the amygdala in processing emotional stimuli and reactions. It may also be related to the borderline hypersensitivity to the state of other people and their uncanny ability to read emotions.[23]

Impulsivity: Anterior Cingulate and Orbitomedial Prefrontal Systems

The second type of dysfunction common in borderlines relates to impulsivity. Borderlines often have difficulty controlling their impulses and behaving in a reasonable and rational manner—especially when they are feeling strongly emotional. One recent study has revealed that borderlines appear to suffer the same problems with impulsivity as those who have suffered damage to their orbitofrontal cortex.[24] To understand the revealing recent imaging studies, you might want to look at the augmented version, shown below, of our neural map. In this expanded version of the drawing, several new areas have been added: the dorsal raphe nucleus and the ventral tegmental area.[25] These two areas produce key neuromodulators—molecules that can boost or dampen the effect of those chemical flares that ferry information across the synapses. The dorsal raphe nucleus produces the neuromodulator serotonin, while the ventral tegmental area produces the neuromodulator dopamine. As the many spidery connections show, the neurotransmitters of the ventral tegmental area and the dorsal raphe nucleus affect many areas of the brain at both conscious and unconscious levels.

The next figure gives a sense of the underpinnings of the strange

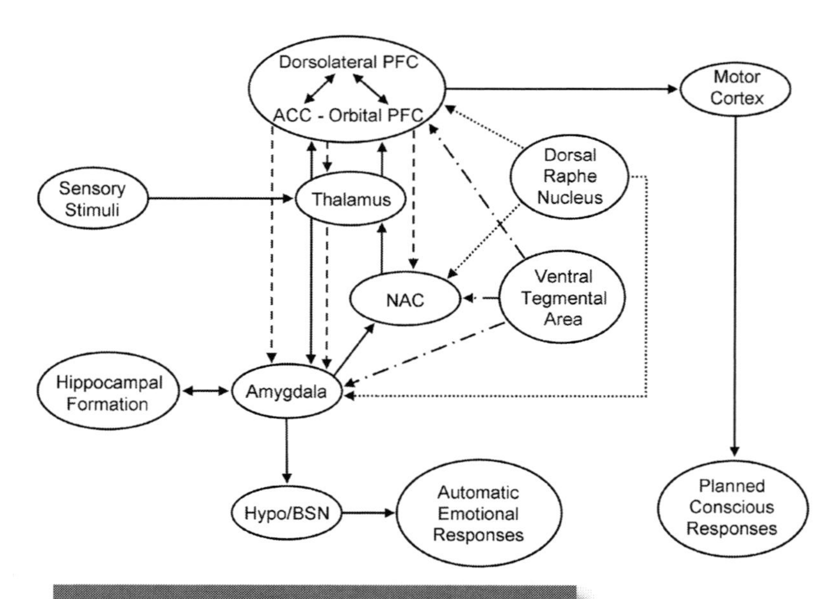

Fig. 8.7. A more detailed sketch of how various areas of the brain connect to one another.

neural behavior of those with borderline personality disorder.[26] Neural PET scans of normal subjects are shown in the top row, while similar scans of borderline subjects are shown at the bottom. ("PET" stands for *positron emission tomography*—which produces images of the brain's chemicals that look almost like color x-rays.) These scans indicate that serotonin levels for borderlines are much lower than normal in many neural regions, including the medial, lateral, and orbital prefrontal cortices—precisely those regions thought to be involved in the increased impulsivity that afflicts borderlines. Other studies have indicated similar problems with imbalances related to neurotransmitters with big-handled names like acetylcholine, norepinephrine, and gamma-aminobutyric acid (GABA). Indirect evidence also points to problems related to dopamine.[27] These neurotransmitter imbalances may well be related to the mood imbalances we discussed in relationship with the amygdala.[28]

The next PET images reveal that glucose metabolism ("feeding" of various areas of the brain), was found to be much larger in borderline patients than in normal people in prefrontal and frontal regions (the patterned area of the images). But glucose metabolism was *decreased* in the limbic regions of

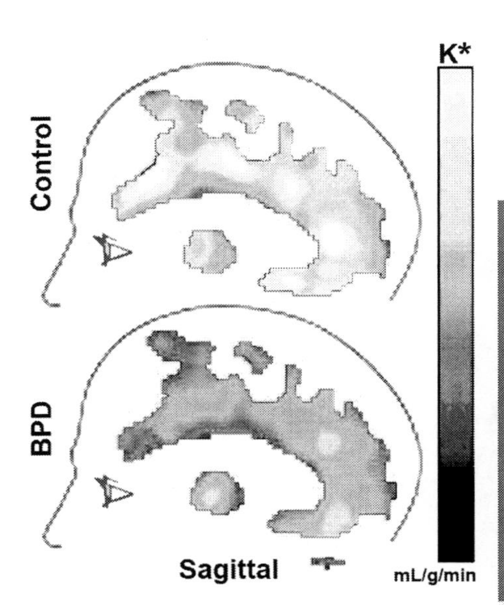

Fig. 8.8. A neural PET scan typical of a normal subject is shown at the top—the lightness indicates serotonin levels are high. A borderline patient, shown below, reveals dark shades that indicate lower serotonin levels in many neural regions, including those affiliated with impulsivity.

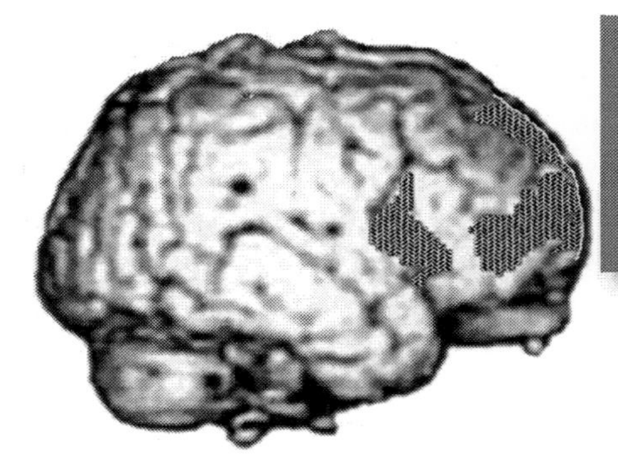

Fig. 8.9. Glucose metabolism ("feeding" of various areas of the brain), was much larger than normal for borderlines in the crosshatched areas shown.

borderlines, for example, the hippocampus.[29] Other studies have confirmed unusual activities in these areas. For example, it has been found that when serotonin levels are increased in the gaps between the neurons, borderline patients, unlike normal controls, show decreased metabolism in the anterior cingulate gyrus and orbitofrontal cortex—important areas for inhibiting impulses.[30]

Yet another study showed that borderline subjects had a significantly smaller right parietal lobe—smaller by 11 percent.[31] (The parietal lobe stretches from ear to ear in a band around the top of the head.) Less is known about the functions of the parietal than the frontal lobes, but the parietal lobes are thought to play a role in integrating information from the various senses, as well as in using objects. It appears that signals to the parietal lobe may arrive out of sync with signals elsewhere in the brain—this may contribute not only to cognitive impairment and impulsivity but also to the borderline's sense of personality fragmentation, as well as

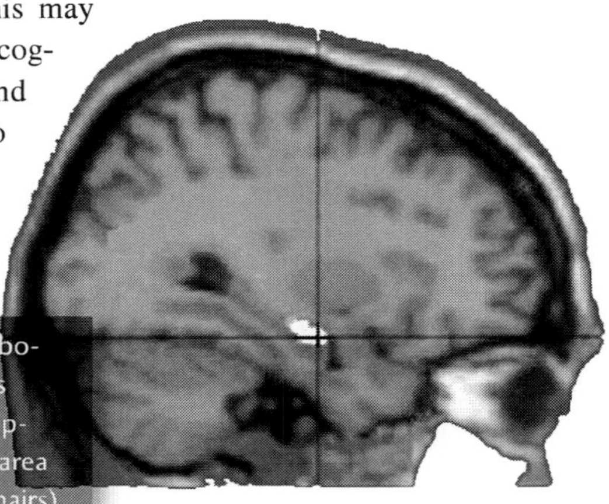

Fig. 8.10. Slow metabolism in borderlines was observed in the left hippocampus (the white area targeted by the crosshairs).

her inability to integrate the positive and negative aspects of herself and the external world.[32]

Interestingly, researchers have found that the smaller the size of the right parietal lobe in relation to the left, the stronger the psychotic symptoms and schizoid personality traits seen in borderlines. Additionally, the volume of the hippocampus has been found to be an extraordinary 17 percent smaller in borderline patients—the smaller the hippocampus, the stronger the borderline symptoms.[33] All these deep-set structural differences in the brain, invisible to the naked eye, appear to be profoundly related to the unusual behavioral "choices" of borderlines.

But there is another important factor to consider. Why do some people show strong indications of borderline personality disorder and, as a result, become clinically diagnosed, whereas others show many of the same symptoms yet are never diagnosed? Oddly enough, the reason may lie beyond mere chance and instead may be related to the nature of their impulsivity.

Impulsivity, if you'll remember, relates to "bottom-up" kicking of the brain into emotional gear by the limbic system. But it also relates to the ability to exert "top-down" control over those emotions once they've been kicked into gear. Top-down control is a function of the conscious control over your sensory and emotional systems that you exert to stop yourself from, for example, swearing in front of your four-year-old after slamming your finger in the car door. The neurological pathways related to conscious control can be seen by looking back at the second flow chart a few pages back—the dotted lines with arrows indicate the top-down conscious control of the body from the prefrontal cortical areas (the anterior cingulate and orbital prefrontal cortices) back down to the thalamus, the amygdala, and the nucleus accumbens. Note that "top-down" executive control is separate from the "bottom-up" pathway.

The capability to exert executive control over emotion may be one of the defining differences between a clinically diagnosed borderline and a person who shows many borderline traits. This ability to avoid emotional meltdown, at least when it is imprudent to melt, can allow a person to avoid personality disorder diagnosis, notwithstanding other emotional dysfunction and cognitive-perceptual impairment. It can also allow for the borderline coping characteristic of "situational competence." As noted psychiatrist Ken Silk has observed, when one of his borderline patients becomes upset, she becomes so swallowed by emotion that she is incapable of logic and is completely unable to assert emotional self control.* This contrasts markedly with what some refer to as a "high-functioning," subclinical borderline—a person with many borderline-like traits but who is able to exert executive control as needed. An example of a high-functioning, nonclinically diagnosed person with borderline-like symptoms might be the manipulative supervisor whose angry tirades, threats, and general malicious behavior toward his subordinates is legendary. Yet this same supervisor, even in the midst of a raging fit, is able to flip a mental switch and slip smoothly into a calm greeting if the company president were to pop in.

A difference in circuits related to executive control capability between those clinically and nonclinically diagnosed individuals with borderline traits has been exposed through a recent set of clever experiments from an interdisciplinary group of collaborators led by psychologist Michael Posner from the Sackler Institute for Developmental Psychobiology in New York.[34] The group hypothesized that there were two differences in temperament between borderline

---

*Many psychologists and psychiatrists prefer not to treat borderline patients, since their manipulative tactics can take a psychic toll. Ken Silk, on the other hand, is that rare therapist who truly *likes* his borderline patients. He understands that their inability to grasp positive emotions, combined with the pervasive and unremitting emotional distress that borderlines experience, can make their lives into a veritable prison, not only for those with the disorder, but for those who are attempting to treat or help them. Silk's ability to deal so compassionately with borderlines may stem in part from his own strong ego boundaries. I once watched a conference audience fall raptly silent as Silk explained how he deals with middle-of-the-night suicidal phone calls, telling his patients: "We are both working very hard in the sessions to decrease your suicidality. But you also need to know that I do not keep special ideas or plans at home that would make you all of a sudden not suicidal. If I had such tools, I would, of course, use them in the office. So if you are really unsafe, then you should go to the emergency room and have the emergency room folks call me. But if you are feeling suicidal but know that you can be safe until the next appointment, then thanks for calling me and letting me know and we can concentrate on this in our next session."

patients and normal controls. The first difference was thought to be in *negative affectivity*, which lies behind the strongly negative mood and volatile anger of borderline personality disorder. The second involved what the group termed *effortful control*—that is, conscious control. Problems here underlie instability in relationships, impulsivity, and difficulties in controlling emotion. (Both negative affectivity and effortful control appear to be strongly heritable traits.)[35] A test of effortful control and negativity, the Adult Temperament Questionnaire, was given to borderline patients and a control group of one thousand students at New York's Hunter College. As expected, the borderlines revealed far higher levels of negativity and far lower ability to demonstrate effortful control than the controls. Then—and this is the clever part—the researchers combed through the large pool of students to find controls who happened to match the borderlines in their scores on negative affect and apparent degree of effortful control. What, the researchers wondered, was the difference between a clinically diagnosed borderline and a nondiagnosed individual who shared the same, often problematic, temperament?

Before they even did any further testing, the researchers noticed that they had difficulty working with the nondiagnosed controls with negative affect and low effortful control. Even though these individuals were prescreened and did not meet the criteria for diagnosis with any personality disorder, they showed unreliability in keeping appointments, made frequent changes of address and phone number, and evinced heightened anxiety and paranoia regarding the experimental procedures. The researchers felt that this subset of the control group contained people whose behavior showed evidence of emotional dysregulation, even though they were functioning in school and did not meet the stringent criteria required for diagnosis of a personality disorder.

Test results surprisingly revealed that there was a single, but very distinct, difference between borderlines and the temperamentally similar controls: clinically diagnosed borderlines had much more difficulty in quickly resolving conflicting information than the temperamentally matched controls and the normal controls.

The ability to resolve conflicting information—an essential aspect of what is known as the executive attentional network—appears to be centered in the anterior cingulate cortex. This region undergoes substantial development relatively early in childhood, between the ages of two and seven years old. Posner and his colleagues hypothesized that "certain individuals possess a genetic propensity related to temperamental characteristics present prior to the disorder. Patients with borderline personality disorder often report incidents of abuse during childhood. These environmental events, together with the genetic propensity, may interfere with development of executive control, which in turn influences the ability to develop clear ideas and empathy for the minds of others as well as the experience of diffusion of one's own identity."[36] In summary then, a very difficult childhood can take a predisposition for borderline personality disorder and turn it into a devastating reality. A decent upbringing, however, can mean that a person with the same predisposition will grow up only to be difficult to deal with, but not necessarily someone who comes to the attention of a clinician.

Interestingly, Posner's work regarding the importance of the attentional network to borderline-like behavior ties neatly with Joseph Newman's work involving psychopathy. Newman, if you'll remember, proposed that psychopathy is actually a disorder related to the attentional network. In fact, just as diseases such as schizophrenia may be found in individuals with a wide variety of underlying personality traits, Newman has found evidence that psychopathy may be found in individuals with many different personalities. The underlying personality appears to shape the expression of the psychopathy. A nonviolent sort might become a con man, while a more violent type might become a hit man.

I can't help but remember Carolyn's brief visit to the store to pick up a few things—during which time she disappeared, to reappear in my life five years later. Once Carolyn's attention was turned to the man she met at the store, it seems, she lost focus on the fact that she was supposed to come back to the cabin and her waiting family.

Cognitive-Perceptual Impairment: Dorsolateral, Ventromedial, and Orbitofrontal Prefrontal Systems

The third type of dysfunction commonly seen in borderlines involves cognitive-perceptual impairment. More subtly, this might show itself as the philandering husband who accuses his faithful wife of cheating on him after he gives her venereal disease. Or the business executive who is unable to recognize that her "brilliant" financing strategy has so many obvious flaws that it will ruin the company. Stronger versions of this might manifest as the stroke victim who believes her husband's body has been taken over by an imposter. These types of cognitive-perceptual impairments relate to the dorsolateral prefrontal system, which underpins our ability to reason, to develop strategies for solving complex problems, to think abstractly, and to maintain a working memory. Strangely enough, people with damage to the dorsolateral and nearby ventromedial areas can have normal intelligence but have no common sense—they are unable to make reasonable decisions.

This phenomenon has been studied in relation to gambling experiments with play money. Players, including brain-damaged patients and normal controls, were given four stacks of cards to draw on—two of which were rigged to give large penalties, while the other two were rigged for small penalties. Unlike normal controls, brain-damaged patients, *even those who became consciously aware that some decks were riskier than others*, continued to play equally from all decks. Their skin showed no change in conductance, which indicated the patients' lack of concern about their risky behavior. It is thought that this dysfunctional pattern of performance was due to the fact that the patients weren't able to develop an emotional "gut" feeling related to the high-risk decks—a theory known as the somatic-marker hypothesis.[37] (*Somatic* is from the Greek word meaning "body.") It seems that conscious, overt knowledge is not enough to ensure common sense decision-making ability; other neural circuits—with the surprising inclusion of those involved in emotion—play a powerful, but hidden, role.

Subtle brain damage has also been affiliated with odd "end justifies the means" behavior. Logically, it might seem rational, for example, to push a hefty man into the train's path to slow a train and save the lives of people farther down the tracks—but normal individuals just can't bring themselves to even *think* about doing it. Those with damage to their frontal lobes, on the other hand, can easily imagine pushing the man in front of a train to impede its motion and thus save other people's lives. (This example may seem a bit contrived, but it's the example that was presented to the test takers.) Brain-damaged individuals, it seems, focus primarily on the consequences, ignoring whatever nasty means might be involved. It may well be that subtly miswired or damaged neural circuits lie behind some of the can't-make-an-omelet-without-breaking-eggs type of behavior seen in dictators and their supporters as they justify the killing of thousands, or even millions, to further their goals.[38]

The medial orbitofrontal cortex appears to be particularly important in suppressing emotional memories that are irrelevant to the current situation. Thus, individuals with borderline personality disorder and its subclinical cousin often seem to respond in "characteristically inflexible and maladaptive ways based not upon current social contexts, but rather according to implicit emotional memories of past interpersonal experiences."[39] This inflexibility may well relate to orbitofrontal cortex dysfunction. Subclinical examples of such behavior might include the supervisor who refuses to see the need for spending money to update equipment despite obvious cost savings, or the father who beats his daughter for being late despite the fact that her car broke down. On a scale of wider importance, it might explain, for example, Hitler's utter inflexibility once he had made a decision.*[40] Borderlines, it should be noted, tend to become particularly irrational when strong emotions are stirred up.[41]

Interestingly enough, substance abuse also appears to produce pre-

---

*Hitler's utter inflexibility regarding decisions he had made could be awesome. Some who tried to counter what Hitler himself called his unshakable obstinacy found that their efforts were in vain and sometimes counterproductive. Toward the end, when it was suggested to him that some things might have been done differently, he exclaimed, "But don't you see I *can not* [*sic*] change!"

frontal dysfunction, which has been associated with various aspects of addictive behavior and impaired decision making in people with antisocial personality disorder.[42] Patients with borderline personality have been found to have similarly impaired decision-making ability and are thought to have dysfunction in the orbitofrontal region.[43] And, of course, some individuals with borderline-like traits attempt to self-medicate with drugs or alcohol, worsening their already impaired prefrontal dysfunction.

Some studies have also shown that volumes of the left orbitofrontal cortex, right anterior cingulate cortex, amygdala, and the hippocampus are all smaller in patients with borderline personality disorder—the shrinkage forming a very distinctive pattern that might help distinguish borderline personality disorder from other disorders.[44] Unfortunately, in a chicken and egg situation, it's not clear whether the smaller size of those neural features causes borderline personality disorder, or whether the disorder itself causes the deterioration. In one form of schizophrenia, for example, brain scans of affected children show a remarkable loss of gray matter in the cerebral cortex between the ages of thirteen and eighteen—the anatomical abnormalities mirrored the increasing psychotic symptoms.[45] And in fact, there is an association between borderline personality disorder and schizotypal personality disorder (often thought to be a mild version of schizophrenia), as well as with schizophrenia itself.[46] One study found reduced N-acetylaspartate (NAA) compounds in the dorsolateral prefrontal cortex in borderline patients. This is significant because NAA depletion, which has been observed in both adults and children with schizophrenia, reflects a state of neuronal damage that often precedes cell death.[47]

But What's the Big Picture?

At this point, it might be nice to paint a bold picture of precisely what is going awry in the neural circuits that handle emotional information processing, impulsivity, and cognitive-perceptual activity in people with borderline personality disorder. But, although researchers are zeroing in on a variety of differences between normal and borderline neural functioning, they still don't know enough about the many different signal pathways, or how defective signal pathways compensate, to be able to state definitely what is going on. In any case, it appears that many borderline features, including poorly regulated emotions, impulsivity, and identity disturbances, are caused by disrupted connections between the prefrontal cortex and other regions of the brain that underlie higher cognitive functions.[48]

Evidence from family studies strongly supports the separate inheritance of impulsivity, moodiness, and cognitive dysfunction—all of which are found in unfortunate confluence in a borderline. Mood and impulsivity traits, for example, are often found in relatives of borderlines, but piecemeal—one relative might have a mood disorder, while another may have problems with impulsivity. Inheritance of such traits means that some of the defects we see so clearly in medical imaging are almost certainly due to problematic genes. Genetic bad luck means getting the whole constellation of a predisposition toward borderline-like personality traits—or, in some cases, outright borderline personality disorder, even without obvious environmental stressors.

In summary, then, it seems that disrupted amygdala function may cause the negative emotions a borderline feels when he is first appraising a person or situation. Disrupted orbitofrontal cortex function may cause impulsivity. And disrupted hippocampal function can cause the typical difficulty a borderline has in ignoring emotional cues that are not relevant to the task at hand. (This would account for the irrational roommate described earlier who was unable to focus on the facts and instead overreacted to her own emotions.) A problematic anterior cingulate cortex may underpin a borderline's inability to resolve conflicting information, while dysfunction in the dorsolateral prefrontal cortex may be involved in the borderline's impaired ability to effectively reevaluate negative stimuli. Ultimately, some of these disrupted activities may be a consequence of reduced activity related to serotonin, particularly in the orbitofrontal cortex.

Underlying all of these issues are the borderline's problems with identity, which, as psychiatrists Katherine Putnam and Ken Silk note, "may be the most profound and damaging result of a chronic state of emotional dysregulation." Problems with identity would include the chameleon-like behavior so often seen in borderlines, as well as the paradoxical mixtures of inflexibility and malleability.[49] (This is much like Milosevic's inability to be swayed by reason or facts, coupled with his utter dependence on his wife, Mira.) Putnam and Silk add: "As emotional experience constitutes the most fundamental part of our selves, it is impossible to know who we are if we cannot identify our feelings, figure out what triggers them, and learn how to modify them to achieve our goals. This enduring frustration, which stems from this primary experience of dysregulation, is an integral part of the experience of BPD."[50]

A Link with the Immune System?

Interesting hints have emerged in recent years that borderline and other personality disorders may be linked with abnormalities in the immune system.[51] These ideas were first proposed by Russian scientists in the 1930s.[52] The linkage the Russians proposed seemed so improbable, however, that it was ignored by the scientific community, even when the ideas were undergirded with more substantive research studies by George Solomon at Stanford in the 1960s. Part of the problem was that conventional research wisdom held that there was no connection between the immune and the neurological (behavioral) systems. Eventually a number of researchers discovered that the seemingly independent nervous and immune systems actually "speak" a similar chemical language that allows the two systems to interact with each other. Finding this connection has given impetus to the small but fascinating body of research that has recently been emerging in this area.

Overlapping Personality Disorders

Borderline personality disorder often shades in with other personality disorders. One study of fifty-nine borderline patients found that all but one was also suffering from an Axis I disorder—almost 70 percent had three or more.[53] The overlap with disorders such as depression, narcissism, and antisocial personality disorder can often make diagnosing the borderline into a complex chess game. If you clear up the depression, for example, will the borderline symptoms go away? Is the patient primarily suffering from antisocial personality disorder with some borderline co-symptoms? Or is he suffering from borderline personality disorder with some symptoms of antisocial personality disorder? Or perhaps the apparent overlap of disorders is just a tool employed by a compassionate doctor to obtain much needed insurance coverage.

But the traits of the different disorders overlap for good neurological reasons. Imaging studies of borderline patients as well as patients with antisocial personality, for example, show that both disorders, which share common attributes of impulsivity and cognitive dysfunction, have diminished regional cerebral blood flow in large areas of the right prefrontal and temporal cortex.[54] Both disorders also show dysfunction in the frontal cortex and limbic systems, which do the heavy neural lifting involved in processing emotions. There are differences, however, which can be more or less distinct, depending on the symp-

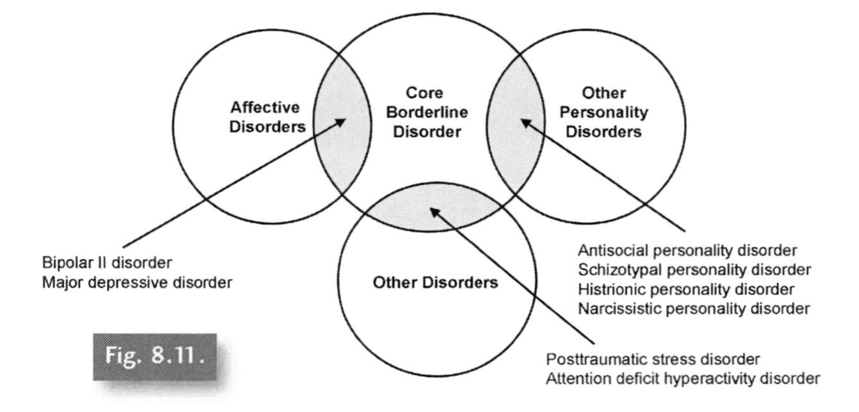

Fig. 8.11.

toms of the patient. Borderlines possess general functional dysregulation, along with hair-trigger amygdalae that respond to even tiny emotional cues. Psychopaths, on the other hand, can sometimes show laid-back responses in the limbic system, including the amygdala, accompanied by revved-up prefrontal areas in response to emotional stimuli. Under fearful situations, however, the limbic and prefrontal areas in psychopaths swap their activation patterns.[55] Notably, many of the "coping characteristics" of both disorders are precisely the same—projection, blame shifting, gaslighting, and narcissistic demands.

In the end, it is clear that human cognition is an extraordinarily slippery creature, which neuroscience is exposing as far less logical and more emotional than we had ever previously realized. This is true even for normal individuals. But for those with even a soupçon of personality disorder, it can be worse—sometimes far worse. Such irrationality can have lose-lose consequences. Not only is personality disorder associated with problematic and even sinister behavior, but it is also related to the naivete that can make relatively normal individuals such easy prey for Machiavellians.

But what is perhaps worst of all is that a person can appear entirely normal, free of any consequential diagnosis of personality disorder, yet have sometimes extraordinary deficiencies in his ability to reason. We saw this with Serbia's Slobodan Milosevic, who, it seems, fooled even himself into believing that the many wars he initiated and people killed were none of his own doing, or were trivial offenses, and that his destruction of the Serbian economy was beside the point.

But there is yet another example of a seemingly normal leader who masked deep cognitive dysfunction. A leader whose charm, duplicity, vindictiveness, and unparalleled desire for adulation, as we shall see, makes Milosevic look gentle by comparison.

CHAPTER 9

THE PERFECT "BORDERPATH"

Chairman Mao

"The hardest thing about being a communist is trying to predict the past."

—Milovan Dilas

Our driver, a small, sprightly man with nary a squeak of English, is named Mao, just like Chairman Mao. Surprisingly, he also looks like Mao, with a "big" forehead (which my Western eyes record as a receding hairline) that bespeaks power and good fortune to the Chinese. Today, riding at eighty miles an hour along a silky-smooth toll road outside Shanghai, I am taking a breather from Machiavellianism. Or at least I think I am.

It's early September 2005. My friend Wenlei and I, along with two bioengineering grad students, are on our way to Dingshan, about three hours' drive northeast from Shanghai. We've just come from the annual conference of EMBS, aka the Engineering in Medicine and Biology Society. The society holds a yearly bash to help stimulate bioengineering research worldwide. A list of previous and planned conference locations reads like a world atlas: Cancún, Istanbul, Lyon, New York City, Amsterdam, San Francisco, and, this year, Shanghai. The meeting is always a treat for me, not only because I have an opportunity to present my own research to the community and visit with many old friends, but also because I'm able to catch a bird's-eye view of the latest breakthroughs related to biomedical imaging. So much is happening in the field of medical imaging that sometimes it causes a bit of a problem—at the Shanghai conference, a quarter of the conference's two thousand or so presentations were related to medical imaging. Everyone, it seems, wants in on the action.

But the conference—a nearly weeklong talkfest of sleepless nights and crowded days—has ended. Now I can indulge in my covert passion: teapots. Not just any teapots: Yixing teapots—those of the absorbent, malleable, unglazed zisha purple clay, with shapes that bring to mind Platonic ideals of geometric perfection. Yixing (pronounced "ee-shing") teapots have been known to inspire a reckless passion akin to a drug addiction. Enthusiasts from Taiwan have sold houses and properties worth hundreds of thousands of dollars to acquire the work of a particular master. Some have given up families and jobs to enable a more complete focus on every detail related to their particular area of focus, perhaps teapots older than 1960, or the works of a particular recent master, or commercial ware.

Yixing teapots are not actually made in the formless regional city of Yixing but rather in nearby Dingshan, a popular place with Chinese tourists, as witnessed by the hundreds of teapot shops that cling to the edges of the crowded streets throughout the town. Western tourists here are rare indeed—glancing at the sweltering sidewalk crowds, I don't see a single other "big nose." Strange, because understanding Yixing ware can lead to surprising insights, not only about China but about people in general.

Our brief stop for lunch at a hole-in-the-wall restaurant immediately after arrival brings us luck—the proprietor, a friendly, open-faced woman of about fifty, takes note of my passion for teapots and agrees to serve as our guide for the afternoon. In the staunch traditions

Fig. 9.1. The author "helps" make a Yixing teapot.

of a true Communist, she refuses any tip. "She's doing this because she has a good heart," says Wenlei.

And indeed, the woman's kindness is real. Rather than take us around to the independent proprietors from whom she could continuously pocket kickbacks for the sale of third- and fourth-rate pots, she takes us directly to the source of all the decent clay and much of the best Yixing pottery currently produced—the oldest Yixing teapot factory in China. Strangely for a civilization that's been around for thousands of years, being the oldest teapot factory is not saying much—the uninspired Yixing Zisha Factory Number 1 was opened in October 1958.

We are taken up an aging elevator on a tour of the potters' studios above the showroom floor, where bleary-eyed craftsmen are luting together flat sheets of clay, shaping round structures with spatulas, and misting spouts. Our arrival sparks enthusiasm—the potters visibly puff under the attention. Soon we are being offered cups of delicious hong cha—black tea—from an exquisite rectangular teapot that bears an elegant sheen from the frequent pouring of the first rinse of tea leaves over its outside surface. Yixing teapots, like fine violins and classical guitars, are meant to be used. In particular, they are meant to be used with black and oolong teas, as well as the half-rotted tea known as pu'er.* A special characteristic of zisha clay is that it holds heat for a long time, thus, it can overcook delicate green teas, which need lower temperature water from the outset. And, unlike more ordinary clays, it is remarkably tough. (Seemingly perfect teapots made of clay from Chaozhou, for example, tend to have spouts that break because of the undesirable presence of sand.) Zisha's exceptional malleability allows for a variety of shapes that other types of ceramics simply can't match. But more than that, Yixing teapots that look the same are not really the same; a slight difference in spout size or angle, for example, can give one teapot a smug look, while another might look radiantly healthy, and yet another might sport a pout. Teapots, like people, can be surprisingly moody.

Arriving at the showroom, I'm faced with row after row of gorgeous pottery. I feel like a kid with a hand in the cookie jar—after all, there are only so many of these singularly beautiful treasures that I can pack in my luggage and stuff into carry-on bags. In the end, I choose a smooth, round teapot with a lovely purple sheen by master craftsman Chen Le Lin, and a rectangular teapot with green and yellow swirls by Fan Yu Lan, along with some less expensive but still exquisite commercial pieces. These pots will add to my caffeinated pleasure collection at home: tea ware purchased at reputable teashops during previous work-related travels to places like Beijing, Harbin, Qiqihar, Dalian, and Kaifeng.

Among all the lovely gems in my cabinets at home is one clunky oddity—a "revolution" teapot I purchased in Hong Kong. This pedestrian teapot was manufactured sometime in the decade following 1966.

---

*A very rare and expensive pu'er is said to be made from the droppings of worms that eat stored pu'er tea bricks. You can enjoy this delicacy with "drunken shrimp"—shrimp that have been doused in alcohol to relax them before they are served, live, in a squirming bowl at your table. A friend helpfully peeled and impaled one for me on a fork the first time I ever tried this delicacy. Just as I raised the fork to my mouth, the shell-less creature managed to leap off the tines and back into the serving bowl. Still, I'd rate the taste of preternaturally raw shrimp above that of stewed donkey meat, another northern Chinese treat, which I think is pretty much on par with rat's ass.

The bottom of the pot is stamped with the characters symbolizing "Yixing, China," after the name of the district. There is no maker's mark.

Revolution teapots are distinctive, in fact, for their lack of distinction. These plain-looking teapots were manufactured during Chairman Mao's Cultural Revolution, when even the faintest whiff of creativity was suspect. Workers were not allowed to put personal marks on individual pieces—it was the factory or the unit, not the individual, that was important. Walls at factories were filled with slogans demanding that literature and art serve the people and the cause of socialism; the state exercised complete control over everything, including arts and crafts.[1] Master potters had no incentive to pour their souls into their craft, since a well-done creative work would attract envy and enemies and in any case wouldn't earn the potter any more money. Many revolution teapots cannot even pour water in a straight line. Some smell of mud from being baked at the wrong temperature, and others have caps that are not level.

Not only did the manufacturing quality of virtually everything suffer during the Cultural Revolution, but nothing was sacred. Nien Cheng, a fluent English speaker who worked for Shell in Shanghai, describes the beginning of her travails during the Cultural Revolution, when youngsters known as Red Guards were encouraged to break into houses and destroy for the sake of an imagined greater good:

> I was astonished to see several Red Guards taking pieces of my porcelain collection out of their padded boxes. One young man had arranged a set of four K'ang winecups in a row on the floor and was stepping on them. I was just in time to hear the crunch of delicate porcelain under the sole of his shoe. The sound pierced my heart. Impulsively I leapt forward and caught his leg just as he raised his foot to crush the next cup. He toppled. We fell in a heap together. My eyes searched for the other winecups to make sure we had not broken them in our fall, and momentarily distracted, I was not able to move aside when the boy regained his balance and kicked me right in my chest. I cried out in pain. The other Red Guards dropped what they were doing and gathered around us, shouting at me

> angrily for interfering in their revolutionary activities. One of the teachers pulled me up from the floor. With his face flushed in anger, the young man waved his fist, threatening me with a severe beating.[2]

Cheng was eventually to suffer six and a half years of cold, hunger, disease, terror, and humiliation in solitary confinement in a Shanghai jail for the crime of having worked for a foreign company. Upon release, she would discover that her only child, her beloved daughter, Meiping, had been beaten to death.

It was Mao who had encouraged the formation of the Red Guards.

> Learning from their fathers and friends that Mao was encouraging violence, the Red Guards immediately embarked on atrocities. On 5 August [1966], in a Peking girls' school packed with high officials' children (which Mao's two daughters had attended), the first known death by torture took place. The headmistress, a fifty-year-old mother of four, was kicked and trampled by the girls, and boiling water was poured over her. She was ordered to carry heavy bricks back and forth; as she stumbled past, she was thrashed with leather army belts with brass buckles, and with wooden sticks studded with nails. She soon collapsed and died. Afterwards, leading activists reported to the new authority. They were not told to stop—which meant carry on.[3]

Mao as Borderpath

Mao was the most Machiavellian leader of the many Machiavellian leaders of the twentieth century. For three decades, he held absolute power over the lives of one-quarter of the world's population. As historian R. J. Rummel writes: "For perspective on Mao's most bloody rule, all wars [worldwide] 1900–1987 cost in combat dead 34,021,000—including WWI and II, Vietnam, Korea, and the Mexican and Russian Revolutions. Mao alone murdered over twice as many as were killed in combat in all these wars."[4] He also killed nearly four times as many as are thought to have died in *four hundred years* of the African slave trade, from capture to sale in Arab, Oriental, or New World markets.[5]

Therefore it is perhaps surprising to find that psychologists, psychoanalysts, and psychiatrists of the time, and even today, rarely make serious efforts to address the possibility of serious mental illness in motivating Mao's behavior. But MIT political scientist Lucian Pye, who published an early biography of Mao in 1976, the year of Mao's death, provides a lucid explanation for the reticence: "[B]ecause I knew that I was already going out onto thin ice by psychologically interpreting the near-sacred Mao, I decided that it would not be prudent, indeed that it would be counter-productive, to use any technical terminology. Therefore, I did not go public in announcing that Mao Zedong was probably a narcissist with a borderline personality, a combination that is not rare. I suspect that if I had stated that this was the case, it would have brought many people's blood to the boiling point."[6]

Indeed, perhaps even more so than Milosevic, Mao showed strong tendencies related to all four of the dimensional symptoms of borderline personality disorder; he also used many borderline-like coping strategies. As Mao biographer Ross Terrill writes: "The evidence . . . that Mao was a borderline personality mounts by the year. It includes his youthful loneliness and fascist ideas, the neurasthenia [nervous system exhaustion] to which his doctor testifies . . . his treatment of family members, his addiction to barbiturates, his lack of give-and-take with colleagues, and his suspiciousness."[7]

But Mao also showed evidence of psychopathy, the deepest form of antisocial personality disorder, which is often diagnosed alongside borderline personality disorder. "Borderpath" does, in fact, seem to be an appropriate term for the fused set of pathologies that propelled Mao on his infamous trajectory, although it is difficult to find any term that truly encompasses Mao's drug-addled mixture of characteristics. In this sense, then, virtually all of the medical studies mentioned about both borderlines and psychopaths so far may shed light on the neurological mechanisms underlying the deeply Machiavellian emotional traits that Mao displayed. And, of course, that little-understood trait of narcissism plays a role as well.

The Early Years

Mao was born in 1893 in the Hunan province—the heartland of China. His ancestors had lived in the same temperate, misty valley for five hundred years. Unlike most of the other villagers, Mao's father, Yi-chang, could read and write well enough to keep accounts. Yi-chang had a quick temper and firm views. Through his hard work and thriftiness, he became one of the most prosperous men in the village. Mao's mother was an exceptionally gentle and tolerant woman from a neighboring village.

Mao was the third son but the first to live past infancy. Later, two more sons—Mao's younger, milder-tempered brothers—were born into the family. Mao had a carefree early childhood living with his mother and her family in their village, where his mother preferred to live, and where his grandmother, uncles, and aunts doted on him. At the age of eight he returned to his native village of Shaoshan to receive an education. Rote learning of the Confucian classics was de rigueur. Mao, blessed with an extraordinary memory, excelled.

Early Antisocial Tendencies

Although a bright student, Mao clashed frequently with his teachers, and at the age of ten, he ran away from his first school. He was expelled from or asked to leave at least three more schools for being headstrong and disobedient—strong evidence of the early behavioral problems often seen in antisocial personality disorder. (Milosevic, in contrast, was a model, if unctuous, child.)

Although Mao's mother remained understanding, the problems with school became one more source of tension between father and son. His father would strike him when Mao would not comply—typical behavior for a Chinese paterfamilias of the time. As with Hitler and Stalin, Mao frequently fought with his father. Mao biographers Jung Chang and her husband, Jon Halliday, report:

> He would tell his father that the father, being older, should do more manual labour than he, the younger—which was an unthinkably insolent argument by Chinese standards. One day, according to Mao, father and son had a row in front of guests. "My father scolded me before them, calling me lazy and useless. This infuriated me. I called him names and left the house. . . . My father . . . pursued me, cursing me as well as commanding me to come back. I reached the edge of a pond and threatened to jump in if he came any nearer. . . . My father backed down." Once, as Mao was retelling the story, he laughed and added an observation: "*Old men like him didn't want to lose their sons. This is their weakness. I attacked at their weak point, and I won!*"[8] (italics added)

Such behavior was unspeakable according to Chinese cultural traditions.*[9] As Francis Fukuyama notes in his book *Trust*, in China, there is "no counterpart to the Judeo-Christian concept of a divine source of authority or higher law that can sanction an individual's revolt against the dictates of his family. In Chinese society, obedience to paternal authority is akin to a divine act, and there is no concept of individual conscience that can lead an individual to contradict it."[10]

Markedly Disturbed Relationships

A common characteristic of both borderline and antisocial personality disorder involves markedly disturbed relationships—Mao could serve as a model in this area. While in his teens, Mao was able to coerce his

*There is little doubt that an abusive father could have contributed to any genetic predisposition Mao might have had for a personality disorder. Yet it should be remembered that many children of problematic fathers grow up without anything like Mao's dysfunctional, deeply sinister personality. For example, Abraham Lincoln's relationship with his father was also "strained by a fundamental conflict." As with Mao's father, Tom Lincoln often beat his son for neglecting his farm work by reading and for other infractions. Despite Lincoln's beatings and hardships as a youngster (unlike Mao, from an early age, Lincoln had no mother to rely on), as an adult Lincoln was so sensitive and caring that he once disappeared from a large group he was riding with to help two baby birds he had noticed were blown from their nest. "'In many things,' remembered Mary Owens, a woman Lincoln courted, 'he was sensitive almost to a fault.'" But Lincoln's extensive problems in adulthood with depression were very likely exacerbated by his early traumatic experiences with his father as well as his mother's early death.

father into paying for further education. But Mao found his education difficult to deal with. He preferred to spend his days reading whatever he wanted rather than pursuing a degree, and his father had to threaten to cut him off to get him to enroll in a professional program. In the spring of 1913, at age nineteen, Mao enrolled in the teacher-training program at the Hunan Fourth Provincial Normal School. He would graduate five years later, but his first years were especially miserable —he liked neither the teachers nor the students.

But Mao's disturbed relationships during this period were not all-pervasive. He came to deeply admire Yang Changji, the head of the philosophy department and an ardent fitness fanatic who organized long hikes and a Spartan physical regime for his students. "When I think of [his] greatness," Mao confided to a friend, "I feel I will never be his equal." Yang felt the same way about Mao, who was one of his favorite students. "It is truly difficult to find someone as intelligent and handsome [as Mao]."[11] Mao began meeting regularly with a small group of other students at Yang's home to discuss "the cultivation of personal virtue, will-power, steadfastness and endurance."[12]

Mao became keen on Yang's daughter, Kaihui. Winning her over with his poetry and letters, they became lovers, but Mao continued to see other girlfriends as he served on the outer circles of the budding Chinese Communist Party. Kaihui was shattered when she discovered Mao's infidelity. But Mao explained the affairs were simply due to his uncertainty about Kaihui's feelings. She believed him, and at the end of 1920, the two were married. (Kaihui was actually Mao's second wife—Mao appears to have rejected his first wife, from an arranged marriage. This first wife died young—long before Mao met and married Kaihui.)

Kaihui was an early feminist and while pregnant with their first child, she worked among the peasants pushing for women's rights and better educational facilities. Meanwhile, again bowing to internal quirks that fueled his disturbed relationships, Mao embarked on several more affairs, one with Kaihui's cousin. Kaihui was again devastated, but because she still loved Mao, she eventually decided to simply let him be. A hint of their tempestuous relationship survives in a love poem Mao wrote to her after a quarrelsome departure following the birth of their second son.

> A wave of the hand, and the moment of parting has come.
> Harder to bear is facing each other dolefully,
> Bitter feelings voiced once more.
> Wrath looks out from your eyes and brows,
> On the verge of tears, you hold them back.
> We know our misunderstandings sprang from that last letter.
> Let it roll away like clouds and mist,
> For who in this world is as close as you and I?[13]

But after the last of their three sons was born, the Communist Party was outlawed by the Nationalists, and Mao, now in jeopardy, left Kaihui in the city of Changsha and headed off to usurp his first armed force. Only four months after having left Kaihui, he married another woman, He Zizhen, who was his local interpreter.[14] Mao told Zizhen that he had not heard from Kaihui and thought she might have been executed—although in a later mention of Kaihui in a letter to a friend, he was well aware of the fact that Kaihui and their children were still alive.[15] Mao's life in bandit country was often quite pleasant—he lived in luxury with his sizable personal staff in the various mansions he commandeered.

Years after Kaihui's execution by the Nationalists for refusing to de-

> ***Lack of Object Constancy***
>
> One common trait of borderlines is "lack of object constancy." Object constancy is the ability we have to soothe ourselves by remembering the love that others have for us. This can be comforting even when the ones we love are far away or no longer living. Unlike most people, borderlines find it difficult to bring to mind the image of a loved one to soothe them when they feel upset or anxious. On an emotional level, if that person is not physically present, they simply don't exist.[16]

nounce her Communist husband, messages that she had secreted in the walls of her house came to light. Kaihui had become so dispirited at learning of Mao having taken another wife that she considered committing suicide. Yet, as her last note reveals, love for Mao still permeated her every breath: "For days I've been unable to sleep. I just can't sleep. I'm going mad. So many days now, he hasn't written. I'm waiting day after day. Tears. . . . He is very lucky, to have my love. I truly love him so very much! He can't have abandoned me. He must have his reasons not to write. . . . Father love is really a riddle. Does he not miss his children? I can't understand him. . . . No matter how hard I try, I just can't stop loving him. I just can't."[17] Mao, however, in his own relationship-challenged fashion, *was* thinking at least occasionally of Kaihui. Despite the fact that his third wife, He Zizhen, had just given birth to their first child, a daughter, Mao wrote a brief note to an old friend, attributing his flat spirits in part to missing his second wife and their little family. Mao asked his friend to find his younger brother in Shanghai to get Kaihui's mailing address so that Mao could write to her. But Kaihui never received a letter.

Mao's new wife, Zizhen, was also to find his womanizing intolerable. She eventually went to Russia, ostensibly for medical treatment, and gave birth to a boy who died of pneumonia after only six months, causing her to sink into inconsolable depression. Despite her attempts to contact Mao, he did not reply. Nearly two years after they had parted, Zizhen learned by chance through a newspaper article that Mao had again remarried. Crushed by ill health and excruciating memories of the children she had been forced to abandon by Mao and the exigencies of war, Zizhen suffered a mental breakdown.

Mao placed emphasis on improving the lot of women in Chinese society by setting aside electoral positions for them in government and promoting women's educational opportunities. But his lofty ideals vanished when his personal life was involved.

Mao's attitudes toward his and Kaihui's son An-ying—his only mentally sound heir—revealed the depths of his continually troubled relationships. Mao's other known son to survive to adulthood apparently suffered from schizophrenia; both known surviving daughters had nervous breakdowns, with one continuing to drift in and out of insanity for years, while the other continued to be troubled by depression.[18] Although a genetic cause for these problematic personalities seems likely, Mao's children suffered unquestionable exacerbating stress as a result of their father's prominence—and neglect. An-ying, Mao's oldest son, had lived with his brother as a street urchin following his mother's execution and then spent most of his adolescence in Stalin's Russia. In his early twenties, he was sent by Mao to China's rural provinces to be hardened by what he would witness. But An-ying, who shared the empathetic nature of his mother, Kaihui, was shocked rather than hardened by the public mass brutality. Not long after this, he was killed in North Korea, a casualty of war.

Mao's dysfunctional relationships were worsened by the fact that he often chose deeply disturbed people with whom to interact. For example, Mao's last wife, Jiang Qing, to whom he was married for thirty-eight years, was even more troubled emotionally than Milosevic's wife, Mira. Jiang Qing's continual demands kept five or six people constantly scurrying in response to her whims. It was considered an honor to work for the chairman's wife, but the level of distress and anxiety among those who served her was high indeed. Jiang Qing's physician was accused of torturing the chairman's wife by pulling the window shades down too slowly when she ordered them drawn. As a consequence, she said, the sun had permanently damaged her eyes. Additionally, she accused her physician of deliberately giving her a chill by lowering the temperature below the eighty degrees upon which she insisted. When her physician showed her the thermometer reading precisely at eighty degrees, she accused him of inflicting mental anguish and had him declared a member of the antiparty group—which subjected him to a terrible ordeal.[19]

Mao's disturbed way of interacting with others revealed itself in other ways. He was deeply anti-intellectual and made a point of flaunting uncouth antics that drew attention to himself, such as "pulling his belt to hunt for lice in his groin as he talked, or pulling off

Fig. 9.2. Jiang Qing in 1954 with Mao. Jiang often tailored the baggy government-mandated clothing styles she wore to draw attention to her slim waist.

his trousers in the middle of an interview as he lay to cool himself down."[20] Unsurprisingly, in the early days of the Chinese Communist Party, when many people had direct, relatively unfiltered knowledge of Mao and his activities, he was an unpopular leader. One official report noted that many felt very bitter toward Mao, seeing him as dictatorial, foul tempered, and abusive.[21] As a consequence, he was ousted on six occasions in his first twelve years as a Communist. But he had powerful friends (especially the Russians), who found him useful; they, along with the plots Mao devised during his moody bouts of nerves at his home or in hospitals, pulled him back up no matter how often he was ousted.

In one of the first mutinies by Mao's men, a circular was sent out describing the men's erstwhile leader:

> He is extremely devious and sly, selfish, and full of megalomania. To his comrades, he orders them around, frightens them with charges of crimes, and victimizes them. He rarely holds discussions about Party matters. . . . Whenever he expresses a view, everyone must agree, otherwise he uses the Party organization to clamp down on you, or invents some trumped-up theories to make life absolutely dreadful for you . . . Mao always uses political accusations to strike at com-

> rades. His customary method regarding cadres is to . . . use them as his personal tools.[22]

The men's attempts to undercut Mao served them no good—ultimately, Mao had them tortured to death.

Years later, after he'd become China's "Great Helmsman," Mao's morning greeting to each member of his staff was always "Is there any news?" Li Zhisui, Mao's doctor, thought it was Mao's way of gathering information and keeping continual check on everyone of importance. "It was his way of controlling us, too," Li wrote in his insightful memoirs. "He expected us to repeat all our conversations and activities and encouraged us to criticize each other. He liked to play one member of the staff off against another."[23] Mao also continually changed his circle of aides, attendants, and bodyguards, all of whom worshiped him, much as Dr. Li had at the beginning of his twenty-two years as his personal physician. Li felt the rare older members of Mao's original staff suffered a similar affliction: "The more one knew of Mao, the less he could be respected. By changing the inner circle and bringing in a fresh crop of worshippers, Mao assured himself continual adulation."[24] This state of affairs was especially sad for the many young women selected for Mao's companionship, who became unbearably arrogant.[25]

The Confusing Façade—Sympathy with Little Empathy

And yet, both despite and because of his disturbed relationships with others, Mao was frequently lonely. Once he gained ultimate power, he led an isolated life. He had no friends, spent little time with his wife, and even less time with his children. He made Dr. Li one of his chief conversation partners, calling him in to chat whenever insomnia struck—no matter how late or inconvenient the hour. But, Li notes, "So far as I could tell, despite his initial friendliness at first meetings, Mao was devoid of human feeling, incapable of love, friendship, or warmth."[26]

Li described the following incident: "Once, in Shanghai, I was sitting next to the Chairman during a performance when a young acrobat —a child—suddenly slipped and was seriously injured. The crowd was aghast, transfixed by the tragedy, and the child's mother was inconsolable. But Mao continued talking and laughing without concern, as though nothing had happened. Nor, to my knowledge, did he ever inquire about the fate of the young performer."[27] Li relayed how

> In 1957, in a speech in Moscow, Mao said he was willing to lose 300 million people—half of China's population. Even if China lost half its population, Mao said, the country would suffer no great loss. We could produce more people.
>
> It was not until the Great Leap Forward, when millions of Chinese began dying during the famine, that I became fully aware of how much Mao resembled the ruthless emperors he so admired. Mao knew that people were dying by the millions. He did not care.[28]

According to Li, this lack of empathy also extended to those in Mao's inner circle. If an aide was no longer useful, he was ruthlessly ejected, even if Mao had worked with him for years. At the same time, Mao was a consummate actor, pretending he was being forced into his actions. In this way, he retained loyalty even as he sent his aide on to hardship and suffering.[29]

Yet Li does recount a few interesting counterexamples. When Mao's unfortunate third wife, He Zizhen, was in her fifties, Mao invited her to visit him.

> Her pallid face burst with delight as soon as she saw Mao. Mao rose immediately and walked toward her, taking her hands into his, and escorting her to a chair as He Zizhen's eyes filled with tears.
>
> He gave her a little hug and said with a smile, "Did you get my letter? Did you receive the money?" He was good to her, as gentle and kind as I had ever seen him. . . .
>
> For a long time after she left, I remained with Mao as he sat silently, smoking cigarette after cigarette, overcome with what I took to be melancholy. I had never seen him in such a mood. I sensed in him a great sorrow over He Zizhen.

Finally, he spoke. He was barely audible. "She is so old. And so sick."

He turned to me . . . "And what is her illness called after all?"

"It is called schizophrenia." . . .

"Is it the same illness that Mao An-qing has?" [Mao's second son by Kaihui who survived to adulthood]

I told him it was . . .*[30]

Schizophrenia and Schizotypal Personality Disorder

It is interesting to note that Mao's younger surviving son, An-qing, seems to have been afflicted with schizophrenia. (Note that An-qing was Mao's son by his second wife, Kaihui—An-qing was *not* related to Mao's schizophrenic third wife, He Zizhen.) Schizophrenia is a strongly hereditary disease or linked set of diseases—relatives of those with schizophrenia have a much greater chance than usual of showing at least mild aspects of the disorder themselves. Schizophrenia is characterized by persistent defects in the perception or expression of reality. Those with untreated schizophrenia may show extremely disorganized thinking and can also experience delusions or auditory hallucinations.

Schizotypal personality disorder is thought by many to be a mild form of schizophrenia. It has been linked with borderline personality disorder through the use of imaging techniques.[31] Patients with coexisting borderline-schizotypal personality disorders have been found to have distinctive reductions in the size of the gray matter in their anterior and posterior cingulates—areas that relate to executive control as well as to mood regulation.

*It is worth noting that Zizhen's shock at seeing Mao probably related to the fact that Mao invited her specifically without letting her know she was going to see him, despite the fact that Mao knew she was in an extremely fragile emotional state. (Zizhen had already relapsed after merely hearing Mao's voice on the radio.) Zizhen suffered a full-blown breakdown as a consequence of her last meeting with Mao, becoming unable to even recognize her own daughter and drooling in her madness. She never fully recovered.

After becoming the leader of China, Mao did perform small kindnesses to a few of those who had known him many years before. He arranged for help for an old classmate who fell into financial straits. Two of Mao's old teachers had written, both in dire need of money. Mao suggested a small subsidy in local party funds for them, as well as for the starving seventy-year-old widow of his revered classical literature teacher, Yuan the Big Beard. He also wrote letters for people he had known, liked, and trusted, "who asked him to spare them the rigors of laboring in the countryside on the Great Leap projects."[32] One such person he excused was the nanny who had looked after his three sons with Kaihui back in the 1930s. The memories of Kaihui appeared to haunt Mao. Although he could have easily rescued her from execution and thus saved their sons from de facto orphanhood, Mao had made no effort to do so. Learning of her death, however, Mao seems to have reacted in genuine grief, writing: "'The death of Kaihui cannot be redeemed by a hundred deaths of mine!' He spoke of her often, especially in his old age, as the love of his life. After he became China's leader, Mao's letters show that he was sending two payments a year to Kaihui's family, each one at least ten times more than a well-off peasant's annual income at that time."[33]

The psychiatrist's words echo: "The borderline is capable of great sympathy and comforting but often may lack true empathy, the ability to put himself in the other person's shoes, in appreciating how others are impacted by his behavior. Additionally, when they are hurt, their rage at those who have hurt them may be intense and cruel and devoid of concern or understanding for the other party."[34]

Impulsivity

Thus far, we've discussed Mao's early antisocial tendencies, as well as his markedly disturbed relationships. But there are other traits that are common to both borderline and antisocial personality disorders—most notably, impulsivity. Although Mao's "top-down" control over his emotions was intact, he clearly used his natural impulsivity and mood swings to great effect. "Getting upset is one of my weapons," he once confided to Dr. Li.[35] As Andrew Nathan writes in the foreword to Li's memoirs:

> The real Mao could hardly have been more different from the benevolent sage-king portrayed in the authorized memoirs and poster portraits that circulate in China today. To be sure, on first meeting he could be charming, sympathetic, and casual, setting his visitor at ease to talk freely. But he drew on psychological reserves of anger and contempt to control his followers, manipulating his moods with frightening effect. Relying on the Confucian unwillingness of those around him to confront their superior, he humiliated subordinates and rivals. He undertook self-criticism only to goad others to flatter him, surrounding himself with a culture of abasement. . . . He understood human suffering chiefly as a way to control people.[36]

Poorly Regulated Emotions

A key trait of borderline personality disorder is poor emotional regulation—this characteristic was noted at virtually every stage of Mao's life. As a young man, for example, Mao was often deeply troubled, racked by bouts of self-questioning and depression; his obstinate and argumentative personality made him a cross to bear for everyone. While attending the teachers' college in his early twenties:

> One moment he was complaining: "Throughout my life, I have never had good teachers or friends." Next minute he wrote intimately to [his close friend] Xiao Yu: "Many heavy thoughts . . . multiply and weigh down on me. . . . Will you allow me to release them by talking to you?" His obstinacy was legendary, even towards those he liked and respected, such as Yuan the Big Beard [his professor of Chinese language and literature], with whom he had a furious row over the title-sheet for an essay which he refused to change. After another dispute, this time with the principal, it took the combined intervention of Yuan, Yang Changji, and several other professors to prevent him being expelled.[37]

Mao would record in his journal his attempts to correct his course: "In the past, I had some mistaken ideas. Now I . . . [have] grown up a bit. . . . Today I make a new start."[38] But six months later, he would again be "starting afresh." He was by turns unhappy, frustrated, enthusiastic, and bitter.

Borderline expert Leland Heller writes:

> Mood swings are a fundamental symptom of borderline. The moods can shift inappropriately from hour to hour, even minute to minute, without apparent justification. . . . A borderline can experience distinct periods of happiness, rage, boredom, mistrust, sadness or joy without anything new happening in the environment. These feelings are often very intense. Some borderlines suffer more mood swings in an average day than most people do in a month . .
>
> Imagine trying to walk down the street blindfolded, never knowing if your next step was one step up, two steps up, one step down, two steps down, or level. It's the unpredictability that makes the situation dangerous. A borderline's emotions are as difficult to overcome as this example of the shifting sidewalk . . .
>
> Borderlines feel only inconsistency—moods that constantly shift.[39]

These mood swings ("affective instability") have been found to be related to borderline symptoms of identity disturbance, chronic emptiness or boredom, inappropriate anger, and the defenses of splitting and projection.[40]

How much of Mao's perceptions were colored by these personality traits? Mao noted:

> I say: the concept is reality, the finite is the infinite, the temporal is the intemporal, imagination is thought, I am the universe, life is death, death is life, the present is the past and the future, the past and the future are the present, the small is the great, the *yin* is the *yang*, the high is the low, the impure is the pure, the thick is the thin, the substance is the words, that which is one, that which is changing is eternal.

> I am the most exalted person, and also the most unworthy person.[41]

The embedded opposites of yin and yang form an ancient part of China's philosophical underpinnings, but Mao's personalization sneaks perilously close to identity diffusion.*[42] "Normally, we experience ourselves consistently through time in different settings and with different people," writes borderline specialist Robert J. Waldinger. "This continuity of self is not experienced by the person with BPD. Instead, borderline patients are filled with contradictory images of themselves that they cannot integrate."[43]

Dr. Li called Mao's condition "neurasthenia," acknowledging that the term is no longer recognized in the United States but explained it as a psychological malaise that manifests itself physically through insomnia, headaches, chronic pain, dizziness, anxiety attacks, high blood pressure, depression, impotence, skin problems, intestinal upsets, anorexia, and bad temper, among other symptoms.

One early example of Mao's neurotic, borderline-like reaction to setback—as well as his deeply Machiavellian lack of ideological commitment—was his response to being asked by the Communist Party to integrate himself into the opposing Nationalist Party. Mao performed the integration with such enthusiasm that he was eventually dismissed from the Communist Central Committee as being too right-wing. The switch demonstrated the low ideological commitment of Christie's archetypical Machiavellian. The dismissal deeply affected Mao, who grew thin and ill; he retreated home for eight months to recuperate. A housemate and colleague said that Mao had "'problems in his head . . .

---

*Mao would later find that these concepts were intimately related to Marx's concept of dialectical materialism. This Marxist interpretation of reality views matter as the sole subject of change and all change as the product of a constant conflict between opposites. The conflict arises from the internal contradictions inherent in all events, ideas, and movements. The fundamentally contradictory basis of this philosophy allowed Mao to fiendishly change his ideological campaigns "to accommodate his political needs, to change direction at will, and to lure real or presumed opponents into exposing their views, the better to strike them down." Such a philosophy also explains how Mao could provide lip service to his followers that told them to think for themselves and not to follow blindly, while also requiring uniformity of thought.

he was preoccupied with his affairs.' [Mao's] nervous condition was reflected in his bowels, which sometimes moved only once a week."[44]

This type of affective instability is well described by Martin Bohus and his colleagues in their research related to the neurobiology of borderline personality disorder.

> Individuals with BPD have profound emotion vulnerability, experiencing accentuated sensitivity to aversive emotional stimuli, intense emotional reactions, and slow return to baseline emotional arousal. In a vicious cycle, the use of dysfunctional behaviors . . . to modulate aversive emotions is negatively reinforced by a short-term reduction in the intensity or experience of these emotions. Emotion dysregulation occurs across the entire emotional system, impacting executive functions and affective learning processes. In turn, the behavioral, physiologic, cognitive, and experiential subsystems of emotional responding are affected. The emotionally dysregulated patient may experience inflexibility in cognitive perspective, along with difficulty controlling impulsive behavior in response to strong positive and negative affect. Further, when emotionally aroused, individuals with BPD may have considerable difficulty organizing and coordinating activities to achieve non-mood-dependent goals and may freeze or shut down under very high stress.[45]

Drug and Sexual Addictions

Those suffering from both borderline and antisocial personality disorders are prone to addictive behavior; Mao was little different in this regard. This related in part to his difficulties with insomnia—his body clock appeared to be set for a twenty-seven-hour day, sometimes even longer, so that he was occasionally unable to sleep for up to forty-eight hours at a time.[46] Dr. Li felt that Mao's biological clock might always have been askew. (Research has uncovered the relationship between mutations in the *clock* gene and lengthened, or even chaotic, sleep cycles.)[47] While trying to resolve his sleeplessness, Mao became addicted to sleeping pills: veronal in the 1930s, then sodium amytal, a powerful barbiturate, which he took more and more, and finally a mixture of sodium seconal with sodium amytal. Ultimately, he discovered chloral hydrate, and became addicted to the euphoria. He took the pills when receiving guests and attending meetings as well as at dance parties. These addictions, as Li noted, could have exacerbated Mao's preexisting symptoms. And, of course, drug addictions frequently accompany personality disorders.

In related addictive behavior, Mao's depraved and unending dalliances with young women eventually brought him venereal disease—for which he refused to be treated. "If it's not hurting me, then it doesn't matter. Why are you getting so excited about it?" he told Dr. Li.[48] Li remarked on Mao's delight in sharing multiple simultaneous women. Men, too, were a target—Mao insisted on nightly groin massages from his handsome young male attendants. At first Li thought Mao might be homosexual—but he later concluded that Mao simply possessed an "insatiable appetite for any form of sex."[49]

Sexual addiction is a disorder that is thought to have similarities to both alcohol and drug addiction as well as to compulsive gambling. Sex addicts are believed to react strongly to the neurochemical changes that take place in the body during sexual behavior—somewhat like becoming hooked on heroin. One researcher in the area, Dr. Patrick Carnes, states, "Contrary to enjoying sex as a self-affirming source of physical pleasure, the sex addict has learned to rely on sex for comfort from pain, for nurturing or relief from stress. This is much like an alcoholic's use of alcohol."[50] Although sexual addiction is little understood, several studies have shown that it appears to be very commonly found in conjunction with substance abuse, anxiety, and general mood disorders. In the few imaging studies that have been done to date, certain areas of the brain have been shown to be deactivated during response to sexual stimuli with arousal. Serotonin levels are also thought to play a key role.[51]

Mao's hypocrisy about sex was breathtaking. "Mao required his people to endure ultra-puritanical constraints. Married couples posted to different parts of China were given only twelve days a year to be together, so tens of millions were condemned to almost year-round sexual abstinence."[52] While ordinary citizens were being sent to labor camps for having had illicit sex, Mao was romping daily with beautiful women—the Chinese Communist equivalent of groupies. Even in his midseventies he was known to invite "three, four, even five of them simultaneously" to share his oversized bed.[53]

Cognitive-Perceptual Impairment

One of the three key dimensional traits of borderline personality disorder is cognitive-perceptual impairment. Mao appears to have displayed dramatic symptoms of this trait. During the Great Leap Forward, from 1958 to 1960, Mao implemented a policy that diverted all human resources into industry rather than agriculture in a misguided and disastrous attempt to catch up with the industrialized West. During the Great Leap, tens of millions of workers were diverted to produce one commodity—steel—in inefficient backyard production facilities. It is estimated that some thirty million died as a result of the ensuing famine—some peasants were reduced to eating each other's children to avoid eating their own.[54] Mao's general reaction to the devastating effects of his policies was to pretend that they weren't happening. His staff, all too aware of what could happen to them if they revealed the truth, served to insulate him even further. But the situation deteriorated so drastically that the truth could not be hidden. Dr. Li describes his insider's perspective on Mao's reactions.

> Mao . . . seemed psychologically incapable of confronting the effects of the famine. When I told him that edema and hepatitis were everywhere, he accused me of inventing trouble. "You physicians have nothing better to do than scare people," he snapped. "You're just out looking for disease. If no one were sick, you'd all be unemployed." . . . I thought Mao was ruthless to close his eyes to the illness that was everywhere around him. But I allowed him his illusions and never mentioned the subject again, behaving in his presence as though hunger and disease had miraculously disappeared.[55]

Dr. Li notes: "Mao was loath to admit his mistakes. His was a life with no regrets . . . I am convinced Mao never really believed he had done anything wrong."[56]

Writing in the late sixties, psychologist Robert Jay Lifton dubbed the policies of the Great Leap, as well as those of the later Cultural Revolution, "psychism." He defined the term as "the attempt to achieve control over one's external environment through internal or psychological manipulations, through behavior determined by intra-psychic needs no longer in touch with the actualities of the world one seeks to influence."[57] Psychologists have since termed this phenomenon "magical thinking"—a trait commonly found in schizophrenia. Magical thinking appears to be associated with unusual features concerning the fusiform gyrus—an area that wraps underneath the brain in the temporal lobe and assists in face, word, and number recognition.[58]

Even Mao noticed and was puzzled by his perceptual impairments—particularly his inability to handle criticism and the input from others. In his youth, he was self-aware enough to tell a friend: "I constantly have the wrong attitude and always argue, so that people detest me."[59] To one of his former teachers he confessed: "I am too emotional and have the weakness of being vehement. I cannot calm my mind down, and I have difficulty in persevering. It is also very hard for me to change. This is truly a most regrettable circumstance!"[60]

Mao's distorted cognitive processes were also apparent in his propensity for saying one thing but doing another. "Mao *did not believe* a lot of what he proclaimed in public," writes Ross Terrill. "No other Communist leader, perhaps no leader in twentieth-century politics, was a match for Mao in this breathtaking insincerity. We find it in small matters and in large."[61] After becoming chairman, Mao, for example, became the premier advocate of traditional Chinese medicine, but as Dr. Li would discover, Mao refused to use it himself.[62] While advocating the "learning-from-the-Soviet-Union" policy, many Chinese studied Russian. Not Mao—who was disgusted by the Russians and instead studied English. Indeed, while telling the Chinese people that America was the apotheosis of evil, Mao privately said good things about America. He said of himself, "My words and my deeds are inconsistent."[63] Some wags may point out that inconsistent words and deeds are always characteristic of politicians. But Mao's inconsistency appears pathological. There is no evidence, for example, that British prime minister Winston Churchill "secretly admired the Nazis or despised Roosevelt."[64]

Mao's decisions to continue with obviously flawed programs such as the Great Leap Forward might be due not only to his innate borderline characteristics but the worsening of those symptoms as a result of subtle, drug-related dysfunction in his prefrontal cortex. This might have reduced Mao's ability to experience "gut feelings" and thus eliminated his ability to make logical, commonsense decisions that were obvious to everyone else.

As Li notes, "To this day, ruthless though he was, I believe Mao launched the Great Leap Forward to bring good to China. . . . The twentieth century was marching forward and Mao was stuck in the nineteenth, unable to lead his country. Now he was in retreat, trying to figure out what to do."[65] Tragically, the cognitive equipment Mao was using to do his "figuring" with was—*different.*

Control and the Purges

One oft-noted characteristic of those with subclinical borderline personality disorder is their extraordinary need for control. This trait was typical of Mao, who had to be in control of every situation—from decisions at the highest reaches of political power to the most mundane details of his everyday life. Nothing that occurred within Zhongnanhai, the former royal garden within the grounds of the old Forbidden City, where Mao lived and worked, happened without Mao's consent, not even the clothing chosen for his wife to wear. Mao expected to be consulted on virtually every major—and minor—decision in China.[66]

Mao's controlling efforts were carried to the ultimate in the form of his purges. His earliest full-blown purge was started in late

November 1930, immediately after he learned that Moscow had decided to support him as their primary Chinese leader in pursuit of Communist hegemony. This purge was designed to eliminate thousands—anyone who had opposed Mao—while simultaneously generating such terror that "no one would dare disobey him from now on."[67] Other purges followed—each building momentum for a still larger purge to come.

One of Mao's favored mechanisms for eliciting purge victims was to urge people to speak out, pledging there would be no retribution. The worst of these deathly false promises occurred in April of 1956, when Mao called for intellectual debate under the slogan "Let a hundred flowers bloom, let a hundred schools of thought contend." Naive intellectuals, encouraged by signs of liberalization, climbed aboard and began criticizing the party. Privately, Mao told party cadres, "This is not setting an ambush for the enemy, but rather letting them fall into the snare of their own accord."[68] After allowing for several months of seeming freedom, Mao cracked down. Quotas ranging between 1 and 10 percent were set of intellectuals to be persecuted and purged. Many who had said nothing were pulled in just to fill Mao's quotas. Mao bragged that one province, Hunan, "denounced 100,000, arrested 10,000, and killed 1,000. The other provinces did the same. So our problems were solved."[69] Mao's Boswell, Dr. Li, remarked on how, although Mao accused others of creating conspiracies, he was the "greatest manipulator of all."[70]

Control slides into vengeance. Mao waited decades, if necessary, to retaliate against those who had differed with him. Thus was born the Cultural Revolution—the parent of all purges, during which at least three million people died violent deaths; another one hundred million are thought to have suffered from the persecutions.[71] Foreign minister Chen Yi was to forthrightly dub the Cultural Revolution "one big torture chamber."[72] Mao closed all schools and encouraged students to join Red Guard units, which persecuted Chinese teachers and intellectuals and enforced Mao's cult of personality. Special enforcers screened millions of officials to explore whether they had

Social Dominance and Control

Social dominance is a key component of both human and non-human primate sociality that is intimately related to neurotransmitters and hormones. The phenomenon seems to walk hand-in-hand with both narcissism and the need for control. Studies have shown that when males climb the social totem pole, their testosterone levels increase, whereas if they slip back down the pole, their testosterone levels also slip. This effect is so widespread that several studies have even shown that male fans of a winning football team have higher testosterone levels after the game than the fans of the losing team.[73]

Many studies have found a strong relationship between serotonin and dopamine levels and a primate's level in its social hierarchy.[74] Manipulating these neurotransmitters can cause a monkey to move either up or down in the social hierarchy.[75] It's reasonable to suppose that Mao's rise to the top of his social structure could have swung his neurotransmitter levels into even higher gear. This in turn would have caused a rise in self-confidence—the same rise in self-confidence observed in primates with purposefully manipulated neurotransmitter levels. Ultimately, this could have helped shut down Mao's already limited abilities to accept criticism from others. As Mao biographer Philip Short writes: "By the mid-1950s, Mao was so convinced of the essential correctness of his own thought that he could no longer comprehend why, if people had the freedom to think for themselves, they would think what *they* wanted, not what *he* wanted."[76]

Perhaps ambition is simply an attempt, in those with certain biochemical predispositions, to self-medicate. This would provide a neurobiological basis to the idea that, at least for those with the proper underlying genetics, absolute power does indeed corrupt absolutely.

ever resisted Mao's orders, even passively. At the top was a special secret group chaired by Zhou Enlai (Chou En-lai), with torturer extraordinaire Kang Sheng as his deputy. This group investigated people personally designated by Mao. Differing from Stalin, who quickly hustled purged victims out of sight via prisons, gulags, or summary liquidation, "Mao made sure that much violence and humiliation was carried out in public, and he vastly increased the number of persecutors by getting his victims tormented and tortured by their own direct subordinates."[77]

Sadism

According to Jung Chang and Jon Halliday, Mao discovered early on a gut affinity for sadism, which is believed by many researchers to be affiliated with the strongest forms of psychopathy. While on a tour of the provinces, he approvingly described the ability of local bosses to toy with and break their victims:

> A tall paper hat is put on [the victim], and on the hat is written landed tyrant so-and-so or bad gentry so-and-so. Then the person is pulled by a rope [like pulling an animal], followed by a big crowd. . . . This punishment makes [victims] tremble most. After one such treatment, such people are forever broken.
>
> The peasant association is most clever. They seized a bad gentleman and declared that they were going to [do the above] to him. . . . But then they decided not to do it that day. . . . That bad gentleman did not know when he would be given this treatment, so every day he lived in anguish and never knew a moment's peace.[78]

In writing his report afterward, Mao waxed enthusiastically about the brutality he saw: "It is wonderful. It is wonderful!" He felt "a kind of ecstasy never experienced before." Soon, he himself was ordering "[o]ne or two beaten to death, no big deal."[79]

Mao's favorite method in the early years of his direct involvement in purges was public execution. Slow killing during public executions

Fig. 9.3. One of the many forms of death by slow execution common in Mao's youth. The prisoners are being asphyxiated as the weight of their bodies stretches their necks.

was a particular favorite. He enjoyed the use of *suo-biao*—a sharp, twin-edged knife with a long handle like a lance. Even local bandits, used to gruesome methods of their own, were frightened by Mao and his organized terror.

Ability to Charm

Those with borderline personality disorder, as well as many psychopaths, are well known for their ability to charm. Mao, along with Russia's Stalin, was well known for his abilities in this area. Mao biographer Jonathan Spence writes: "When American advisory groups came to Yan'an and began to explore the possibilities of using the Communists more systematically against the Japanese, Mao was able to charm a new constituency with his earthy ways and his easy laugh. He also knew how to lobby skillfully for supplies and aid, posing his 'democratic' peasant society against the landlord tyrannies of Chongqing. And always his reach and his mandate spread."[80]

Dr. Li observed Mao's methods for obtaining loyalty from his retainers. He writes: "Incapable himself of affection for others, Mao expected no such feelings toward him. Repeatedly in my years with Mao I watched him win loyalty from others in the same way he had won it from me. He would begin by charming people, winning their trust, getting them to open up, to confess their faults . . . Mao would then forgive them, save them, and make them feel safe. Thus redeemed, they became loyal."[81]

Mao also deliberately chose an influential American journalist, Edgar Snow, who wrote for the *Saturday Evening Post* and *New York Herald-Tribune*, to charm for the Western world. Snow swallowed Mao's fabrications wholesale, calling Mao and other party leaders "direct, frank, simple, undevious."[82] Mao's journalistic charm campaign had long-term payoffs for both Snow and Mao. Other prominent figures joined Snow in praising Mao and his regime. Harvard professor John K. Fairbank "returned from a visit to China and remarked: 'The Maoist revolution is on the whole the best thing that has happened to the Chinese people in centuries.' Feminist philosopher Simone de Beauvoir excused Mao's murderous regime by arguing that 'the power [he] exercises is no more dictatorial than, say, Roosevelt's was.' Jean-Paul Sartre, de Beauvoir's consort, celebrated Mao's 'revolutionary violence,' declaring it to be 'profoundly moral.'"[83]

But Mao and the Communists' ability to fool the press extended further, in unexpected ways. Because the Nationalists had a much freer press, where frank complaint and discussion could take place, the Nationalists' own atrocities and blunders were magnified in people's minds. The contrast with the carefully controlled positive press coming from the Communist camp allowed many people to come to the conclusion that the Communists were the lesser of two evils. Nationalist captain Hsu Chen provides one example of an individual who had witnessed the terrors of Communist rule firsthand, becoming strongly anticommunist as a result. Coming home to Ningbo, near

Shanghai, he found that people were in denial and did not want to hear his views: "I talked to every visitor, til my tongue dried up and my lips cracked . . . I told them about the heartless and bestial deeds of the Communist bandits. . . . But I was unable to wake them up from their dreams, but rather aroused their aversion."[84]

Narcissism

Narcissism is frequently seen in both borderline and antisocial personality disorders. Mao's narcissism, however, was so extraordinary that it bordered on religion—he had near-mystical faith in his own role as leader. "He never doubted that his leadership, and only his leadership, would save and transform China. . . . He shared the popular perception that he was the country's messiah."[85] Mao's personality cult, which he strongly encouraged behind the scenes, began to emerge in the 1940s. The new preamble to the Constitution of the Communist Party was ultimately written to affirm that "[t]he Chinese Communist Party takes Mao Zedong's thought . . . as the guide for all its work, and opposes all dogmatic or empiricist deviations."[86] As Mao himself said: "The question is not whether or not there should be a cult of the individual, but rather whether or not the individual concerned represents the truth. If he does, then he should be worshipped."[87]

It's difficult for someone with a Western upbringing to understand how completely a national personality cult can overtake people's minds. In Mao's China, toddlers chanted, "We are all Chairman Mao's good little children." Mao wasn't venerated—he was indeed worshiped: "The 'Little Red Book' of his aphorisms was ascribed the power to work miracles. Chinese newspapers reported how medical workers armed with it had cured the blind and the deaf; how a paralytic, relying on Mao Zedong Thought, had recovered the use of his limbs; how on another occasion, Mao Zedong Thought had raised a man from the dead."[88]

Unusual insight into Mao's brand of narcissism comes from a set of commentaries he wrote while in his mid-twenties: "I do not agree with the view that to be moral, the motive of one's action has to be benefiting others. Morality does not have to be defined in relation to others. . . . People like me want to . . . satisfy our hearts to the full, and in doing so we automatically have the most valuable moral codes. Of course there are people and objects in the world, but they are all there only for me."[89]

Reading Mao's own words gives insight into an earlier issue. If Machiavellianism is most closely related to borderline personality disorder of all the personality disorders, why is the correlation only a 0.40 instead of close to a perfect 1.0? The answer lies in the wording of some of the questions on the Mach-IV test. By looking at Mao's own thoughts as he described them in his commentaries, it's clear that Mao—that most Machiavellian of men—would have answered the question "One should take action only when sure it is morally right," for example, with the seemingly low-Mach "strongly agree."[90] Mao's commentaries clearly stated his feeling that his own moral codes were the most valuable. Given his profound narcissism, we can reasonably assume Mao felt morally justified in doing whatever he wanted.

Remember the "wisdom of crowds" and the value of using different framing lenses to come to the optimal solution to complex prob-

Fig. 9.4. In this 1966 photograph, *Little Red Books* are waved as Mao clearly relishes the adulation.

Narcissistic Personality Disorder

The DSM-IV diagnostic criteria for **narcissistic personality disorder** consist of a pervasive pattern of grandiosity (in fantasy or in behavior), need for admiration, and lack of empathy. It has been found to be the most strongly heritable of all the personality disorders.[91] The disorder begins by early adulthood, and is indicated by at least five of the following:

- a grandiose sense of self-importance (patient exaggerates own abilities and accomplishments)
- preoccupation with fantasies of beauty, brilliance, ideal love, power, or limitless success
- belief that personal uniqueness renders the patient fit only for association with (or understanding by) people or institutions of rarefied status
- need for excessive admiration
- a sense of entitlement (patient unreasonably expects favorable treatment or automatic granting of own wishes)
- exploitation of others to achieve personal goals
- lack of empathy (patient does not recognize or identify with the feelings and needs of others)
- frequent envy of others or belief that others envy patient
- arrogance or haughtiness in attitude or behavior

Subclinical narcissists, on the other hand, are often happy people who take stress in stride. *Psychology Today* writer Carl Vogel writes: "Mild narcissism also seems to help people recover from accidents or other trauma—it gives them an unrealistic sense of their own invulnerability, and they believe that they will be able to handle whatever else life throws at them. As one researcher put it, being somewhat narcissistic is like driving a huge SUV: you're having a great time, even while you hog the road, suck up extra resources and put other drivers at higher risk."[92]

lems? By stifling dissent and ensuring all eyes were turned only to him for all decisions, Mao effectively made a hundred-million-fold cut in the effectiveness of planning and governing in revolutionary China. It's no wonder the Chinese economy stuttered into reverse.

Paranoia

Another of the key traits the DSM-IV uses to define borderline personality disorder involves transient, stress-related paranoid thinking.*[93] Paranoia, in fact, became an increasing problem for Mao as he grew older. Dr. Li first noticed it in relation to a special swimming pool designed especially for Mao, which Mao became convinced was poisoned (although he happily encouraged others to use it).[94] By 1965, in conjunction with his excessive consumption of sleeping pills (more than ten times the usual amount), Mao's paranoia began to tighten its grip. He began to feel that his guesthouse was poisonous, and he had to move. But, as Li noted, "the only poison was political, the intrigue and backstabbing at the highest levels of communist power."[95]

In his final days, Mao spent hours going over diatribes he had written thirty-three years before against one of his most loyal and zealous followers, Zhou Enlai. Zhou was dying of bladder cancer—Mao had refused permission for an operation during the early stages, when it was still curable. Mao relished the old insults he had written, which he had never felt wise to publish, and cursed Zhou afresh, along with other people who had crossed him over the years.

Forensic psychologist J. Reid Meloy, expands on the type of behavior Mao displayed:

---

*Paranoid behavior appears to be affiliated with the neurotransmitter dopamine. People with psychotic—that is, paranoid and hallucinatory—tendencies seem to show increased activation in the right hemisphere of the brain. Increased levels of right hemisphere activation have also been found in healthy people with high levels of paranormal beliefs or in people who report mystical experiences. Creative individuals also show a similar pattern of brain activation.

> [A person with paranoid personality disorder has] a remarkable capacity to remember slights or insults, perhaps delivered inadvertently, many years after they occurred. . . . [He is] much more deeply wounded . . . than normals, because he is much more narcissistically sensitive to such slights. He is very thin skinned, because the underbelly of narcissistic pathology, which the paranoid individual has in abundance, is covered by an emotional epidermis of shame and humiliation. Hurtful words or actions of others, quickly forgotten by normals, will become a source of preoccupation for this individual. He thinks about them again and again, and as he ruminates, feelings of anger, even hatred, are stirred, and begin to constellate around the memory of the insult. As time passes, these insults, now consciously mixed with unpleasant feelings, begin to fester.[96]

Mao's Manipulation

Mao could be exceptionally subtle in his attempts to manipulate and split his colleagues. One of his favorite techniques was to refuse to issue a categorical ruling—allowing a measure of doubt to subsist. In this fashion he "placed his colleagues before a situation where they had to make a choice, and then stood back and waited to see which way they would jump."[97] He could then in his paranoid fashion find anyone guilty, often flip-flopping his interpretation midstream to catch people on both sides of the fence. Supporters of Mao's hand-selected successor, Lin Biao, drew up a document with a devastatingly accurate assessment of Mao's political tactics:

> Today he uses this force to attack that force; tomorrow he uses that force to attack this force. Today he uses sweet words and honeyed talk to those whom he entices, and tomorrow he puts them to death for some fabricated crimes. Those who are his guests today will be his prisoners tomorrow. Looking back at the history of the past few decades, is there anyone he supported initially who has not finally been handed a political death sentence? . . . His former secretaries have either committed suicide or been arrested. His few close comrades-in-arms or trusted aides have also been sent to prison.[98]

By the end of his life, Mao had vilified nearly every one of his closest colleagues, even those who had been at his side for decades. "After 1971," writes biographer Philip Short, "general cynicism prevailed. Only the young (and not all of them), and those who had profited from the radical upsurge, still believed in Mao's revolutionary new world."[99] As psychiatrist Vamik Volkan points out: "[T]he narcissist in power has special psychological advantages in terms of sustaining his grandiose self-image. He can actually restructure his reality by devaluing or even eliminating those who threaten his fragile self-esteem."[100]

Did Mao Believe in Communism?

Mao's early willingness to ride whatever horse would allow him power and his later ebb and flow of political ideology make his actual belief in conventionally defined communism unlikely. Psychiatrist Jerrold Post astutely notes:

> It is hard to identify the narcissistic personality with any consistent beliefs about the world because his beliefs tend to shift. More than any other personality type, what the narcissistic personality says should be viewed as "calculated for effect." The only central and stable belief of the narcissist is the centrality of the self. What is good for him is good for his country.
>
> The conscience of the narcissist is dominated by self-interest. Unlike the sociopath, who is without an internal beacon, without an internalized body of scruples and principles, the narcissist does indeed have a conscience, but it is a flexible conscience. He sincerely believes himself to be highly principled and scrupulous, but can change positions and commitments rapidly as "circumstances change." The sincerity of his beliefs is communicated, so that the unwary may be completely persuaded of the trustworthiness of the narcissist; and indeed, *at that moment*, he is. The righteous indignation with which he stands in judgment of the moral failure of others often stands in striking contrast to his own self-concerned behavior, which seems hypocritical to the outside observer.[101]

Endgame and Aftermath

After causing the deaths of so many millions, Mao was to die quietly in his bed, aged eighty-two, his body ravaged with an incurable, and ultimately fatal, motor neuron disease—amyotrophic lateral sclerosis—Lou Gehrig's disease. "I felt no sorrow at his passing," noted Dr. Li.[102]

After more than a quarter of a century of Mao's rule, seventy million Chinese had died needlessly, often gruesomely, at his behest. China's economy was in a shambles; its agriculture based on inefficient, backbreaking hand toil; its industrial base worthless, turning out heaps of defective equipment. As a military power, China's only claim to fame was a multitude of warm bodies. There were entire fleets of planes that could not fly and a navy that could barely navigate the seas. (Of course today, with the free-market reforms initiated by Mao's successor, Deng Xiaoping, things are quite different.)

Mao's many pathologies that led to China's sad deterioration were almost certainly rooted in his genetic predisposition. If his obstreperous, argumentative temperament was present early on, which is natural to suppose, it would have had a particularly poor fit with his father's brutal authoritarianism. The resulting stress could have worsened the worst of Mao's already problematic character traits. But Mao's less savory characteristics were also undoubtedly worsened through the years as his neurochemistry was thrown further awry with drug usage and the intoxicating positive feedback of the cult of personality he himself had created.*[103]

But early on, at least, Mao had a few redeeming virtues—perhaps

---

*The idea that a personality disorder could grow from too much attention and adulation is a very old one—the Greeks dubbed it *hubris*, while Carl Jung called it *inflation*. Robert B. Millman, professor of psychiatry at Cornell Medical School, has put a modern spin on the concept, calling it *acquired situational narcissism*. This is the type of narcissism that celebrities, for example, can develop after they've achieved success and are followed everywhere by fans and an entourage. Keeping in mind the physical changes in the brain that can occur from practicing a musical instrument or memorizing the streets of London, it's easy to suppose that acquired situational narcissism might actually be related to physical changes in the brain. It would be fun to stick some Oscar winners in a scanner to find out.

the very virtues that allowed his wife Kaihui and his mother to love him so deeply. Judging from the evidence of Mao's occasional acts of real kindness, Dr. Li, the man who knew him best, was probably right to think that at least a few of Mao's programs were conceived with a vague sense of decency behind them. As biographer Ross Terrill writes:

> The "late Mao" described by Dr. Li was different from the early and middle Mao. He had not always been vain, insincere, vindictive, arrogant, or duplicitous. In middle age, to take examples from the testimony of his most impressive secretary, Tian Jiaying, Mao was not immodest. When commended for his opening speech at the Eighth Party Congress, he said, "Do you know who wrote my speech? A young scholar—Tian Jiaying." The Mao of 1949–50 read all of the many letters that reached him from the general public, but the later Mao lacked a sense of the general public's problems and views. The Mao of 1950 could cry over the suffering of individuals from the grass roots, but the Mao of the Cultural Revolution did not. The Mao of 1951 called the sending of gifts and silk banners in homage to the government a "waste" and an "error," but the later Mao could not get enough homage. In 1962 at the 7,000-cadres conference, after words of outrageous flattery of Mao were uttered by Lin Biao, Mao said, "Lin Biao's words are always so clear and direct. They are simply superb. Why can't the other party leaders be so perceptive?" . . . Tian Jiaying could testify that Mao changed, for the change wrecked Tian's career and ended his life.[104]

As Mao grew older, his increasingly disturbed neurological underpinnings damped out his already poor ability to take in and act upon critical feedback. In point of fact, Mao, like Milosevic, seems to have had little neurological capability of processing criticism. And perhaps in the same fashion that memories can be purposefully suppressed, Mao seemed perfectly capable of suppressing what little empathy he was able to feel. Like Joseph Newman's violent—and nonviolent—psychopaths, Mao had a strangely warped attention; one that allowed him to leave an inconveniently distant wife and family, or ignore the fall of a child acrobat, or discourage attempts to bring his focus to the plight of those suffering from his policies.

Mao used his native intelligence and dysfunctional qualities to great advantage in making his way to the top. He was aided by a society in turmoil as well as by often inept opposition. And he was given cover by an unchecked Communist political party that gained followers through its unabashed idealism and ability to provide the poor and working classes revenge against those they worked for, owed money to, or were simply jealous of. (Of course, many a right-wing dictator has been able to create similarly unchecked powers under cover of "democracy.")

Even taking maximum advantage of every dollop of favorable circumstance, however, Mao's rise was never certain. At any number of junctures, particularly early on, he might have been caught and summarily executed. Here Mao's paranoid tendencies, perhaps rooted in his hints of schizotypy, served him in good stead. In fact, virtually every borderline, psychopathic, and narcissistic tendency in his arsenal proved useful: Mao's supernal gaslighting abilities were used to throw his opponents into disarray; his sadism and vindictiveness shocked his opponents to silence; his black-and-white thinking kept people scrambling to prove themselves as being on Mao's good side; while his explosive temper became a key tool for manipulation. Projection kept the spotlight off him when the problems he created became too widely known to suppress. Eventually, his narcissism would lead the great bulk of the population to literally worship him. As Terrill writes: "The imperial-plus-Leninist system acted as a magnifying glass, giving huge dimensions to each of Mao's personality quirks."[105]

In a capitalist economic structure, Mao might have made his way to the top of a business enterprise. There, like a surprising number of managers today, he would have run roughshod over colleagues and subordinates while devising unreasonable programs even as he took out anyone who objected. If he would have chosen a religious vocation, his smooth charm would have found him a ready pulpit to snare both men and women to feed his sexual drives even as he preached chastity—much as he did in real life as China's chairman. If he had chosen democratic politics, his dictatorial style and ease with corruption, coupled with his modicum of idealism, might have emulated Louisiana's political "Kingfish," Huey Long, who President Roosevelt himself labeled one of most dangerous men in America. Yet in any of these Western-style roles, most of which are constricted by a modicum of checks and balances, Mao's worst traits would have likely been kept in check, or possibly even exposed. An American-born Mao would have had to worry about a judicial system not entirely under his control and an open society that, although far from perfect, provided for some degree of illumination and accountability. Mao might have also had to worry about appeasing political constituents—spending time deflecting reporters and pesky law enforcement officers, and, if he became a major public figure, misdirecting reporters from *Time*, the *Drudge Report*, and gossip magazines. The higher he might have climbed, the more obvious any illegal activities could become—a pattern of suspicious deaths, for example, would inevitably have begun attracting attention. (Though cutthroat, American CEO types have been able to dispatch their adversaries without actually murdering them.)

It's customary for those in Western societies to point fingers at their own undeniable problems in leadership and its corruption, malfeasance, incompetence, and malign intent. But these problems truly pale in comparison to those under Mao's China. In fact, the unfettered political structure of the Communist Party allowed Mao to give full play to every Machiavellian trait in his disordered psyche, feeding his innate sexual, sadistic, and narcissistic proclivities while allowing him to rise to the top. Once at the top, he could write and rewrite rules for his own personal benefit. This in turn provided him with vast powers to implement his dysfunctional pet projects and to give full vent to his paranoid ideas, exposing millions to the consequences of his pathologies.

But today, Mao's personal characteristics and his active role in the deaths of millions have been suppressed in China through the self-serving dictates of his successors. The benevolent image he cunningly portrayed to his people lives on: his portrait still dominates Tiananmen Square in Beijing and the shrine containing his corpse is an object of pilgrimage. Mao's photos are used as charms by street vendors and as icons in the homes of peasants. "I can't pin down why it is I worship Chairman Mao so much," taxi driver Ma Junjing says. "But if the Chairman were alive today and called on us youth to go to war, I'd be the first to register."[106]

CHAPTER 10

EVOLUTION AND MACHIAVELLIANISM

"Genetics explains why you look like your father, and if you don't, why you should."

—Anonymous

The Web page of psychologist Linda Mealey rides ponderously in cyberspace—weighted down with listings of dozens of awards, journal papers, presentations, and reviews. Often, professors ease off teaching once the gravitas of international renown sets in, but Mealey apparently made her own rules. While serving as president of the International Society for Human Ethology, for example, she kept up not only with her extensive scholarship but raised the professional bar still higher with her teaching standards. "I will go over rough drafts in detail if you wish to turn them in," she notes in one of her class syllabi, conveniently downloadable from her site. She is asking for an extra burden most professors would refuse.

The Web page is a bit old-fashioned as Web design goes—but then, since its first deployment in 1990, the Web itself has evolved more quickly than anyone might have thought possible. Still, despite, or perhaps because, of the quaintness, Mealey's Web page communicates boundless energy. A banner scrolls enthusiastically across the bottom of the page: *See Announcements for Information about My New Book!!!!!!!* (The book itself, *Sex Differences: Developmental and Evolutionary Strategies*, provides a tour de force evolutionary explanation of differences between the sexes.) To the right, the Web page shows a dour image of a beetle-browed Darwin. On the left is a photograph of Linda herself, with flashing dark eyes and a mischievous smile.

The Web page is that of a woman with enormous gifts and plans. But a closer look shows something askew. It's not just that the style of the Web page is quaintly obsolete—other issues are also oddly out of sync. The list of publications in Mealey's curriculum vitae, for example, show an explosion of work beginning in the late seventies—up to a dozen or more papers, book chapters, and book reviews each year, backed up with dozens of presentations given at all points of the compass—Salamanca, Perth, Montreal. For over two decades, the vita is swarming with productivity.

And then, suddenly, nothing. Nada. Zip.

At forty-six, Linda's brilliant life was cut short. But her remarkably prescient body of work remains, clearly outlining a perspective that has, over the past dozen years, helped reshape and inform dialogue about the origins of Machiavellian behavior.[1]

For centuries, people have viewed psychopathy and related antisocial behavior as an emotional impairment or disorder. Mealey's contribution to research in this area, outlined in the fundamental paper "The Sociobiology of Sociopathy: An Integrated Evolutionary Model," was to pull together in one massive review the many pieces of evidence suggesting that psychopathic and Machiavellian behavior resulted from something quite different. Instead, she suggested, these were often viable personality traits flowing from the selective forces of evolution—hardly the result of a dysfunctional environment or an accident of biology.[2] Reaching broadly across disciplines, Mealey knit together related ideas of researchers from a variety of disciplines, all pointing toward the fact that psychopaths or Machiavellians can obtain long-term benefits by acting in me-first fashion that hurts others. In fact, the more sinister among us can reproduce and live quite nicely by taking advantage of others, thereby perpetuating any genes that might have played a role in their Machiavellian characteristics. It might not be nice, for example, to steal food from your baby brother's mouth during a famine—but which of the two of you has a better chance of surviving? In a more pointed example, it might violate profound social mores to insinuate yourself into your neighbors' life, and then rape their daughter. But even if you might be caught and killed, the pregnant girl could very well end up giving birth to a child who will possess your genes—the same genes that contributed strongly to your committing the deeply antisocial behavior involved in the rape in the first place. (If the girl was exceptionally kind and caring, her child would have an even better than usual chance of surviving. Kind-hearted naivete puts one at risk of being taken advantage of by a psychopath, which is perhaps why sweet-tempered Laci Peterson, brutally murdered by her husband while eight months pregnant with their first child, had previously dated a man who eventually received a fifteen-year prison sentence for shooting another girlfriend in the back.)[3]

Just as the cuckoo has found an evolutionary niche laying its eggs in the nests of other birds (taking advantage of their nurturing instincts), psychopaths and Machiavellians have found their evolutionary niche in taking advantage of the natural altruism of other humans.*[4] Such variation in human emotional outlook is bred into our very genes.

Research has progressed since Mealey wrote her seminal paper in 1995. But the essential idea she reviewed and synthesized is unchanged—that is, congenitally deceptive individuals—cheaters—

---

*In this section, psychopaths, borderlines, Machiavellians, and the "successfully sinister" are often alluded to in virtually synonymous fashion. To some, this may seem an unfair blurring of phenotypes. To clarify matters, it might help if you were to think of psychopaths and borderlines as extreme examples, in slightly different but often overlapping fashion, of Machiavellian tendencies. The successfully sinister might be thought of as a less extreme form ("intermediate phenotype," or subclinical manifestation) of these personality types. In a sense, then, all four of these terms blend into one another.

can thrive and reproduce in society. How much these cheaters succeed depends on how many of them there are. If their numbers are tiny, they can easily find victims to dupe, and so they thrive. If their numbers grow large, however, the surrounding population grows more wary. In this more savvy population, it's harder to find a gullible target, and so the duplicitous have a more difficult time being successful—and being able to reproduce successfully. Thus there are fewer cheaters in the subsequent generation. And so it goes in a seesaw of counterbalancing activities, much like a predator-prey relationship.

Tit for Tat

Over the past four decades, a great deal of research has centered on the concept of altruism—that is, the desire to help others without any expectation, or at least conscious expectation, of help in return. After all, a person can gain so much by simply being self-serving and greedy that it's tough to figure out how helpful people could have ever evolved. Any time and energy spent helping others' genes takes away from one's own. To solve the conundrum of altruism's evolution, a number of different propositions have been put forward. Altruistic behavior toward your own relatives, for example, would obviously help preserve the same genes that run in you. Sexual selection might also play a role—a man or woman who behaved altruistically might be more attractive as a mate. Other possible reasons include growth in the neurological apparatus that allows us to acquire a conscience; the sheer reason involved in knowing that someone will probably help me later if I help him now; and religion or ethical philosophies, which often promote altruistic behavior.

But there is yet another reason that altruism may have arisen. It may be related to a strange, semi-kind, semi-vindictive strategy known as reciprocal altruism (first proposed by groundbreaking biologist Robert Trivers in the early 1970s). This strategy alternately cajoles and browbeats people into forming a win-win way of working together. A memorable set of competitions organized over a quarter century ago by political science professor Robert Axelrod showcased just how successful the strategy could be.

A version of the game "Prisoner's Dilemma" was used as the basis for Axelrod's competition. In Prisoner's Dilemma, two suspects are jailed separately and given a choice as to whether or not to squeal on their partner. If one partner gives in to the temptation, that partner goes free, and the other—the sucker—is left with an extended jail sentence. If neither partner squeals, however, both serve short jail sentences and then are set free. Obviously the best mutual strategy is to both stay silent and take the significantly reduced jail time. But a misjudgment of the partner's loyalty can prove devastating. What's a prisoner to do?

Axelrod invited game theorists, computer scientists, psychologists, teenage game freaks, and the general public to submit strategies for winning an iterated version of Prisoner's Dilemma in a massive tournament. The submitted strategies were then played off against one another. One parameter that Axelrod was able to tweak in his simulations proved crucial—whether the players expected to meet again. (This was set up mathematically with a discount weighting factor, w, related to the relative importance of the next move.) Axelrod found that if the players did expect to meet again, they were much more likely to cooperate successfully than if they were dealing with an opponent they expected to meet only once.

Long-term, the most consistent winning strategy turned out to be the simplest—"tit for tat." Tit for tat starts out with a first move of cooperation. Then, after the other player makes a move, the strategy involves simply mimicking that move. If the other player defects, defection is in order; if the other player remains loyal, then both stay loyal. Perhaps most importantly, Axelrod's study (which became one of the most cited articles ever published in the journal *Science*) showed that nice guys with a willingness to retaliate when necessary can finish first.

The mechanism that reinforces this strategy is becoming apparent in medical imaging related to our ability to form moral judgments. It turns out that if we intentionally violate a social norm, our amygdala (the "fight-or-flight" processor deep in the limbic system) becomes activated. If we *accidentally* transgress a social norm, however, our amygdala isn't activated. It's believed that the amygdala activation occurs as part of a punishment system that is a consequence of one's own immoral behavior. The expectation of a reward, on the other hand, is thought to bring positive feedback through the brain's natural reward system, the nucleus accumbens. It's easy to see how variations in these systems might be genetically based and thus subject to natural selection.[5]

One can also imagine variants of tit for tat that involve deceit—people who signal they will remain cooperative but who then defect. Such cheaters often have fine-tuned abilities to hide their intentions, almost as if they were bred with such abilities. Indeed, a number of researchers have posited that the presence of cheaters among human populations has led to a sort of evolutionary arms race over time, "in which potential cooperators evolve fine-tuned sensitivities to likely evidence or cues of deception, while potential cheaters evolve equally fine-tuned abilities to hide those cues."[6]

But how applicable is the tit for tat strategy to real human dealings? As it turns out, very—it's just that we use our emotions to do our strategizing. President Lyndon Johnson, for example, was famous for the "Johnson treatment," a mixture of flattery (that's the kind part of the tit for tat), pressure, and, when all else failed, threat of severe sanctions (those are the retaliation tactics), all of which won him high marks as a legislative leader and allowed him to pass much of his "Great Society" legislation.[7] Sumo wrestlers sometimes allow opponents to beat them in bouts when they are comfortably ahead in their overall rankings—later, their opponents are quite likely to return the favor.[8] Nomadic hunters worldwide generally share their kills within their group. But a lazy nomadic hunter could, unless retaliated against, constantly mooch off the kills of others.

Axelrod's study sparked evolutionary psychiatrist and psychologist Randolph Nesse to posit an evolutionary model focusing on the emotional behavior that underlies tit for tat. Nesse concluded that emotions such as those involved in kindness and retaliation strategies are "specialized modes of operation, shaped by natural selection, to adjust the physiological, psychological, and behavioral parameters of the organism in ways that increase its capacity and tendency to respond to the threats and opportunities characteristic of specific kinds of situations."[9] Basically, Nesse proposed that emotions can internally reward, punish, or motivate an individual to either perform or not perform certain types of social interactions and also communicate one's intentions to others.

Evolution involves a pattern of spreading into unfilled niches by using different strategies. People with different emotional makeups use different strategies, and so, in a sense, emotions themselves are subject to evolutionary pressures. Cooperative strategies, which are often beneficial only in the long-term, can coexist with alternate strategies based on cheating, deception, and "rational" short-term selfishness. Adrian Raine, whom you met earlier regarding his research on psychopathy, notes: "An essential component of [a] successful cheating strategy *must* be a gene machine that lacks a core moral sense. One way to create such a cheating machine would be to engineer individuals lacking the neural circuitry essential for moral feelings and behavior."[10] In the end, a mixture of different types of people—and different personality-related genes—fills out a population, with most people being largely cooperative. But a "frequency dependent" group of cheaters rides along. Indeed, some have proposed that the problem of detecting cheaters is what lies behind the dramatic leap of human intelligence.[11] After all, being continually snookered doesn't usually lead to having a long, healthy life.

Although tit for tat forms perhaps the optimal yin-yang set of strategies, people realize a variety of strategies by using the different emotions that are inherent in their personalities. Some people, such as psychopaths, are constant defectors. Others, like the kindly woman who won't divorce her husband even when he beats her, are congenital cooperators. Most people, however, have an emotional makeup that allows—or provokes—them to use both cooperation and retaliation strategies. Indeed, imaging results have shown that we feel disgust (as evidenced by significant activation of the anterior insula) when faced with the behavior of cheaters, and very real satisfaction (that is, activation of the caudate nucleus), when we punish those cheaters.[12]

Throwing Away the Steering Wheel

There is another game that can enlighten our understanding of altruism and antisocial behavior—the game of "Chicken."

In this parental nightmare of a pastime, two drivers square off and make a run toward each other in souped-up cars. The first driver to swerve loses the game and also loses the respect of his adolescent peers. But if neither driver swerves, both drivers lose the game in catastrophic fashion.

What's the best way to win such a game?

In a book called *The Strategy of Conflict*, economist and game theorist Thomas Schelling suggested the ultimate winning strategy: in full view of the other driver, toss your steering wheel out the window. Once the other driver knows that you will never, ever swerve, he has every reason, at least if he's rational, to swerve himself.[13] By eliminating all your other options, you've just won the game.

How does this apply to real life? In straightforward fashion, it shows how successful seemingly irrational emotional strategies can be. For example, if you are known for losing it when hearing something you don't want to hear, or being ruthlessly vindictive when crossed, or going on the attack at even the slightest of imagined provocations, you've essentially torn off your steering wheel in front of those around you. Your own seemingly irrational behavior can work as a completely rational strategy for getting your own way. The only rule for such a strategy to be effective is that you must be known to have torn off your steering wheel. To truly coerce others into steering clear, your ruthlessness and combative nature should be complete and obvious—you must be regarded as being willing to pursue utterly vindictive strategies regardless of the consequences. The reputation you garner by possessing such traits can often provide enough advantages to make up for the occasional disadvantage.

Blink and You'll Miss It—The Quickness of Evolution

If Machiavellians exist and reproduce because sometimes those traits can be advantageous, you can't help but wonder—is the percentage of Machiavellians in modern society changing because our modern-day technological environment is very different from that of our nomadic and farmer ancestors? At first glance, you might think not. After all, evolutionary psychologists have long held that people's evolution took place at a tortoise's pace. If human genes are still nearly identical to a chimp's, and if it took 1.5 million years to evolve from the near-human *Homo habilis* to today's *Homo "laptop,"* people could hardly have been expected to change even a tiddle in the last ten thousand years, much less the last few hundred.

That snail's pace perspective of human evolution has, however, changed dramatically. After all, virtually the only constant in human environments is *change*. Rains can stop falling, as with the once green Sahara, or their effects can increase greatly, as with the recent spate of floods in Bangladesh. Disease can reshape the genetics of an entire population, as when the black death killed a third of all Europeans and when smallpox decimated the indigenous population of the Americas. And, because ancient humans were capable travelers, they could find themselves, within the span of thirty to forty generations, having moved from a warm and humid to a dry and frigid environment. Those peoples with the natural genetic toolkit to help them adapt quickly to whatever they encounter could be expected to survive better than those without such a genetic knack.

Archaeologists are now fairly confident that anatomically modern humans made their break from Africa some seventy thousand years ago, not long after the ability to speak had stabilized in a population that had originated from a bottleneck of a mere five thousand anatomically modern humans. The rapid development of speech may have given the critical edge that early humans needed to break through the menacing ring of other hominids that blocked Africa's exits. In the seventy millennia since that African breakout, humans worldwide have evolved dramatically, with variations in skin color being just the tip of the iceberg.

American Indians, for example, have been found to have a genetic shift that allows for more effective use of the cellular heat engines known as mitochondria. This adapted the Indians to the cold and allowed them to more easily live in, and cross, the frigid regions of Siberia and Beringia on their way to the Americas.[14] Tibetans and inhabitants of the Andes and high African plateaus have acquired differing suites of genetics that allow them to breathe more easily at high altitudes.

The agricultural revolution, beginning some ten thousand years ago, has fueled even more extraordinary changes. Once we humans took charge of growing our own food, we didn't need such a razor-sharp ability to sniff out prey or poison. Consequently, we've begun to lose the genes that gave us olfactory abilities.*[15] And since we now don't generally forage among potentially dangerous plants, we are also losing the genes that detoxify natural plant poisons—a change that means some people can still quickly clear medications from their bodies, and others seem to have lost that ability.

Even our brains are still continuing to evolve. Most people in Europe and East Asia carry an allele called "microcephalin," which appears to have some relationship with cognition, and in any case, spread with contagious quickness after it first arose thirty-seven thousand years ago. Another allele related to cognition, a variant of ASPM, emerged only six thousand years ago but also appears to be spreading quickly.[16] Perhaps,

---

*Wine connoisseur Robert Parker has made a living from his ability to discern and remember flavors and odors—perhaps a rare gift he received from our distant ancestors. Parker has an extraordinary taste memory, and is known for recognizing wines on blind tests that he hasn't tasted in over a decade. One wine reporter noted that Parker "has got a nose like a dog. He sees wine like other people see color." The canine olfactory comparisons run deep—as a child, Parker's father, an avid hunter who kept bluetick hounds, mentioned that his son could tell a breed of dog by the smell alone. Only later did Parker realize his father shared the same abilities.

as David Sloan Wilson has pointed out, "rather than marvelling at the antiquity of our species, we should be asking what kinds of evolutionary change can be expected in 10, 100, or 1000 generations."[17]

Overall, we seem to be evolving rapidly—far more quickly than had been previously thought. Part of what may be allowing these changes to take place so speedily is the rapidly changing nature of some of the junk DNA. Rapid changes had been previously noted in those locations. That's part of why it was thought to *be* junk—because any old change could occur and it didn't matter.[18] But now it appears that the ability to quickly change regulatory sequences is akin to being able to swap out a new motherboard on your computer—you can easily get an upgrade.

Baldwinian Evolution

The laws of genetics state that physical changes you make in your body, such as pumping iron to get a sculpted physique or lying in a tanning booth to get toasted skin, are not passed on to your children. But there are ways that physical changes you make in yourself and your environment *can* be passed down to your children. For example, some eight thousand years ago, a few groups of humans stumbled onto the trick of keeping cows, goats, and sheep around and drinking their rich, vitamin D–laden milk. This was terrific—except for one problem. Some humans suffer from ghastly, flatulent side effects when they drink milk because they have stopped producing the enzyme required to digest lactose once they become adults. This meant that those adults who had milk available and could drink it without side effects tended to be healthier—and have healthier kids. Thus people with milk-drinking genetics became far more common in societies where milk was an important food source. In Denmark, for example, a Neolithic dairy hotspot, 98 percent of all adults today can drink milk. In Zambia, however, where tsetse flies haven't allowed for widespread use of dairy cattle, virtually no one past infancy can drink milk.[19]

Thus, a change in people's behavior made changes in the environment. These environmental changes in turn made changes in people's genes. This type of indirect evolution is called Baldwinian evolution, after American psychologist Mark Baldwin, who first proposed this concept as a variant of Darwinian evolution.[20] There are many other examples of Baldwinian evolution. For example, slash-and-burn agriculture, along with animal husbandry, has allowed malarial mosquitoes to flourish in the new ponds and puddles that replaced the forests. This in turn allowed for rapid spread of a number of different genetic mutations that help humans to cope more easily with malaria. These mutations aren't entirely beneficial, as attested to by their role in causing various sickle-cell anemias and thalassaemias. These mutations would never have spread (been "selected for") if it hadn't been for their beneficial effects in warding off the side effects of malaria.

Can Culture Create Machiavellians?

We've seen how a cultural change as seemingly simple as adoption of slash-and-burn farming methods can cause a widespread, unintended consequence—malaria—that alters people's genetics. This leads to a new question: could cultural changes cause increases or decreases in the numbers of Machiavellians in a society? More pointedly—can cultural practices cause an increase in frequency of some or many of the genes that underpin Machiavellian behavior?

This is not an off-the-wall question. Harvard anthropologist Richard Wrangham has proposed that humans have tamed themselves over the past fifty thousand years, a process he feels is still at full throttle. Violent and aggressive males are, after all, more prone to dying young, leaving fewer offspring than their milder brothers. Wrangham sees the evidence for our domestication in our thinning skulls and smaller jaws and teeth.[21] But Wrangham's thesis hinges on nuanced parsing of the word *tame*. Perhaps instead, in a yin-yang scenario, we are making ourselves less overtly physically aggressive while increasing the percentages of the more subtly aggressive traits of the successfully sinister.

After all, was Mao tame?

Prior to the advent of agriculture, human groups were small—perhaps made up of fifty or fewer, and perfectly capable of "voting with their feet" to escape unfair treatment. Psychopathic or self-serving Machiavellian behavior would be obvious in such a restricted environment and would be difficult to tolerate long-term. There is evidence that when such behavior arose in those small, ancestral nomadic groups, it was eliminated in straightforward fashion. Harvard anthropologist Jane Murphy, for example, notes that the Yupic-speaking Eskimos of northwest Alaska have a word, *kunlangeta*, which means "his mind knows what to do but he does not do it." This word

> might be applied to a man who, for example, repeatedly lies and cheats and steals things and does not go hunting and, when the other men are out of the village, takes sexual advantage of many women—someone who does not pay attention to reprimands and who is always being brought to the elders for punishment. One Eskimo among the 499 on their island was called *kunlangeta*. When asked what would have happened to such a person traditionally, an Eskimo said that probably "somebody would have pushed him off the ice when nobody else was looking."[22]

Murphy goes on to describe a similar word, *arankan*, used by the Yorubas of Africa. It is applied to a person who always goes his own way regardless of others, who is uncooperative, full of malice, and bullheaded. Interestingly, neither *kunlangeta* nor *arankan* were thought to be curable by native healers. Psychopathy is rare in those settings, notes psychologist David Cooke, who has studied psychopathy across cultures.[23]

But what about more urban environments? Cooke's research has shown, surprisingly, that there are more psychopaths from Scotland in the prisons of England and Wales than there are in Scottish prisons. (Clearly, this is not to say that the Scottish are more given to psychopathy than anyone else.) Studies of migration records showed that many Scottish psychopaths had migrated to the more populated metropolitan areas of the south. Cooke hypothesized that, in the more crowded metropolitan areas, the psychopath could attack or steal with little danger that the victim would recognize or catch him. Additionally, the psychopath's impulsivity and need for stimulation could also play a role in propelling the move to the dazzling delights of the big city—he would have no affection for family and friends to keep him tethered back home. Densely populated areas, apparently, are the equivalent for psychopaths of ponds and puddles for malarial mosquitoes.

We have agriculture to thank for those increased population densities. The first settled communities centered around farming began to gradually form about ten thousand years ago. With these settlements, the egalitarianism of the nomadic groups disappeared and a hierarchical society with chiefs and underlings took its place. Becoming more sedentary wasn't all peachy, either—one tradeoff for a steady food supply was that it made the settlements into honeypots for raiders. It also attracted vermin and allowed the newly compacted groups of people to serve more easily as disease reservoirs. (Indeed, since humans are the only natural host of poliovirus, the disease may have arisen and spread with the increased population densities of agriculture.) Settled community living also required new ways of interacting with one another. "Settlement," notes science writer Nicholas Wade, "would have created a quite novel environment, to which people probably adapted by developing a different set of behaviors, including a range of intellectual skills for which there was no demand in hunter-gatherer societies."[24] Property, value, number, weight, measurement, quantification, commodity, money, capital, economy—all these, Wade goes on to note, are concepts that would rarely have been necessary to mobile foragers. It may well be that the modern mind has emerged in a gradual process, "operating in several phases and stages, and perhaps independently in different parts of the world."[25]

Agriculture not only allowed for exponential leaps in the numbers of people, it also provided for enormous increases in the potential of those in power to have sex and children. Recent studies have shown that a remarkable sixteen million men, one in every two hundred men worldwide alive today, are direct male descendents of thirteenth-century Genghis Khan—a nomad who commandeered virtually all the many sedentary peoples within his long reach. From the Great Khan's six Mongolian wives, as well as the many daughters of foreign rulers that he also took on as wives, and the great numbers of beautiful women he demanded as his due from conquered territories, the Great Khan is thought to have sired an enormous number of children, although the exact number is unknown. Genghis also brought his relatives to the party: "It's likely that some brothers and male cousins of Genghis Khan who shared his Y-chromosome enjoyed heightened reproductive success in his enormous wake," notes controversial writer Steve Sailer, "rather like how it is said that some of the sex appeal of the rock band Led Zeppelin rubbed off on its lucky roadies."[26] The fact that so many men are direct male descendents of either the Great Khan or one of his near paternal ancestors indicates that virtually everyone on the Asian steppes is, through some line of descent, carrying the DNA of the Great Khan's family. Even Queen Elizabeth II is thought to be a descendent, through Mary of Teck and her Basarab dynasty ancestors. Physicist turned evolutionary theorist Gregory Cochran observes: "This disproves the theory of history promoted by Marx and Tolstoy that says only social forces matter, not individuals. This shows that one man can make a difference."[27] The difference, of course, lies not only in the virtually single-handed creation of a vast empire, but also in a genetic imprint that leaves competing males in the dust.

Research on this type of genetic stamping is still in its infancy. Recently, for example, Giocangga, progenitor of China's Qing dynasty, has been found to be the probable ancestor of approximately 1.6 million men living today.[28] The average man from Giocangga's era should have only about twenty living male descendents—which means that Giocangga has left a genetic imprint eighty thousand times larger than that of the typical man of his time. Geneticist Bryan Sykes notes, "Mini-Genghises were probably all over the place in medieval times."[29] Niall of the Nine Hostages, for example, an early medieval Irish king once thought to be apocryphal, is now understood to be a direct paternal ancestor of one in five men from northwest Ireland—and an ancestor to virtually every Irishman.

These recent discoveries reinforce the findings of anthropologist Laura Betzig. Her 1986 *Despotism and Differential Reproduction* provides a cornucopia of evidence documenting the increased capacity of those with more power—and frequently, Machiavellian tendencies—to have offspring. The *Guinness Book of World Records*, for example, has long cited Ismail the Bloodthirsty as the most prolific man ever—siring 888 children. (In reality, the total was probably closer to three hundred.)[30] In any case, Ismail was one of the greatest leaders in the history of Morocco. He was also a man of legendary cruelty who inflicted barbaric punishments on slaves and subjects alike, torturing and beheading on a whim.

Ismail the Bloodthirsty's successfully sinister behavior was shared by despots across a broad range of agriculture-based societies. "Under complex preindustrial hierarchies, despotism appears clearly to have been a general phenomenon," Betzig writes, defining despotism as "an exercised right to murder arbitrarily and with impunity."[31] As Machiavellian researcher Richard Christie and his colleague Florence Geis aptly note: "[H]igh population density and highly competitive environments have been found to increase the use of antisocial and Machiavellian strategies, and may in fact foster the ability of those who possess those strategies to reproduce."[32]

Betzig's book is filled with example after example, such as that of the Byzantine emperor Justinian, who literally gave *despotism* a bad name. Among other memorable deeds:

> Within days of coming to power, he executed the head of the palace eunuchs for making an "injudicious remark" against an arch-priest; he accused a member of the opposition faction of committing "offenses against boys," and had him dismembered, . . . and he took immense wealth from the grandson of a former emperor when a will

> openly rumored false was produced upon his "unexpected" death. . . . These sorts of accusations, punishments, and false documents were liberally brought . . . against anybody "who happened to have come up against the rulers in some other way," and especially, against those exceptionally well off. . . . He and Theodora eventually asked subjects to prostrate themselves flat on their faces in front of them, kiss one of each of their feet, and address them as "*despotes*" and "*despoina*," master and mistress, making themselves, by implication, their slaves.[33]

Betzig's ultimate point is not that the corrupt attain power but that those corrupted individuals who achieved power in preindustrial agricultural societies had far more opportunity to reproduce, generally through polygyny, and pass on their genes. The more Machiavellian, that is, despotic, a man might be, the more polygynous he tended to be—grabbing and keeping for himself as many beautiful women as he could.*[34] Some researchers have posited that envy is itself a useful, possibly genetically linked trait, "serving a key role in survival, motivating achievement, serving the conscience of self and other, and alerting us to inequities that, if fueled, can lead to escalated violence."[35] Thus, genes related to envy—not to mention other more problematic temperaments—might have gradually found increased prevalence in such environments.

Incan kings, for example, are recorded as keeping more than seven hundred women with whom to take their pleasure. Further thousands of women were sequestered throughout the kingdom for the king's sole use if he happened by. These women "lived in perpetual seclusion to the end of their lives, and preserved their virginity; and they were

*You may wonder why modern despots don't seem to have many children. While it's true that Hitler had none, in general, the relatively recent spread of venereal diseases unheard of in Genghis Khan's day may play a role. Additionally, many despots seem to have had a number of children who are unacknowledged or often simply not mentioned. Mao, for example, is thought to have had from seven to ten children before becoming infertile, probably due to venereal disease. Kim Jong Il has far outdistanced Mao's relatively tame sex life by systematically recruiting the most beautiful girls in the country to join the thousands of women in his "Happy Corps." As biographer Bradley K. Martin wryly comments—"Eat your heart out, Hugh Hefner." Both Kim Jong Il and his father, Kim Il Sung, are thought to have scores of well cared for but publicly unacknowledged children.

not permitted to converse, or have any intercourse with, or to see any man, nor any woman who was not one of themselves."[36] Below the rank of king, reproductive rights were precisely prescribed—caciques were given fifty women, vassal leaders thirty, provincial leaders twenty, and still lower leaders were allocated fifteen, twelve, eight, seven, five, or three, depending on their positions. The remaining women were given to whomever was left.[37] Betzig's work provides example after example of similar societies.

Questions we might reasonably ask are—has the percentage of Machiavellians and other more problematic personality types increased in the human population, or in certain human populations, since the advent of agriculture? And if the answer is yes, does the increase in these less savory types change a group's culture? In other words, is there a tipping point of Machiavellian and emote control behavior that can subtly or not so subtly affect the way the members of a society interact? Certainly a high expectation of meeting a "cheater," for example, would profoundly impact the trust that appears to form the grease of modern democratic societies and might make the development of democratic processes in certain areas more difficult. Crudely put, an increase in successfully sinister types from 2 percent, say, to 4 percent of a population would double the pool of Machiavellians vying for power. And it is the people in power who set the emotional tone, perhaps through mirroring and emotional contagion, for their followers and those around them.[38] As Judith Rich Harris points out, higher-status members of a group are looked at more, which means they have more influence on how a person becomes socialized.[39]

An example of how a leader's tone might contaminate an entire society was related by Ceausescu biographer Edward Behr, who recalled his attempt to interview a senior ex-minister who was a brilliant, British-trained medical specialist with a reputation as an Anglophile.

> I found myself in a hospital consulting room face to face with a frightened but at the same time angry individual who, instead of

answering my questions about the Ceausescus, about whom he knew a great deal, seemed solely preoccupied in finding out how I had come to track him down. He clearly intended to report on this intolerable invasion of privacy to the appropriate authorities. His harsh, inquisitorial tone was such that I could not refrain from pointing out, somewhat curtly, that while I would have expected such conduct during the Ceausescu era, I was amazed that the changes in the regime since December 1989 had had so little effect on him. It seemed, I said, that Ceausescu lived on—at any rate, in the hearts and minds of all those who had had anything to do with him. Visibly shaken, he paused. There were tears in his eyes as he said, in a completely different tone of voice, "I don't know what came over me. I don't know why I started asking you these questions. There's no reason for it. But there's something of Ceausescu inside me that will never go away."[40]

Gold Diggers, Stable Sinister Systems, and the Slow-Motion Implosion of Empires

But it is not only despotic, Machiavellian men who might pass along more than their share of "evil" genes. Machiavellian women can also play that game. As it happens, there is enough documentation passed down surrounding the intrigues of the Ottoman Empire to make a good guess at how the game might have been played.

In the early 1500s, a fifteen-year-old girl—probably captured in one of the many Tartar slave raids on the Ukraine—was purchased from the open-air market in Istanbul. The girl's buyer was Grand Vizier Ibrahim, the best friend of Sultan Suleyman the Magnificent. The girl's name was Roxalena; her new home was to be the Ottoman sultan's harem. Much of what is now known about Roxalena comes from widespread gossip and hearsay—yet the compelling story that has come down to us parallels that of many other successfully sinister individuals.

Arriving at the harem, the quick-witted Roxalena soon surpassed dozens of other concubines to become a favorite of the sultan—she

Fig. 10.1. Roxalena, the wife of Sultan Suleyman the Magnificent. She lived from around 1500 to 1558.

was "so full of light that Suleyman seemed blind to her dark side. He named her Hurrem, 'the laughing one,' because of her crystalline laughter and freedom from inhibition."[41] Roxalena soon bore Suleyman a son, Mehmed, which elevated her to third-highest woman in the harem.

Both Western and Ottoman sources of the time viewed Roxalena as an extraordinarily dark and malevolent influence on the throne. A contemporary remarked: "For myself I have always heard every one speak ill of her and of her children, and well of the first-born and his mother."[42] (Recent attempts to rehabilitate Roxalena's reputation involve putting a positive, feminist spin on her more unsavory, power-hungry attributes.)[43] In any event, it appears Roxalena forged a letter that made it seem that crown prince Mustafa—Suleyman's favorite son—was scheming with the shah of Persia to dethrone his father. Mustafa was subsequently strangled on Suleyman's orders. Roxalena also orchestrated Grand Vizier Ibrahim's death, taking advantage of every bit of gossip and information to inflame Suleyman's mind against his stalwart commander and best friend since childhood. Suleyman had Ibrahim strangled as well.

Roxalena had five sons with Suleyman. Three, through inconvenient deformity or death, rendered themselves ineligible for the throne. This left Beyazit, who was "able but cruel" (echoes of his mother), and Selim—a drunkard. Selim, clearly the weaker-willed of her two sons, was Roxalena's choice to succeed her husband. And indeed, Roxalena's genes lived on through this son, "Selim the Sot," and his descendents, each generation's heir mixing his genes with those of the most beautiful—and often, the most Machiavellian—slave girls in the harem. Descendent after descendent proved to be squirrelly or outright mad. Roxalena's great-great-great grandson, Murad IV, enjoyed haunting the streets incognito, the better to catch

The House of Hilton

Those on the lookout for a modern version of the Ottoman dynasty might be satisfied with the warped characters and antics of the Hilton dynasty, as chronicled by respected biographer Jerry Oppenheimer, from whose work this account is drawn. Conrad Hilton, Paris Hilton's great-grandfather, was a hotelier and womanizer of no mean talent. But it is Paris Hilton's gold-digging maternal grandmother, "big Kathy," who seems to have set the tone for her modern-day descendents. Big Kathy was an emergenic stage-mother-from-hell who was so vindictive that she put a screw in her stepdaughter's cheeseburger in an attempt to ruin her perfect teeth. The stepdaughter later noted: "I was so afraid of [big Kathy] that I just finally withdrew into a complete shell. She has a violent disposition and she intimidated me and she *knew* she intimidated me and she enjoyed it. She had a sadistic streak. She needed help."[44] Big Kathy was also extraordinarily materialistic and obsessed with marrying rich. Big Kathy's daughter, little Kathy, was viewed by those in the know as a clone of big Kathy—one noted, "Little Kathy and big Kathy loved hurting people. . . . They are very bizarre."[45] Both women realized their dream when little Kathy snagged a Hilton heir as a husband. One observer of the time noted that "she nailed him with her fake personality, her false way of being."[46]

The problematic personalities carry on. "[Little] Kathy Hilton's *very* selfish and *very* spoiled and *very* self-centered, and that absolutely carries through to [her daughter] Paris."[47] Of Paris Hilton herself, one close observer has noted, "She's bright about three things—money, men, and how to get attention—and those are the only things she really *cares* about; she's basically classically self-involved and narcissistic . . . I've spent an enormous amount of time with her, and after she leaves a meeting I always find myself wondering, where did this creature come from? What are the genetics happening here? Who's responsible for turning out a persona like Paris? Where'd she get her values and ethics and morals?"[48]

people and kill them for some small infraction. (Corpses, it was said, hung at every street corner.) Murad managed to kill his brave and handsome younger brother, Beyazit, leaving only an insane youngest brother, Ibrahim. On his deathbed in 1640, Murad ordered Ibrahim's death, telling his mother "it would be better for the dynasty to end than to continue with insane royal seed."[49] But Ibrahim lived, and the problematic personalities continued through his progeny. Five generations further removed, in the 1830s, an outside observer unaware of Ibrahim's dissolute predecessors would write of Sultan Abdul Mecit: "[H]e has the taint of madness which has existed in the family since Sultan Ibrahim, who was known as 'the madman.'"[50]

Recollecting our discussion of "evil genes," we can easily imagine how a trickle of troubled serotonin receptors and transporters here; a brilliant but neurotic splash of BDNF or COMT there; and a pinch of inefficient MAO-A alleles rounding off the already troubled upbringing of many of the sultans' sons might contribute, generation by generation, to the dissolute character of the Osmanli line. Some of each sultan's inherited alleles were undoubtedly associated with the charismatic, temperamental, and often deeply Machiavellian personalities of the women who clawed their way to the top of the harem to charm the sultan and destroy their competitors. These genes would have been affiliated with some of the best and brightest, but also some of the most troubled of offspring. Unfortunately, the best were usually killed by their more Machiavellian rivals or by Machiavellian powers-behind-the throne who needed a weak and easily manipulated front man. Like a slow-motion train wreck, the Ottoman Empire—the Sick Man of Europe—continued on with its troubled system of inheritance through the seraglio, losing territory and careening from one weak, cruel, or unhinged sultan to another. The real question is how such a clearly dysfunctional system of governance could have lasted for so long.

We'll get to that.

It may be a little tough to segue from the Ottomans to the Hiltons to the Roman Empire, but the parallels are there, if you're willing to dust off your ancient *National Enquirer* equivalent: the *Historia Augusta*.

Marcus Aurelius, for example, who lived from 120 to 180 AD, was the last of the five "good" emperors who were selected for merit rather than direct blood propinquity. Marcus married his cousin Faustina, who, although apparently possessing a "lively" personality, was also characterized as a woman who enjoyed whoring around with gladiators and sailors—she may also have actively worked against her husband's interests. Although questions remain as to whether Faustina was really as bad as she was made out to be,* there is little question that her son Commodus, ostensibly fathered by Marcus Aurelius, was a disastrous character who could charitably be described as a neurotic megalomaniac. As the (admittedly spotty and unreliable) *Historia Augusta* notes:

> Marcus tried to educate Commodus by his own teaching and by that of the greatest and the best of men. . . . However, teachers in all these studies profited him not in the least—such is the power, either of natural character, or of the tutors maintained in a palace. For even from his earliest years he was base and dishonourable, and cruel and lewd, defiled of mouth, moreover, and debauched. . . . In the twelfth year of his life, at Centumcellae, he gave a forecast of his cruelty. For when it happened that his bath was drawn too cool, he ordered the bathkeeper to be cast into the furnace . . .
>
> The more honourable of those appointed to supervise his life he could not endure, but the most evil he retained, and, if any were dismissed, he yearned for them even to the point of falling sick. When they were reinstated through his father's indulgence, he always maintained eating-houses and low resorts for them in the imperial palace. He never showed regard for either decency or expense.[51]

When Marcus Aurelius appointed his son Commodus rather than a more deserving candidate as his heir, the results proved devastating for

*It's been pointed out that Marcus Aurelius appeared to really love Faustina, so she likely wasn't all that bad. Still, evidence abounds of decent people who truly love their borderline spouses—trying endlessly to please them and happily lapping up the moments of wickedly funny humor and deliciously uninhibited activities. No one will ever know definitively whether Faustina—or Roxalena, for that matter—had borderline or borderpathic traits, but it's worthwhile to keep in mind that the same personality traits and disorders seen worldwide today were undoubtedly present during Roman and Ottoman times as well.

Rome. Instead of taking serious interest in matters of state, Commodus showed himself to be interested only in staged gladiatorial events, for which he charged Rome extortionate sums, and sex, indulging himself with hundreds of female concubines and young boys. He surrounded himself with sycophantic fellow Machiavellians who ran the government for their personal profit rather than the empire's benefit—which, after an extended period of extraordinary malfeasance, eventually drove the Romans to kill both Commodus and his claque.*[52] The creepy Severan dynasty followed, marked by incompetence, brutal rivalries, and marriages with boundlessly ambitious women. After the last of the Severans was butchered, the empire was left in churning chaos. It was kept afloat only with the occasional inspiring rule of emperors like Claudius III (who fell unfortunate victim to the plague), Aurelian, and Constantine the Great.

Historians of the Roman Empire note that dynastic successions were often given to the most plausible heir rather than simply the oldest male descendent of the emperor—although the latter was generally preferred if he looked like a reasonable choice.[53] But many of the most successfully sinister emperors, including Caligula and Nero, portrayed themselves, chameleon-like, as having pleasing personalities. Thus, they seemed like credible candidates when they came under consideration. Often, these decadent and malevolent sorts were the sons of social climbing mothers with reputations for troubled personalities—clearly women with propensities for "evil genes" (although one still can't discount acquired nastiness).

---

*Stanley Bing's delightfully entertaining *Rome, Inc.* points out the many similarities between the Roman Empire and the modern multinational business environment. As he notes: "Sometimes the emperors were just kids, sons of somebody with some marquee value, thrust into the corner office by one special-interest group or another. These had to be 'helped' by powerful second and third bananas, as is often the case with any weak chief executive. This decayed culture fosters henchmen and elevates toadies and other forms of reptilian life to very senior roles indeed. I've been there, and I can tell you that it's worse than working for a straightforward despot any day."

Bing also notes: "Before 100 B.C. the incidence of gigantic, pathological, preening, egotistical and thoroughly modern nutcases in the ranks of senior management is relatively rare, and you have many examples of noble Romans who lived to serve the state. After, you begin to get moguls. In the absence of a strong center, and the corruption of daily life in the Republic, these competing moguls have free rein to marshal their forces and make a run at the top slot, which they may occupy for a time, but never own."

For insight into the phenomenon of powerful men (who often come with their own set of psychological quirks), attracting charismatic, troubled, sometimes deeply sinister women, it's useful to hop back to the present and take note of modern psychiatric findings. Jerold Kreisman, a psychiatrist and leading expert on borderline personality disorder, uses a description of Princess Diana to introduce his most recent book on borderlines: *Sometimes I Act Crazy: Living with Borderline Personality Disorder.* Kreisman writes: "The fractures in Diana's personality became more prominent during her adolescence. She could be charming, charitable, and remarkably empathic with friends at times, but on other occasions she exhibited an unpredictably cruel rage when these same friends disappointed her. Sometimes, during stressful periods, she appeared calm and stoic, but at other times she became irrationally emotional, alternating between inconsolable grief and ferocious anger."

Kreisman tellingly adds: "Typically, the borderline seeks partners who are in a position of power. The most common scenario involves the younger, attractive, borderline woman and the older, narcissistic man: the secretary embarks on an affair with her older, married boss; a student becomes involved with her professor; a patient with her doctor."[54] Marriage into the aristocracy or royalty, in times of old, or for Diana in modern times, offers the readiest route for borderlines to partners in positions of power.

In short, it seems hereditary aristocracies—not to mention the decidedly wealthy—can attract mates with ambitious, manipulative, controlling, chameleon-like, semi-neurotic personalities, who in turn are more prone (although not guaranteed) to have children with a genetic predisposition for similarly idiosyncratic personalities. And indeed, there is a growing body of research literature that reveals how people selectively seat themselves into positions that suit their personalities.[55]

There is little question that the unrest and decay surrounding the most sinister emperors and sultans, as well as the corruption they and their Machiavellian minions fostered, contributed to internal weakness of the empires. It was only a matter of time before increasingly powerful external forces were able to leverage this weakness to their advantage.[56] It seems that the longer an empire is in existence, the more time the successfully sinister have to find ways to subvert the system and insert themselves into positions of power. But even as empires—be they political, religious, or business—begin gradually to founder, they can still muddle on, year after dysfunctional year. Sometimes, especially in political or religious enterprises, they can slither through century after dysfunctional century.

Pockets of such "stable sinister systems" are apparent not only in the ancient world but in modern-day social systems as well: a cubicled group led by an unscrupulous, chameleon-like director of sales; a city government controlled by a corrupt longtime mayor; a religious group that rewards and promotes even the most unsavory types as long as they obey; or—the ultimate in free rein to rewrite rules—the repressed country led by a dictator. By taking advantage of their own dysfunctional but simultaneously advantageous traits, as well as the compliant characteristics and emote control reasoning of others, Machiavellians can build tightly interlocked systems that keep naysayers in check and allow themselves to remain in control.

A Prototypical, Modern-Day, Stable Sinister System—Texas Southern University

In August 2006, flamboyant Texas Southern University president Priscilla Slade, along with three board members, was indicted for "misapplication of fiduciary responsibility" in relation to millions of dollars of misspent, misused, and disappearing funds. Many of Slade's apparent accomplishments were ultimately shown to disguise a sordid reality. For example, TSU's doubling of enrollment brought a dangerous element to campus even as the tuition helped fund her flamboyant lifestyle—Slade was eventually caught illegally spending $260,000 to landscape and furnish her home, $10,000 for limousines, and $9,000 for a bed. Meanwhile, only 6 percent of TSU students graduated in four years—one of the lowest rates in the nation.

Freshman class president Justin Jordan and his friends Oliver Brown and William Hudson—the "TSU 3"—were motivated to investigate the school after the death of a student bystander who died when a firefight erupted in a campus parking lot. Their investigation uncovered rampant corruption on the TSU campus. Christina Asquith, a reporter for *Diverse Issues in Higher Education*,[57] related how the TSU 3 discovered a paper trail of evidence revealing that associates of campus administrators were being paid thousands each month even when they didn't work for the university. State representatives were paid by TSU to be "guest lecturers." Two highly publicized parking garages were built for tens of millions of dollars over budget. Administrators at many levels appeared to be stealing state funds.

Through their diligent efforts, the TSU 3 built a slam-dunk case against TSU's administration that immediately provoked indignation from the board and state authorities and resulted in the immediate firing and indictment of the guilty parties.

JK, as the instant messengers say. Just kidding.

Instead, despite the increasingly squalid nature of the material the TSU 3 was uncovering, the administration responded by offering semesters abroad and other bribe-like inducements to the trio of would-be whistleblowers. When the TSU 3 brought their evidence of corruption to the university's board, board members responded with a vote of confidence for TSU's corrupt president—neatly shifting blame for the problems on lack of funding from Republicans. When the students met with Texas state governor Rick Perry to provide evidence for criminality, the governor simply referred the matter back to the TSU board—who ignored it. The young men were harassed by campus police officers and ultimately arrested on trumped-up charges. Then, as Asquith relates:

> In late Spring 2005, administrators brought the students before the Student-Faculty Disciplinary Committee on charges that included "inflicting mental harm," "insubordination, vulgar lan-

guage" and "disturbing a meeting." They say they were denied legal representation and told to write a letter to Gov. Perry saying that "everything was OK now" at TSU. One of the TSU 3, William Hudson, was suspended for a year and required to take anger management classes in order to return. He was also fired from his campus job in the office of enrollment management. Each of the TSU 3 was forced out of his role in student government. . . . By the fall of 2005, the three were feeling demoralized and ready to give up. "Every time we took information to someone, we ran into a brick wall," said Jordan.[58]

Finally, luck turned their way—a sympathetic DA took on the case and the goings-on at TSU came under legal scrutiny. The indictments came down, and Slade lost her job, after a fashion. She was a tenured professor, so she was simply moved to a teaching position.

"With corruption, everyone pays," Jordan says. "Now the faculty has to teach more classes, the students have had a tuition increase, the taxpayers—they're sick of paying more money, and people in the administration are going to jail. We are all paying somehow." Adds Jordan: "Dr. Slade and the administration did a wonderful job of charming the board. They were mesmerized by her. People were mesmerized by her."[59]

One can easily imagine that, if the charismatic Slade had had friends in the DA's office, Jordan and his friends would have been further harassed until they had no psychic resources remaining. The lives of the TSU 3 would have been derailed, and corruption at TSU could have gone unchecked for decades to come.

DEFINING "MACHIAVELLIAN"

Earlier, I defined a Machiavellian—which I've used interchangeably with the term *successfully sinister*—as a person who is charming on the surface, a genius at sucking up to power but capable of mind-boggling acts of deceit for control or personal gain. I've also stated that Machiavellians are unscrupulous and self-serving and therefore are capable of deeply malign behavior. But, given the research results we've covered, it's now possible to refine the definition. Ultimately, a Machiavellian, as I use the term throughout this book, is a person whose narcissism combines with subtle cognitive and emotional disturbances in such a fashion as to make him believe that achieving his own desires, and his alone, is a genuinely beneficial—even altruistic—activity. Since the Machiavellian gives more emotional weight to his own importance than to that of anyone or anything else, achieving the growth of his preeminence by any means possible is always justified in his own mind. The subtle cognitive and emotional disturbances of Machiavellians mean they can make judgments that dispassionate observers would regard as unfair or irrational. At the same time, however, the Macchiavellian's unusual ability to charm, manipulate, and threaten can coerce others into ignoring their conscience and treading a darker path.

I recognize it's problematic to use a single term, *Machiavellian* (or my synonym, *successfully sinister*), for disturbed individuals who in all likelihood come by their disturbances from a variety of neurological quirks and environmental influences and who vary significantly in their dysfunction. But psychologists use inherently vague terms like *antisocial personality disorder*, which has a similar nomenclature-related problem, all the time. The trouble is, it's difficult to conceive of an alternate, more refined shorthand that conveys the same sense of phenotype as *Machiavellian*. In point of fact, as with many personality types and disorders, there are almost certainly hundreds—perhaps thousands—of genotype configurations and resulting differences in brain function that could underlie Machiavellian behavior, all of which could vary further, depending on slight differences in how one might define the term *Machiavellian*. And the effect of the environment on those with a potentially Machiavellian genotype is not necessarily as straightforward as it might seem. For example, a talented boy with an underlying set of problematic genes might, as a result of abuse, descend by adulthood into obviously pathological behavior—borderline or psychopathic—that could result in his incarceration and removal from society. However, the same Machiavellian-oriented child with a mild upbringing might flower into a full Machiavellian as an adult—a charismatic man whose sinister influences could ultimately affect millions.

Perhaps psychiatrist Regina Pally puts it best in her description of borderline personality disorder:

> Neuroscientists agree with Darwin's assumption that variation is healthy for a species as a whole, even if some variations may be maladaptive for a given individual. Darwin's theory of natural selection argues that, in order for a species as a whole to survive and to adapt to changing environments, the individuals of that species need to exhibit a wide variety of physical traits and capacities. . . . Since what is maladaptive in one environment may be adaptive in another, evolutionary pressures have resulted in the retention of maladaptive variations. . . . It can be conceptualized that for evolutionary reasons such as these the biological impairments of BPD [borderline personality disorder] have been retained in the human species. What Darwin's theory implies is that normal is relative. Normal exists as a range. Every biological factor, whether it be height, eye color, blood type, serotonin level, cortisol level, or autonomic reactivity, exists on a continuum—a bell-shaped curve—in which some variations are more common (i.e., in the middle range of the curve) and other variations are less common (i.e., at the tails at either end of the curve). I stress Darwin's theory of natural selection to emphasize that every symptom of BPD exists somewhere on a bell-shaped curve of the traits found in humans, albeit on the statistically less common tails of the curve.[60]

Ultimately, of course, the genes that predispose borderline traits, when combined with genes that predispose psychopathy and narcissism, could be thought to be the foundation for Machiavellian behavior. For some with a naturally gentle, giving, and loving character, almost inconceivably great environmental influence would be needed to provoke Machiavellian behavior.*[61] For others with a particularly unfortunate confluence of traits, Machiavellian behavior seems to come naturally.

LINDA MEALEY

We've covered a great deal here, tossing emotional steering wheels out the window on our way to visit "evil" genes, playing tit for tat as we drank milk and lingered with royalty. But it might be nice to end this chapter back where we first began, with psychologist Linda Mealey. Despite Linda's disappearance from the Web half a decade ago, she still affects the studies and careers of those in her field and beyond. Linda's father, George Mealey, has worked with the International Society for Human Ethology to establish an in-perpetuity fund to maintain the Linda Mealey Award for Young Investigators. In contrast to the sociopaths she sought so diligently to understand, Linda continues to serve as a model and inspiration for the generations to come. Her Web site hangs suspended in cyberspace, never updated, in tribute to her life. Even as she lay dying, she faxed copies of notes and handouts to friends at the International Society for Human Ethology meeting in Montreal, enduring her illness with great courage, grace, and dignity. In the end, however, Linda was no match for the tiny, mutant genes inside her. She passed away on November 5, 2002, of colon and liver cancer.

---

*Scottish anthropologist Colin Turnbull described a kindhearted little girl named Adupa who was born into the Ugandan tribe known as the Ik, which was just beginning a nightmare descent into famine. By Ik standards, Adupa was insane:

"Her madness was such that she did not know just how vicious humans could be, particularly her playmates. She was older than they, and more tolerant. That too was a madness in [the world of the Ik]. Even worse, she thought that parents were for loving, for giving as well as receiving. Her parents were not given to fantasies, and they had two other children, a boy and a girl who were perfectly normal, so they ignored Adupa, except when she brought them food that she had scrounged from somewhere. They snatched that quickly enough. But when she came for shelter, they drove her out, and when she came because she was hungry they laughed that Icean laugh, as if she had made them happy."

Adupa was eventually locked away by her parents and allowed to starve to death.

CHAPTER 11

SHADES OF GRAY

"He is absolutely untrustworthy, as was his father before him."
—Lord Derby on Winston Churchill[1]

It is people like Linda Mealey who remind us that not everyone is a self-serving Machiavellian. Far from it. Ordinary people by the thousands have raced into burning buildings to save children from certain death, thrown themselves onto hand grenades to protect their squad mates, or leapt into icy waters to help strangers who were drowning. And, just as even the most sinister of the successful have their good traits, it seems even the most angelic among us have bad traits. Christopher Hitchens, for example, describes Mother Teresa's buttering up of despotism in his *The Missionary Position.*[2] Altruistic Christian pacifists have given their lives for their beliefs—yet so have Muslim terrorists. In loop-the-loop fashion, one man's Machiavellian could be another man's Messiah. You can't help but wonder—are Machiavellian traits *necessarily* evil? If there are Machiavellians, couldn't there also be quasi-Machiavellians? Would quasi-Machiavellians do quasi-good?

What are we to make of people such as Salvador Allende, a socialist with the best of intentions who set himself above the law and sent the Chilean economy into a tailspin? Allende's successor, brutal martinet Augusto Pinochet, was responsible for torturing and killing thousands who opposed his regime. Yet most observers agree that Pinochet's economic policies have left Chile the envy of South America—arguably saving tens of thousands of lives, and improving the lives of millions more, by providing better possibilities for economic sustenance. At the same time, many far less able, but sometimes even more brutal, right-wing dictators have waved the anti-communist flag to maintain their power and simultaneously drive their country toward ruin, as with Haiti's "President for Life" François Duvalier, or Nicaragua's Somoza regime. Along similar lines, fascist Benito Mussolini had a reputation for making the trains run on time and built magnificent monuments, but at what price—particularly for the hundreds of thousands of Ethiopians killed in Mussolini's grand imperialist schemes? Yet Turkey's Ataturk and Poland's Pilsudski could also be ruthless, with no compunction against using dictatorial methods, but each inarguably left his country far better off than it had been before he took power.

Still further back in history was Catherine the Great, empress of all the Russias, who seized the throne in a coup d'état, while her eccentric husband and other claimants to the throne conveniently died around the same time. However Catherine achieved power, she was the epitome of the enlightened despot, doing much to improve the lot of her subjects. Even further back was Genghis Khan, who conquered vast territories in an often horrifically brutal fashion. Yet once these lands were under his control, the Great Khan proved himself to be a surprisingly benevolent and visionary ruler whose new political system, based on talent rather than nepotism, helped tie East to West.[3]

Machiavellian-*cum*-altruist and altruist-*cum*-Machiavellian—it's enough to make your head spin as you try to tease a clearer picture from the crazy jigsaw we call personality. But, perhaps surprisingly, there *is* a clearer picture. The fuzzy shades of gray that at first seem so confusing can, in the end, fill in a far more nuanced portrait, not only of the successfully sinister but of people in general. More than that, these shades can lead us to a much more complete understanding of why seemingly "evil" genes persist.

NARCISSISM, DECEIT, HUMBLENESS, AND CONSCIENCE

We can first see how shades of gray fit into the picture by harking back to narcissism—that vain self-fascination and inordinately high self-esteem that has received so little attention in hard science research. This trait forms a hallmark of those with borderline, antisocial, and narcissistic personality disorders. As mentioned previously, suspicions abound that it has a strong genetic component.[4] Virtually every nasty dictator has shown the worst of narcissism's ugly features. Hitler, for example, deigned to share his feelings with an interviewer at Berchtesgaden: "Do you realize that you are in the presence of the greatest German of all time?"[5] Romania's Ceausescu told his health minister in the early 1970s, "A man like me comes along only once every five hundred years."[6] Turkmenistan's Saparmurat Niyazov, frequently criticized in the West as one of the world's most authoritarian and repressive dictators, had images of himself in virtually every public place and a gold-plated statue in the capital that rotated so it always faces the sun and shines light into the capital city. Niyazov modestly noted: "I'm personally against seeing my pictures and statues in the streets—but it's what the people want."[7]

But what happens to talented people when narcissism is *not* present? Such individuals help form the backbone of society—the superb secretary whose adept business skills make her boss look good, or the guy who never even sees fit to mention to his family that he had won the Bronze Star for his cool heroism under fire.

One such brilliantly talented, non-limelight-hogging person was Gregor Mendel, the man now known as the "Father of Genetics." Mendel was an inordinately neurotic individual who spent his teenage years in bed with a mysterious illness that now appears to have been akin to acute anxiety.[8] In keeping with his neuroticism, Mendel suffered so badly from test anxiety that he twice failed the examination to become a high school teacher. But Mendel, who loved both plants and mathematics, was a curious character. In his happy hideaway at the monastery, he spent eight years and raised thirty thousand pea plants figuring out why variations in heritable traits occur.

Mendel did attempt to communicate the results of his remarkable studies, but his pedantic lectures, paltry published study, and bashful attempts at correspondence with other scientists went ignored. Ultimately, although Mendel suspected his results were of supreme importance, his lack of confidence led him to give up and turn away from science altogether.

If Mendel had had the ego, self-esteem, or sheer, untrammeled narcissism* to repeatedly trumpet his findings to the world, researchers would have been clued in to the central ideas underpinning genetics some thirty-five years earlier than they did.

Mendel makes an interesting contrast with Charles Darwin and Alfred Russel Wallace, co-discoverers of evolution and variation with natural selection, who superficially appeared to share Mendel's lack of self-esteem. Darwin was an inhibited man with a reputation for integrity and a pride so well veiled that Wallace admired him from afar for being "so free . . . [of] egotism."[9] After a five-year, round-the-world voyage on the *Beagle*, Darwin returned to publish his findings related to zoology and geology. Secretly, however, he also embarked on a never-finished five-hundred-thousand-word masterwork (the equivalent of two thousand double-spaced manuscript pages) that was to summarize the theory of and evidence for evolution.

Much of the twenty years Darwin spent tucked away at his country estate preoccupied with puzzling out the secrets of evolution, Alfred Russel Wallace spent puzzling at the same problem in his adventures

---

*I hope the reader can forgive my loosely interchangeable use of narcissism, ego, self-esteem, self-importance, conceit, arrogance, and the like. Clearly these concepts are related to one another, but not identical. In the end, *I'm* fuzzy because hard science *research* in this area is fuzzy.

studying and collecting the flora and fauna of both the Amazon River basin and the Malay Archipelago. Like Darwin, he published his findings. Unlike Darwin, however, Wallace also began publishing articles related to the origin of species—poaching on evolutionary turf Darwin had thought was his alone. In a moment of feverish malarial brilliance, Wallace conceived a comprehensive theory of evolution and, in his enthusiasm, wrote it up and sent it to a man he knew would appreciate its importance—Charles Darwin.

Biographer-physician Ross Slotten notes: "Whatever the reason for [Darwin's delay in publication]—failure of nerve, a passion for perfection, periodic debilitating illness—it was not until the unexpected appearance of Wallace's essay that the issue of priority suddenly reared its ugly head."[10] Darwin, wringing his hands at the thought of his research being relegated to a footnote, wrote to his friend Charles Lyell, "I rather hate the idea of writing for priority, yet I certainly should be vexed if anyone were to publish *my* doctrine before *me*."[11] With Darwin's tacit encouragement, his friends arranged a neat sleight-of-hand joint publication of the theory, with ever-so-slight seniority accorded to Darwin's efforts, and Wallace's more complete work used to bolster Darwin's claim.

If Wallace had sent his results directly to a journal, rather than to Darwin, he would have unquestionably have laid claim to the theory of evolution. But Wallace never worried over issues of priority. In truth, Wallace hadn't a drop of self-aggrandizement in his body—he was happy his work was recognized at all. (As science historian Michael Shermer notes, Wallace was "agreeable to a fault.")[12]

Darwin, with his curiosity, brilliance, and well-concealed egotism, became canonized. Wallace, on the other hand, with the same curiosity and brilliance, coupled with an utter lack of egotism, became an impoverished footnote. Granted, neither of these men were flaming narcissists, but Darwin did have just enough ego to trump Wallace's hand.

Far more flagrantly egotistical than either Darwin or Wallace, however, was the sublimely arrogant James Watson, the misogynistic codiscoverer of the structure of DNA. Watson had no qualms about using data pilfered from scientist Rosalind Franklin to make his seminal discovery—and then writing a book describing "his" discoveries that mocked virtually everything about Franklin.[13] Later, Watson would try to block the development of the computerized approach to gene sequencing. Instead of hailing Craig Venter's automated sequencing machines at a senate meeting about the Human Genome Project, Watson derisively cracked that the machines "'could be run by monkeys.' Venter, sitting next to him, turned pale. 'You could see the dagger go in,' a witness later recalled. 'It killed him.'"[14] Later, of course, Venter's sequencing machines would help decode the human genome years ahead of the government's desultory schedule.

Want ego? Science alone provides plenty of examples. Narcissistic Nobelist William Shockley, the inventor of the lucrative junction transistor, was goaded into his discovery by jealousy of his colleagues' invention of the point contact transistor (which, indeed, used the underlying theory that Shockley had developed). Despite his genius, Shockley's arrogance and heavy-handed style alienated those who worked with him—he butted into everyone's business, sadistically blocking the careers of those he disliked. (Nobel co-laureate John Bardeen would leave Shockley's group in high dudgeon and go on to win a second Nobel Prize for the superconductivity research that Shockley had tried to prevent him from completing.)[15] When founding his own company, Shockley deliberately hired the brightest men around, but he could become unhinged, pounding the table in rage, during the rare occasions they accidentally outshone him.[16] Willing to do anything to keep in the spotlight, he took up controversial theories of eugenics, which undoubtedly assuaged not only his need for publicity but also his obsession with his own superiority. Ultimately he was left with racist allies whom "no moral, thinking soul would ever be associated with."[17] Even Shockley's own children became estranged—not surprising, considering that he publicly announced they had "regressed" from his own intelligence because of their mother's inferior standing.[18]

All of this doesn't even begin to do justice to the myriad of other cutthroat battles for glory surrounding the sciences. There was the mean-spirited Jonas Salk, with his continual public humiliation of Alfred Sabin. (Sabin had developed a far better polio vaccine that Salk did everything in his power to block and discredit.)[19] And the bitter feud between Newton and Leibniz over the invention of calculus.[20] And neuroscientist Solomon Snyder's blithe usurpation of credit from his doctoral student, Candace Pert, for the discovery of the opiate receptor.[21] (Snyder had, in fact, tried to stop Pert's research in this area because he thought it was a waste of time.) And brilliant Edwin Armstrong's invention of FM radio, which was hijacked by the unsavory Lee de Forest.[22] Armstrong would eventually leap to his death in despair over the legal imbroglio that left him destitute.

***

Why does it so often seem necessary for there to be at least a smidgeon, if not a heaping helping, of narcissism to get one's just (or unjust) due in this world?

A big part of the problem seems to lie in the fact that so many people are so darned *creative*, not only in science but also in thousands of different areas. Simply walking into a corner bookstore and thinking about the billions of hours of imaginative work encapsulated there can make you gasp with astonishment. And that's not to even mention the ongoing creativity swirling worldwide in software, music, cinema, science, art, sports, and contraptions of all sorts. No matter what creative enterprise one might undertake, there are frequently so many other people doing something similar that it's difficult to stand out. The Beatles, who'd floundered for three years with no recognition (there were over three hundred rock groups in Liverpool alone), used their manic-depressive "drama queen," Brian Epstein, to get them off the ground.[23] There would never have been a Motown without Berry Gordy, who has been dubbed a Jekyll-and-Hyde "thief of dreams" as well as a monstrous manipulator.[24] Madonna, with her ego and me-first sense of ethics, purportedly found whoever she needed to boost her up and then cut them out.[25] She, like many another superstars, understands that being nice when competing against those who use their elbows is likely to leave you in the shadows. (Darwin was lucky to have had a sweet-natured competitor, and he knew it, writing Wallace that "[m]ost persons would in your position have felt bitter envy and jealousy. How nobly free you seem to be of this common failing of mankind."[26] The fundamentally decent Darwin worked hard to arrange a civil list pension for Wallace. Even so, Wallace was still forced to continue publishing in his frail, final years in the hopes that his royalties would sustain his children.)

In the arts, it is difficult enough for an individual or group to stand out, even with the assistance of world-class, in-your-face promoters. But many modern-day creative concepts in other spheres—such as sequencing the human genome, building an assembly line and creating the automobile, coding a "killer application" for a computer system, or designing a high-definition TV—require an even more complex interweaving of innovation, tenacity, flexibility, and resourcefulness in order to be successful. To make matters worse, virtually all new innovations contain hard-to-protect creative concepts, either in execution or marketing, that other researchers or businesspeople love to emulate—or steal. This is where a spearhead person—a visionary who "gets it" yet also has a protective cloak of narcissism—is invaluable. And when the rewards of the enterprise are large, competition by those visionaries can become ruthless—a veritable clash of egomaniacal titans. As Roy Kroc, "the founder of McDonald's, once said of competition in the fast food industry: 'This is rat eat rat, dog eat dog, I'll kill 'em, and I'm going to kill 'em before they kill me.'"[27]

Narcissism can be a crucial asset not only in art, science, and business but also, understandably enough, in politics. Winston Churchill's sense of self-importance can be gleaned from an early letter to his mother from the battle lines: "I am so conceited I do not believe the Gods would create so potent a being as myself for so prosaic an ending."[28] In the dark days of 1940 and '41, when the Nazis seized the bulk of Europe and the lonely little islands of Britain were the next target, it was Churchill's convincing, egotistically certain manner that rallied the troops and the populace around the idea of standing fast rather than continuing with fruitless appeasement—as Lord Halifax, Churchill's competitor for the prime ministership, was wont to do. (Churchill once said: "Halifax's virtues have done more harm in the world than the vices of hundreds of other people.")[29] Where would England have been without Churchill's hyperinflated ego—coupled with his cunning intelligence and rapier wit?*[30] We might do well to listen to Churchill's own admonition: "Megalomania is the only form of sanity."[31]

Shades of narcissism might be needed to get your music heard, your ideas out, your innovations noticed—or your country saved, for that matter—but as people slide into the darker shades of that gray area, we find successful characters among us truly willing to hurt others to benefit themselves. As one former close associate of billionaire CEO Martha Stewart observed: "Martha often got involved with highly creative women whom she could dominate, manipulate, use, and abuse, women who wouldn't fight back."[32] Stewart's one-time business partner Norma Collier, whose ideas were cribbed during Stewart's me-first climb to the top, says of her former best friend: "I hope I never hear that woman's name again in my life. She's a sociopath and a horrible woman, and I never want to encounter her again or think about her as long as I live."[33]

Interestingly enough, one of the few lawsuits Stewart has filed was one against the *National Enquirer* for an article characterizing her as having many of the traits of borderline personality disorder. In 1997, reporter and celebrity biographer Jerry Oppenheimer published *Martha Stewart—Just Desserts*, a meticulously researched book that characterized Stewart as a narcissist of "almost diabolical dishonesty,"

*Unfortunately, egotism alone does not do the trick—as Churchill's talentless son Randolph revealed. Churchill biographer Gretchen Rubin summarizes: "[Randolph] was universally considered an over-bearing, egotistical snob—in fact, one club's constitution stipulated, 'Randolph Churchill shall not be eligible for membership.' Drunken arguments, broken marriages, and unfulfilled ambitions marred his life."

who suffered from fits of depression, had threatened suicide, possessed a mercurial and explosive temper, and was capable of profoundly abusing those around her.[34] In the *Enquirer* article, borderline expert Leland Heller maintained that traits such as those described in Oppenheimer's book were consistent with borderline personality disorder. The *National Enquirer* didn't take Stewart's lawsuit lying down. After two years of wrangling, Stewart dropped the suit.[35] Subsequently, of course, she was convicted of insider trading and sentenced to five months in jail.

Individuals like Martha Stewart can be tempted to run with "cutting-edge" remunerative ideas that are ill-advised or frankly illegal (although in Stewart's case, there's evidence of prosecutorial bias in her jailing for a relatively minor offense).[36] As biographer Christopher Byron relates in *Testosterone, Inc.*, Sunbeam's "Chainsaw" Al Dunlap, the Turnaround King, used channel stuffing—which entailed reporting shipments as revenue when the revenue hadn't actually been received—to fool people into thinking that Sunbeam had achieved a stunning surge in profitability when it was actually going broke.[37] Sunbeam eventually went bankrupt. (Executive Jerry Ballas, who had worked with Dunlap at Scott Paper Company, said, "It's terrorizing working for the man. What you do is you avoid, at all costs, getting near him . . . avoid contact with him.")[38] In yet another selfish sleight of hand, Dennis Kozlowski, CEO of Tyco, was convicted of misappropriating company funds to support a lifestyle that included a one million-dollar birthday party for his wife on the island of Sardinia that included an ice Statue of David urinating Stolichnaya vodka.[39]

And then, of course, there's Enron.

Enron—The Power of Unchecked, Mutually Supportive Machiavellians

Cursed with a tag team of Machiavellian leaders who shunted away or fired underlings with ethics, Enron Corporation followed the money deep into the dark side. ("We don't need cops," said Enron

CEO Jeffrey Skilling, when asked to explain why he was moving a manager who was beginning to question some of the transactions.)[40] Chief financial officer Andy Fastow was perhaps the key instigator of Enron's mythmaking flameout. Fastow was a smooth, deeply corrupt chameleon, able to charm his superiors into allowing him to supervise personally remunerative deals that were steeped in blatant conflicts of interest.

An overweening narcissist, Fastow told Enron's head of corporate communications Mark Palmer, "I ought to be CFO of the Year. I've seen it in *CFO* magazine . . . I want it to be me. Could you do that, get them to write a nice article about me?" Palmer was repulsed and became further appalled when he'd watch Fastow turn from tiger to pussycat in front of chairman Ken Lay. "It was like something out of a movie, with Fastow in the role of the obsequious yes-man."[41] In a set of stunningly adroit Machiavellian coups, Fastow would ultimately get his wish and be declared CFO of the Year, while *Fortune* would dub Enron America's best-managed company.

Fastow was not a genius—his ignorance of fundamental issues involving finance could at times be jaw-dropping. *"Is this guy for real?"* wondered one financially astute colleague. "How could someone making a play for the CFO job have such a fuzzy understanding of the basics?"[42] But in Enron's top-down mandated culture of greed, traits such as competence and integrity were given short shrift. In any event, Fastow's temper served as an excellent guard to keep people from knowing his incompetence—or his dark secrets. Ray Bowen, a finance officer who had questioned Fastow's suspicious-looking partnerships, once received a late-night phone call from Fastow that quickly degenerated into a screamfest:

> "I'm doing this because it's good for Enron, not for me!" Fastow shouted.
>
> "Goddamn it! I am sick and tired of people attacking this! It's good for you, it's good for your business! So fuck you guys!"
>
> Bowen hadn't said a word.
>
> "I'll tell you what!" Fastow yelled, careening out of control.

> "We'll shut it down! And you *fucking* guys won't be able to get your *fucking* deals done because you won't have the *fucking* capital. So just figure it out on your own!"
>
> Bowen held the phone away from his ear as the screaming escalated.[43]

If it had been just Fastow and his duplicitous schemes, the damage could have been caught and contained early on—Fastow's lack of ability alone would have seen him ushered to the door in most companies. Yet, with equally Machiavellian, cognitively dysfunctional CEO Jeffrey Skilling averting his eyes as necessary, Fastow's every incompetence and illegality was overlooked or somehow explained away. After all, whatever the means, Fastow was able to magically produce the profits that Skilling and others on the management team were so eager to see.* CEO Skilling presented an even richer level of Machiavellianism:

> Skilling thrived on confrontation and had a perfect command of the minutiae of deals. In interviews he could stun financial writers with his grasp of details, but that same superiority made corporate meetings enervating for his colleagues. His vision was messianic. . . . From the beginning, colleagues say, Skilling's pattern was to scapegoat others without leaving a trail that could lead back to him. In meetings that Ken Lay chaired, Skilling was often silent, letting Lay believe that he was completely in control. But at other times Skilling could be very volatile. . . . He would often blurt out astonishing remarks in public—he once, famously, called a stock analyst an asshole during a conference call—and the public-relations staff worried each time he gave an interview.[44]

It was Skilling's egotistical, charismatic, almost borderpathic ability to convince listeners that he was creating a new vision for business rather than recycling a de facto pyramid scheme that led whistle-

---

*Enron's culture of rewarded incompetence was the antithesis of Microsoft's. Whatever Microsoft's sometimes cutthroat business tactics, Gates's own dazzling technical and business acumen underpins virtually every major decision. A willingness to argue intelligently with Gates's ideas is prized.

blower Sherron Watkins to openly declare during meetings that "[t]his is a circle jerk."[45] But others were swept into rapt agreement with Skilling. (After Enron's demise, Skilling would become more obviously delusional, suffering a nervous breakdown on the streets of New York City, "running up to people in bars and on the street, pulling open their clothes, and claiming that they were undercover FBI agents.")[46] Overseeing Skilling and Fastow was glad-handing chairman Ken Lay, a man so dysfunctionally clueless that whatever the evidence presented to the contrary, he believed the entire issue was simply a PR problem that could be solved with a press release.[47] Those who looked the seemingly gullible, self-serving Lay in the eye and told egregious lies were forgiven—their "minor" sins excused.

In short then, encouragement from Enron's semi-delusional Machiavellian top fostered development of a dim-witted Ponzi scheme. This was coupled with hiring, retention, and reward practices that selected for the unethical or their willing codependents, Intimidation of those who might have spoken up ensured that dissent was kept to a minimum. All this was bolstered by an oblivious chairman who refused to take firm action no matter what was brought to his attention. This sinister system was so obscenely and delusionally corrupt that it bent the lax rules for sinister stability past breaking. In the end, as the company went bankrupt, thousands of Enron employees lost their jobs and retirement savings, and some investors their life savings.

***

In short, then, narcissism—like many of the Machiavellian attributes—can be a double-edged sword. Too little of it can allow even the most talented and intelligent of individuals to pass by unnoticed. Too much of it—well, it seems there can never be too much of it. Extreme narcissism combined with even a modicum of talent can be a recipe for success on a grand scale. But when narcissism finds itself combined with intelligence, charisma, a too-easy glibness with truth, chameleon-like identity diffusion, and a Mao-like ability to manipulate mood with frightening effect, it can lead to individual success at high cost to others. When abetted by other Machiavellians and the oblivious dysfunctionality of blind optimists, Machiavellian narcissism can lead to the worst sorts of social disasters. On an organizational level, it can lead to Fastow, Skilling, and Lay's Enron. On a broader historical level, it can lead to Hitler's Nazi Germany, Stalin's Soviet Union, or Mao's China.

Temper, Temper, Temper

But there is another seemingly dysfunctional trait with a positive flip side, perhaps best shown by George Washington—the "Foundingest Father of them all."[48] Gouverneur Morris got right to the heart of the matter, eulogizing Washington as a man of "tumultuous passions" who was capable of terrible wrath.[49] At the Battle of Monmouth, Washington tracked down a commander who had allowed his troops to retreat and, as one observer noted, "swore that day till the leaves shook on the trees. Charming! Delightful! Never have I enjoyed such swearing before or since."[50] Years later, Thomas Jefferson dryly noted Washington's reaction to a provocation at a cabinet meeting: Washington became "much inflamed; got into one of those passions when he cannot command himself."[51]

Yet, despite—and possibly related to—his passion and sometimes overwhelming efforts to master it, Washington managed to control and resist a temptation to remain in power that Julius Caesar, Oliver Cromwell, Napoleon, Lenin, Mao, and thousands of other leaders, great and small, have been unable to resist. An anguished Napoleon commented on his deathbed, "They wanted me to be another Washington." But he wasn't.

Washington wasn't alone in harboring a volatile side that he attempted to control even as he performed noble deeds. Spiritual master of nonviolence Mahatma Gandhi shared the same characteristic. (Beyond their shared temper, Gandhi, like Washington, wasn't above rewriting his own history to burnish his legend.)[52] Biographer Louis

Fischer, who knew Gandhi personally, reported, "He had a violent nature and his subsequent mahatma-calm was the product of long training in temperament-control."[53] Early on, it was Gandhi's wife who felt the brunt of his temper. "Once," Fischer reports, "they quarreled so fiercely he packed her off from Rajkot to her parents in Porbandar." But where Washington made a virtual religion of self-control, Gandhi made it an actual religion. He took the Hindu ascetic practice of *brahmacharya* to its broadest interpretation to include "restraint and control of all of the senses, including diet, emotions, speech, and actions."[54]

It's enlightening to contrast Gandhi's combination of hot-blooded emotion and generally tight control with Hitler's emotional makeup, as described in a secret analysis written by Dr. Walter Langer in 1943 for US intelligence:

> [Hitler] shows an utter lack of emotional control. In the worst rages he undoubtedly acts like a spoiled child who cannot have his own way and bangs his fists on the tables and walls. He scolds and shouts and stammers, and on some occasions foaming saliva gathers in the corners of his mouth. [An eyewitness observer], in describing one of these uncontrolled exhibitions, says: "He was an alarming sight, his hair disheveled, his eyes fixed, and his face distorted and purple. I feared that he would collapse or have a stroke."
>
> It must not be supposed, however, that these rages occur only when he is crossed on major issues. On the contrary, very insignificant matters might call out this reaction. In general they are brought on whenever anyone contradicts him, when there is unpleasant news for which he might feel responsible, when there is any skepticism concerning his judgment, or when a situation arises in which his infallibility might be challenged or belittled.[55]

Hitler, in other words, had an extraordinary temper—with only a rare desire to put a damper on it. No doubt Washington's and Gandhi's abilities to control their sometimes overwhelming emotions was in part abetted by their conscious decision to exert control—just as Hitler's perceived lack of desire to control his emotions was abetted by his realization that he could get his way more easily through temper tantrums. (In fact, architect Albert Speer, one of the few who was close to Hitler, argued that "self-control was one of Hitler's most striking characteristics."[56] Biographer Ian Kershaw agreed that Hitler's rages and outbursts of apparently uncontrollable anger were in reality often contrived.[57]) Hitler clearly had a passionate temper—which he was perfectly capable of switching on and off as he needed to manipulate others.

Passionate emotions, as evinced by impulsive, angry outbursts—sometimes, but not always, kept under control—are found surprisingly often in a great number of high-achieving individuals, good or bad (or good *and* bad). A random list of those who have been said to possess such a temper might include Microsoft's Bill Gates; designer Ralph Lauren; opera singer Maria Callas; France's prickly Charles de Gaulle; "Iron" Mike Ditka; and a broad slew of US presidents, ranging from Bill Clinton to Richard Nixon to fiery nineteenth-century battle hero Andrew Jackson—a brawler who killed a man in a duel for casting aspersions on his wife.[58] And of course, impulsive tempers are found widely in the less talented, or less fortunate, run-of-the mill population: the friendly florist pulled over for road rage, the mother with an acid tongue, the landlord with an attitude.

In the end, impulsivity and temper may form part of Machiavellianism, but they also form a part of the broader spectrum of human behavior. If there is a difference between normal and sinister behavior, it is that the successfully sinister often appear to use their temper in a more consciously manipulative fashion for malevolent ends. As the perceptive Abigail Adams would write of George Washington, "[I]f he was really not one of the best-intentioned men in the world, he might be a very dangerous one."[59]

Cognitive Function and Dysfunction

But temper and ego aren't the only double-edged traits. Perhaps surprisingly, cognitive dysfunction can also carry good as well as bad aspects. After all, it was the near-delusional idealism of another founding father, Thomas Jefferson, that lay behind the inspiring opening words to the Declaration of Independence: "We hold these truths to be self-evident; that all men are created equal." Jefferson's flight of rhetorical hyperbole, and the well-intentioned mindset it sprang from, inspired a nation to recognize the principles of individual rights and freedom that have since spread from "men" to women and people of all backgrounds. Indeed, Jefferson's extraordinary affinity for "idealized or idyllic visions, and the parallel capacity to deny evidence that exposed them as illusory [was] a central feature of [his] mature thought and character."[60] Jefferson declared, for example, that "all men are created equal," even as he owned slaves and bedded at least one of them. But despite Jefferson's Mao-like tendencies for duplicitous behavior (George Washington endorsed a characterization of Jefferson as "one of the most artful, intriguing, industrious and double-faced politicians in America"),[61] Jefferson retained a very un-Mao-like mental flexibility. He had a sincere aversion for conflict and carried a lifelong willingness to absorb advice from his many friends. As a consequence, most of Jefferson's more lunatic ideas—such as canceling all debts every nineteen years—were pruned before ever reaching public discourse.

It is that ability to listen and, at least on occasion, to change one's views in response (perhaps echoes of the ability or inability to resolve conflicting information that Posner's group was studying), that appears to be the key difference between inflexible tyrants such as Hitler and Mao, and vastly more effective, although still tough, leaders such as Turkey's founder Kemal Ataturk; Britain's Winston Churchill, and, in other fields, business executive Jack Welch, basketball coach extraordinaire "Red" Auerbach, and Manhattan project director J. Robert Oppenheimer (a probable polio survivor).[62] After all, as James Surowiecki has shown in *The Wisdom of Crowds*, although groups don't always converge on the right answers, they can frequently get pretty close. One smart but inflexible person will always be wrong part of the time—and sometimes about crucially important decisions. But a critical thinker who accepts the best of surrounding input, instead of tuning out what he or she doesn't want to hear, can obviously do far better than any one inflexible thinker acting alone.

Delusions

It's worth lingering a bit on the dark side of our shades of gray to discuss outright delusional thinking, which can sometimes be found even in high-functioning, seemingly rational individuals.[63] Recently, the editors of *Popular Mechanics* saw more than their share of such thinking as a result of their book *Debunking 9/11 Myths: Why Conspiracy Theories Can't Stand Up to the Facts*. The editors' conclusions? Conspiracy theorists, it seems, are often completely incapable of assimilating facts that counter their claims.[64]

Research in delusional thinking is still in its infancy, but it seems clear that delusions must involve a fairly complex process. After all, it isn't just that the delusional person makes a mistake when perceiving something. Instead, a delusion can be adopted and maintained as a belief despite convincing contradictory evidence, and in the face of the fact that it is completely implausible. Delusions can be held with great conviction and defy rational counterargument. Such delusions often also involve jumping to conclusions: a negative bias in the way facts are absorbed; a way of processing information so that it becomes focused on the delusional individual herself; and biased recall that can actually seem emotionally richer to the delusional person than real memories. Indeed, one of the strongest characteristics of delusional thinking is the unwillingness to admit to any evidence that would refute the belief. Such thinking is reminiscent of the inflexible thinking of many a nefarious dictator—or difficult college roommate.

Researchers who have studied deluded patients have taken care to point out that these patients aren't delusional about *everything*. But such patients often do show a personalizing bias—a tendency to blame other people when things go wrong (in a word, projection). Interestingly, it seems that there is a separate neural circuit for threatening information that pertains directly to the "self" as opposed to anything else. Delusional patients, it seems, are not able to tone this circuit down, which means excessive attention is paid to "self-referential" information. This inclines the delusional person to think in a self-serving fashion. Research in delusions may help to provide a neurological-based understanding for the sometimes incomprehensibly self-centered behavior found in extreme narcissism.*[65]

---

*I can't help but wonder whether this same set of circuitry might be involved in Hannah Arendt's "banality of evil." As the Rape of Nanking, the murderous actions of the Nazi Einsatzgruppen, and savage onslaughts of the Huns show, virtually anyone can be taught to kill. However, careful examination shows that these killers have often been taught, sometimes through lifelong indoctrination, that other people aren't *people*. (This practice has a long tradition: It appears nomadic human bands generally refer to themselves as *The People*—other bipedal types, of course, being just humanoid imitations straight out of *The Thing*.) It may be that such killers have often been reared or trained to use a "non-self" or "not one of us" referential circuit that is less able to activate feelings of empathy. This would explain why, for example, savagely brutal WWII Japanese soldiers, taught since childhood to believe that the Chinese were worse than dogs, could return home from their rape and slaughter with little or no feelings of guilt and show themselves to be decent, upstanding family men. And this would explain Goldhagen's thesis that pervasive and violent German anti-Semitism lay at the heart of the Holocaust. Other types of killings may be based on development of a social frame that sparks neural circuitry related to morally justifiable actions. Cambodian refugee Youk Chhang, for example, is haunted by his teenage memories of heckling a couple as they were beaten and buried alive for the crime of falling in love without official permission.

Those with borderpathic traits, however, would need little training to commit their sometimes heinous crimes. Such sinister individuals could serve as ideal shock troops to inflame and train ordinary people. For example, researcher Paul Brass points out that Indian riots are generally fomented by "riot specialists"—somewhat sinister types ranging from scruffy young hooligans to university professors who specialize in converting what is often a minor local incident into a major regional or even national problem. (In borderline-speak, you might call these "splitting" specialists.) These riots are often ordered up by either the Moslem or Hindu elites to keep people on edge and make sure focus is maintained on Hindu-Moslem relationships. When the time is right for the full-scale riot, lumpen elements, including criminals, hooligans, and willing students, are brought in to get the ball rolling.

In some sense then, most of us are indeed capable of horrendous acts, but it may be that people with different neurological underpinnings would be induced to commit those acts much more easily and for very different reasons.

In relation to these ideas, Philip Zimbardo, a former president of the American Psychological Association, recently published *The Lucifer Effect*—a "penetrating investigation" of his famous 1971 Stanford Prison experiment involving college students who proved themselves capable of becoming sadistic prison guards or abjectly submissive prisoners. Zimbardo drew sweeping conclusions to the effect that it was the situation alone that drew these "good people" into doing "evil."

Zimbardo's understanding was that he had gone out of his way to select "young men who seemed to be normal, healthy, and average on all the psychological dimensions we measured." However, as pointed out by astute researchers Thomas Carnahan and Sam McFarland,[66] that does not at all appear to have been the case. On one test related to authoritarian attitudes, Zimbardo's volunteers scored higher than every standardized comparison group except San Quenton prisoners. Moreover, Zimbardo and his group did not indicate which version of Christie's Machiavellian test was used as another of the tests to determine normalcy. This makes it impossible to be certain what the scores

One hypothesis relates delusional thinking to defects in the regulation of dopamine and perhaps other neurotransmitters. This could lead to a person improperly assessing the importance of the information she is receiving, because dopamine helps a person figure out whether whatever she is perceiving is either good or bad. It's thought that there may be two very different types of delusional thinking—one that is driven by emotion, and one that seems to have no relation to emotion at all. Interestingly, treatment of mood disorders seems to reduce delusional thinking that is based on emotion. No one knows the cause of many of the nonemotion-related delusions.

The Delusions of Dictators

Dipping again into the darkest shades of gray, we find that Hitler's borderline-like thought processes followed the emotion-driven pattern of delusion—his thinking was observed to "proceed from the emotional to the factual instead of starting with the facts as an intellectual normally does. It [was] this characteristic of his thinking process that [made] it difficult for ordinary people to understand Hitler or to predict his future actions."[67] (This is an eerie echo of Milosevic, who, if you'll remember, decided first what was expedient to believe, and then believed it.)[68] As early psychoanalyst Walter Langer pointed out, Hitler was so clever at finding facts to prove his emotions correct that he appeared to be making rational judgments when that was actually far from the case. This was particularly true in discussions, where

---

reported by Zimbardo's group actually mean. In fact, it appears that, by most interpretations of the data, the Machiavellian scores of those involved in the experiment were far higher than normal. This logically implies that Machiavellian individuals tend to be attracted to prison-related situations.

Carnahan and McFarland tested this idea by writing two different newspaper advertisements for study volunteers. One ad was virtually identical to Zimbardo's original, which referred to "prison life." The other was also virtually identical—except it was missing the words "prison life." Testing of those who responded to the ad revealed there was indeed a dramatic difference in Machiavellian scores between the two groups of respondents—prison-related work apparently is a magnet for Machiavellians. Moreover, those who volunteered for Carnahan and McFarland's study were higher not only in Machiavellianism, but also in narcissism, dispositional aggressiveness, authoritarianism, and social dominance—and lower in dispositional empathy and altruism.

Ultimately, then, it is probable that Zimbardo's sweeping conclusions about Abu Ghraib, genocide, human nature, and evil itself are based on a fundamentally flawed study.

Hitler was "unable to match wits with another person in a straightforward argument. He [would] express his opinion at length, but he [would] not defend it on logical grounds." One observer noted: "He is afraid of logic. Like a woman he evades the issue and ends by throwing in your face an argument entirely remote from what you were talking about."[69] Hitler's near-schizophrenic magical thinking led him to believe "that his 'will' could accomplish what others thought impossible, [he would thus] brook no contradiction from lesser souls. The absolute power he in fact obtained served then to reinforce his idea that his will was magical."[70]

"No matter how impulsive, bizarre, destructive, or lawless his actions were, Hitler rationalized them as legitimate."[71] And, like each of the other dictators we've discussed, Hitler was particularly gifted at the borderline trait of gaslighting—that supernal technique of denying reality that can so throw an opponent. Particularly disconcerting in light of Hitler's phenomenal memory (about which more will be said later) was his capacity for "forgetting." He would say something one day and then, several days later, say something that would completely contradict the first statement. If the inconsistency was pointed out, Hitler would fly into a rage, demanding to know whether the other person thought he was a liar. Leading Nazis took to mirroring Hitler's trick (shades of the emotional contagion seen in Ceausescu's Romania—and in Skilling's Enron). As former Nazi leader Hermann Rauschning observed: "Most of the Nazis with Hitler at their head, literally forget, like hysterical women, anything they have no desire to remember."[72] He noted further that Hitler was "capable of entertaining the most incompatible ideas in association with one another."[73]

The Delusions of Madmen

But, on the other hand, could humankind do without utterly inflexible, sometimes almost delusionally visionary people? Could we have done without determined teachers such as Socrates who, rather than accept exile, cheerfully drank hemlock as punishment for refusing to recognize

Fig. 11.1.

the gods and for "corrupting" youth with his teachings? Or brilliant, tragic Joan of Arc, whose visions inspired her countrymen to fight off the yoke of the English? Or archly inflexible Galileo (*Eppur si muove*—"and yet it moves")? Or the mysterious man of China's Tiananmen Square, courageous enough to stand for a just cause in front of massed tanks?

It can sometimes be difficult to know whether a political, religious, business, or scientific leader is cognitively disturbed or instead an avant-garde visionary who sees the truth others are missing.*[74] Or

*As I tell our kids—you can always find a distinguished scientist who backs up your views, whether you believe that US government agents destroyed the Twin Towers or that smoking cigarettes is good for you. People are often surprised to learn that a person can be simultaneously both an intelligent scientist or public personality and a crackpot. An immunologist friend once spoke with preeminent scientist Peter Duesberg—principal proponent of the idea that HIV (human immunodeficiency virus) is harmless, and that AIDS is actually caused by noninfectious factors, such as the very drugs being used to treat AIDS. Duesberg's work has inspired people like Christine Maggiore, an engaging, articulate, well-to-do HIV-positive woman who heads up a group that denies standard treatment is necessary or effective for AIDS. After Maggiore's three-year-old daughter died of AIDS-related pneumonia, Maggiore still leads the movement.

While conversing with Duesberg, my friend asked him why people with AIDS showed a certain well-studied pattern where certain cells in the immune system were killed, while others were left alone—an unlikely pattern if chemicals are the cause of the disease. Duesberg, although friendly, dismissively waved her question off with "I haven't seen that data." In point of fact, it is mind-boggling that Duesberg would not be aware of, and obviously uninterested in looking at, that very well-known perhaps both. Churchill, for example, was rightly characterized as depression-prone and at times dependent on alcohol. But he was correct to see Hitler's menace when other British politicians settled for a groupthink of appeasement. As in Churchill's case, will history prove correct those who now see similar menace in the Machiavellians who have found purchase in fundamentalist Islam? Will well-intentioned policies of cultural relativism, in the long run, prove equivalent to Chamberlain's similarly benign, seemingly rational, and humane policies of appeasement—policies that led willy-nilly to genocide?

PERSONALITY UNDERLIES IDEOLOGY

In the end, illusion, delusion, happy optimism, or other forms of cognitive dysfunction or seeming dysfunction may be good or bad—depressives have often been found to be more realistic—but they are certainly not necessary to outstanding leadership. George Washington, for one, was highly respected and effective in large part because he was a supreme realist, "temperamentally incapable of tilting at windmills or living by illusion" and carrying an "instinctive aversion to sentimentalism and all moralistic brands of idealism."[75] Washington, as biographer Joseph Ellis reminds us, was "that rarest of men: a supremely realistic visionary. . . . His genius was his judgment."[76]

---

and relevant data. Incidentally, the foreword to Duesberg's book, *Inventing the AIDS Virus*, was written by Nobel Prize–winner Kary Mullis, who has also written of his abduction by aliens from his California forest hideaway.

As Richard Feynman noted in a commencement address to CalTech: "The first principle is that you must not fool yourself—and you are the easiest person to fool. So you have to be very careful about that."

The reason it's so hard to be sure that "crackpots" aren't actually all they're cracked up to be is that occasionally, a seeming eccentric is proven correct. Dr. Barry Marshall, for example, came up with the idea that bacteria were the cause of most stomach ulcers. He was ridiculed by an establishment that had long held ulcers were caused by stress, spicy foods, and an overly acidic stomach. Scientists felt that bacteria simply could not live in such an acidic environment. "Everyone was against me," said Marshall, "but I knew I was right." Marshall eventually proved his theory by drinking a petri dish of bacteria and giving himself gastritis (and to his wife's dismay, bad breath). A dose of antibiotics cured him. Marshall—and his stomach—eventually won the Nobel Prize for his groundbreaking work.

Perhaps surprisingly, Ellis cites as the cause of Washington's judgment his lack of schooling—his "mind was uncluttered with sophisticated intellectual preconceptions." And there may be some truth to Ellis's notion. But a number of leaders with minimal schooling—Zimbabwe's Robert Mugabe, Stalin, and Ceausescu spring immediately to mind—taught themselves with entirely different outcomes. While these ultimately evil dictators voraciously absorbed the idealistic teachings of Marx, which they then used to mask for their self-interested behavior, Washington was busy bringing himself to the opposite conclusion that "men and nations were driven by interests rather than ideals."[77] Ideology, it seems, whether liberal, conservative, Communist, capitalist, or religious, is often seized by people of certain temperaments, for their own purposes, whether for good or for ill.

Although pinned to opposite ends of the political spectrum, how very similar Hitler's fascist Germany was to Stalin's Communist Russia.

And how very different Washington's moral and rational leadership was from Mao's.

Ambition and Control

We've talked about the mixed advantages of narcissism, temper, and cockeyed cognition. But what about the desire for control and its often conjoined twin, ambition? Certainly we know that an obsession with power and control is one of the most common traits of tyrannical dictators. For example, after wrecking Zimbabwe's strong economy and relatively sound human rights record, tyrannical dictator Robert Mugabe "made no attempt to deal with any of the calamitous economic and social issues facing his government. All that mattered to him was the exercise of power. . . . Whatever the cost, his regime was dedicated towards that end."[78]

Likewise, biographer Robert Waite summarizes apropos Hitler: "[T]he most basic single characteristic of both his personal life and his system of government can be reduced to one overriding need: to force others to do his will."[79] Attaining and maintaining control was also a central tenet of Stalin's existence. His closest circle and top generals had naturally devolved to those with a gift for acquiescing and brown-nosing. Stalin's lackeys studied Stalin "like zoologists to read his moods, win his favour and survive."[80]

But, like so many other traits, controlling behavior, when mixed with, for example, sensible cognition, can be a winning combination in a much more positive sense. "Iron Lady" Margaret Thatcher, for example, the first female prime minister of Great Britain, has earned both love and loathing for her controlling revamp of British economic policies. Much like Stalin, Hitler, Mao, and Washington, Thatcher's ambition, not to mention her egotistical certitude, was boundless—she could hold her own with anyone.*

> Her style was built on domination. None of her colleagues had ever experienced a more assertive, even overbearing, leader. That had always been her way of doing business, and it became much more pronounced when, having defeated all her male rivals in 1975, she needed to establish a dependable ascendancy over them. With her command of facts and figures and her reluctance ever to lose an argument, she seemed so damnably sure of herself that nobody could suppose there lurked much uncertainty anywhere in her makeup.[81]

Unlike Hitler and many other despots, and much like George Washington, Thatcher used "a 'thinking' or 'rational' solution to problems rather than a 'feeling' or 'emotional' response."[82] And, unlike the typical chameleon-like Machiavellian, Thatcher's identity was stanchion-solid—as she said, "We don't change our tune to whoever we are talking."[83] (An echo of Ataturk's in-your-face public law-breaking in Islamic Turkey as he quaffed fiery raki and proclaimed, "Hypocrites and frauds of old used to drink a thousand times more, secretly in hovels as they indulged in all sorts of nastiness. I am not a fraud. I drink to my nation's honor!")[84] Perhaps most importantly, the policies

*Anyone, that is, except Deng Xiaoping, who easily bested the "Iron Lady" in negotiations for Hong Kong. But then, the brilliant, wily Deng had trained by surviving for decades under Mao.

that Thatcher chose to pursue ultimately emphasized moving *away* from unitary control by any one person or group in government—including herself. It seems that Thatcher's desire for control was nuanced by practical cognition that, Churchill-like, saw through the groupthink of economic appeasement of labor unions as well as idealistic but ultimately detrimental government handout policies that had overgrown their original beneficial purposes.

On the other hand, labor unions and government handout policies have proven themselves at times to be of vital importance. Franklin Delano Roosevelt, characterized by one British politician as having the "cunning of a schemer and the ambitions of a genuine altruist,"[85] used his cool intellect and keen desire for control to ram through work and relief programs that helped ease millions of destitute families through the crisis of the Great Depression. And there is no doubt that the labor unions Roosevelt helped champion, although imperfect, helped serve as counterweights to some of the more outrageous offenses of employers. Different times and situations call for different solutions—but under almost any circumstances, ambition and desire for control play a role.

The Surprising Attributes

As is perhaps becoming clear in our discussion of shades of gray, Machiavellian attitudes alone do not necessarily make for an individual's success (or *sinister* success). Many individuals, Machiavellian or not, and no matter their flavor of political or religious orientation, compete for positions that provide power or gain. Those who make it to the top of any given social ladder often have a number of *non*-Machiavellian traits that give them significant advantages over their competitors. Only a few of these non-Machiavellian traits will be described here, even so, these examples give a sense of the advantages certain personality traits can provide in achieving either the fame of a Gandhi, or the notoriety of a Hitler.

Native intelligence obviously lies among those advantageous

traits, as, perhaps more surprisingly, does a good memory. Selecting examples from the fascist side of politics, "Papa Doc" Duvalier, who enjoyed watching torture by peeking through an eyehole from a neighboring room, had "a reputation for great intelligence."[86] Paraguayan Nazi-sympathizer Alfredo Stroessner was "darkly brilliant" in his ability to profit as a dictator from the mistakes of others, as were any number of other right-wing dictators, many of whom retained power because of US support for their anticommunist stance.[87]

Underlying Hitler's extraordinary speaking skills, which are often cited as the key to his rise to power, was his memory. Biographer Robert Waite explains:

> Everyone who knew Hitler was struck by his incredibly retentive memory and the extraordinary range of his factual knowledge. He could remember the trademark and serial number of the bicycle he had used in 1915; the names of the inns where he had stayed overnight 20 years previously; the streets down which he had driven during past political campaigns; the age, displacement, speed, strength of armor, and other data of every capital ship in the British and German navies; the names of the singers and their roles in the operas he had seen in Vienna as a youth; the names of his commanders and their precise armaments down to the battalion. . . .
>
> Hitler also used his data bank as a defensive weapon to ward off displeasing arguments. When field commanders on the Eastern Front pointed out the strength of the enemy, Hitler would either dismiss their argument as irrelevant—his steel-like will would overcome all problems—or overwhelm the doubter with production statistics and precise weaknesses in the armament of the enemy. Or he would undercut and embarrass commanders by demanding from them information that they simply could not remember. If, for example, they raised objections to a tactical plan, he would bombard them with questions such as the name and rank of each of their subordinate commanders or the military decorations each was entitled to wear. When a field commander admitted ignorance of these matters, Hitler would provide the answer triumphantly and announce that he had more knowledge of their sector than they had.

> He had little interest in coming to grips with difficult intellectual problems, and had the habit of repeating the same question about a complex historical event without making an effort to investigate the answer.[88]

In general, people associate a steel-trap memory with high intelligence, although the two qualities don't appear to be necessarily linked (witness idiot savants). A sharp memory can easily be used as a manipulative tool by the less bright, but more Machiavellian among us. Thus, for example, Mussolini used his prodigious memory to fool people into thinking that he had an exceptionally wide knowledge of science and philosophy. In reality his knowledge was often limited to what he'd happened to skim a few pages of—but which he could recite practically verbatim.[89]

On the Communist side, repressive dictator Fidel Castro early on showed far more interest in sports than in academics, but he caught the attention of his teachers with his remarkable memory, which he used to easily memorize entire books.[90] And if communism's grand progenitor, Stalin, was different from many dictatorial wannabes, it was only in his intellect and, perhaps most importantly, his "rolodex of a memory."[91] One railways commissar who had reported to Stalin hundreds of times pointed out, "One felt oppressed by Stalin's power, but also by his phenomenal memory and the fact that he knew so much. He made one feel even less important than one was."[92] Mao and Milosevic were similarly blessed with remarkably good memories.

A retentive memory also plays a surprisingly important role in such factors as charm and charisma. Who is not delighted to discover that he is important enough that his name is remembered, even after a meeting that lasted only seconds many years before? Teachers with extraordinary memories can hold students mesmerized with their ready command of facts and be endeared for their ability to remember student names. (As Philip Wankat notes in *The Effective, Efficient Professor*, "The most important single activity you can do to show students that you are interested in them is to learn and use their names.")

Famed Antarctic explorer Ernest Shackleton, popular with his crew (and consequently disliked by fellow explorer Robert Falcon Scott), delighted everyone with his wonderful memory and "amazing treasure of most interesting anecdote."[93] Many top political leaders with good or great reputations—and remarkable memories—include Mahatma Gandhi, Winston Churchill, Charles de Gaulle, Franklin Delano Roosevelt, Margaret Thatcher,*[94] and Chinese president Hu Jintau, as well as business leaders such as Warren Buffett, Jack Welch, and Bill Gates. Other top business leaders with a different sort of reputation—but no less remarkable a memory—include indicted Hollinger CEO Conrad Black, convicted former Enron CEO Jeffrey Skilling, convicted CEO Martha Stewart, and (if you consider mob bosses to be business leaders), dreaded *capo di tutti i capi* Toto Riina.[95]

President Bill Clinton, with his marvelously retentive memory, could cover gaffes such as being given the wrong speech for his first State of the Union address through recollection and ad-libbing—no one ever guessed what was going on until later. And President Ronald Reagan was a near-professional raconteur, with ready quips always at hand to loosen tension; he could quote long passages from memory of books that had impressed him. It was Reagan's extraordinary memory that underpinned his moniker of "The Great Communicator" (which makes his later Alzheimer's all the more tragic). Charismatic, straight-from-the-heart speeches are easier for someone who doesn't necessarily need to look down at papers or slightly askew at teleprompters for reminders.

One other useful attribute shared by many top leaders, Machiavellian or not, is an indefatigability that hints of hypomania. (Hypomania is a mild manic state that, in more extreme forms, can shade into bipolar disorder.) Gandhi, for example, could display almost supernatural endurance, walking enormous distances with little food or rest, or embarking on lengthy, well-publicized fasts of self-purification to

---

*As with Reagan's Alzheimer's, Thatcher's case is particularly poignant. Her daughter noted: "She had such a brilliant memory—like a website. She could quote inflation statistics going back years without reference to a single note." But a series of mini-strokes left Thatcher with the frustrating inability to remember, by the end of a sentence, what the beginning of a sentence had been.

bring attention to causes he believed in. Margaret Thatcher would say, "I've never had more than four or five hours sleep. Anyway, my life is my work. Some people work to live. I live to work."[96] Kemal Ataturk could stay up all night reading a book he found interesting or partying with friends and then still be on top of his duties the next day.

Virtually every "evil" dictator who founded his own regime shared a similar hypomanic intensity, including Hitler, Mussolini, Stalin, Mao, Ceausescu, Castro, Papa Doc Duvalier, and Robert Mugabe. Each of these individuals was capable of almost superhuman efforts—as long as the work pertained to the all-consuming desire to achieve control.

So, Are the Successfully Sinister Really Different?

Quite commonly, people appear to succeed—not only in politics, but also in the arts, the sciences, business, and even in religious leadership—because of a healthy dollop of "evil genes." As sociologist Daniel Chirot notes: "The competition for power is rarely won by those who are considerate of their enemies, who lack self-confidence, or who think they do not have something important to contribute to the problems of the group they want to rule."[97]

But is deeply sinister, Machiavellian behavior just an extreme version of "regular" dysfunctional traits—the tail end of the Gaussian curve for nastiness? Or is being successfully sinister one of those emergenic qualities—something that, like genius, springs forth as far more than the sum of its parts?

It seems the answer may involve the combination of extreme traits and emergenic qualities. Churchill and Ataturk displayed broad indications of narcissism, temper, and mood disorders—yet both went out of their way to process and act responsibly on information whether or not it was something they "wanted" to hear. In other words, whatever dysfunction they might have had, had little relationship with that of the conventional evil dictator who manipulated underlings to bring facts that would only make them happy.

Moreover, it appears to be individuals with a widespread dysfunction that involves not only narcissism, impulsivity, and mood disorders but also identity disorders, cognitive dysfunction, and, sometimes, sadism who emerge with the markedly different personalities seen in the successfully sinister. Such a borderpathic constellation of personality characteristics can be as distinct in its own Machiavellian way as bipolar disorder, autism, and schizophrenia are in theirs. Hitler, Stalin, Mao, Milosevic, and other nefarious dictators shared many borderpathic characteristics, as undoubtedly do some present-day leaders in a variety of fields. (One must be wary of naming names, however, as the *National Enquirer*'s episode with the tyrannical Martha Stewart made clear.) Such distinctive Machiavellians can leave a trail of fear, loathing, and lawsuits even long after their deaths. Woe betide those, for example, who delve too deeply into the life of—well, my publisher won't let me tell you who.

NAIVETE

Credulity involving deceptive, deeply pathological behavior crops up everywhere. Indeed, it's difficult for many people to understand how emergenically different the successfully sinister can be—we just can't believe these people can be that different from us. Historian Robert Waite describes the wonderful impression Hitler made on others: "To the sophisticated French ambassador, he appeared as 'a well-balanced man, filled with experience and wisdom.' An intellectual found him 'charming,' a person with 'common sense' in the English sense. The British historian Arnold Toynbee came away from an interview thoroughly 'convinced of his sincerity in desiring peace.' The elegant and precise Anthony Eden was impressed by Hitler's 'smart, almost elegant appearance' and found his command of diplomatic detail 'masterful.'"[98]

Sadly, we have plenty of current examples as well. George W.

Fig. 11.2. Hitler in 1938. Nothing beats a photo op with kids for portraying kindness.

Bush initially thought subtly devious Russian president Vladimir Putin was "straightforward and trustworthy." Media mogul Ted Turner agreed, dropping in to spend an hour with his "old friend" Putin during a visit that was heavily covered on CNN. (CNN *didn't* cover the nearly simultaneous armed raid on Putin's nemesis—news source Media Most Group—by masked men armed with automatic weapons claiming to be "tax inspectors." The many recently liquidated critics of Putin's regime would also testify to Putin's chameleon-like nature.)[99]

President Jimmy Carter, arguably a decent man, befriended and feted Nicolae Ceausescu, handing a propaganda coup to one of the world's nastiest dictators. Carter has also made a post-presidential habit of being conned: befriending career terrorist Yasir Arafat, singing the praises of brutal North Korean dictator Kim Il Sung, and certifying as fair and aboveboard many a questionable third world election.

President Clinton was just as gullible regarding Saddam Hussein: "I think that if he were sitting here on the couch I would further the change in his behavior. You know if he spent half the time . . . worrying about the welfare of his people that he spends worrying about where to place his SAM missiles . . . I think he'd be a stronger leader and be in a lot better shape over the long run."[100] Dan Rather was similarly mushy with Hussein, allowing him the commercial airtime to speak, without rebuttal, about how much he loves peace and humanity.[101]

60 Minutes stalwart Mike Wallace was charmed by Mahmoud Ahmadinejad, the deeply fanatical president of Iran, effusively describing him as "very smart, savvy, self-assured, good looking in a strange way . . . infinitely more rational than I had expected him to be."[102] Wallace had been similarly obsequious with Syrian tyrant Hafez Assad as well as Palestinian leader Yasir Arafat, prompting journalist David Bar-Ilan to note: "Had [Wallace] treated American . . . politicians this way, he would have been drummed out of the profession."[103]

Naivete about people's motives, especially from well-known figures, often allows Machiavellians a public stage to work their confusing, deceptive wiles. But sometimes, Machiavellians achieve this publicity by covertly appealing to their interlocutor's narcissism rather than naivete. When mediators and interviewers interact with a well-known sinister character and bring out the seeming best in him, it provides an opportunity to flaunt their own character: "See. He may seem evil, but he's really not so bad. At least not to intelligent, nice, right-thinking people like me. The guy's just a pussycat when *I* talk to him."

Warren Buffett—Multifaceted Genius

But there is yet another way that emergenic traits can combine to shape extraordinary, radically different behavior that has a potent influence worldwide. Investor Warren Buffett's life provides proof you don't need to be Machiavellian to succeed. But Buffett's success also shows the importance of recognizing the possibility of darker motives in others.

Warren Buffett is widely acknowledged to be the greatest investor the world has ever known. As biographer Roger Lowenstein summarizes: "Starting from scratch, simply by picking stocks and companies for investment, Buffett amassed one of the epochal fortunes of the twentieth century. Over a period of four decades—more than enough to iron out the effects of fortuitous rolls of the dice—Buffett outperformed the stock market, by a stunning margin and without taking undue risks or suffering a single losing year. This is a feat that market savants, Main Street brokers, and academic scholars had long proclaimed to be impossible."[104] How does he do it?

Buffett merges so many extraordinary abilities in one ebullient fireball that it's impossible to single any one trait out and say *that's* the key. He possesses an Einstein-like ability to focus (that is, a Lamborghini of an anterior cingulate cortex) and a photographic memory (making it easy to posit *val/val* BDNF with a deluxe option package of accompanying alleles). Buffett doesn't need a computer because, as he told one interviewer, "I am a computer."[105]

Perhaps surprising in light of Buffett's extraordinary ability to discern patterns from numbers is his talent for intuiting people's ability and motives—an underappreciated tool in the investment toolkit. If a would-be manager is more interested in the money than the business, Buffett either settles on a different manager or pulls up financial stakes and instead invests elsewhere. One whiff of Enron-style leadership, one suspects, and Buffett would make a speedy getaway.

Buffett himself writes:

> I would agree that I have been pretty good at sizing up people. Not perfect, of course—I've certainly made a few mistakes in selecting managers at Berkshire. I would say that part of the reason for the success I've had is that I only take the easy cases. In other words, if you gave me 100 people to evaluate on a scale of 1–10 in terms of how they would work out at Berkshire, I would be pretty good at selecting a few 10s. I would also miss a few other 10s in my screening and I would be terrible at differentiating between the 3s and the 7s. This is similar to my method of selecting stocks where I

> only have to be right on a few decisions and can put most of the rest into the "too hard" pile.[106]

Buffett's practical ability to tell the difference between genuinely outstanding managers and their dark, chameleon-like doppelgängers has shaped his approach to business and propelled his efforts to steer Wall Street toward an ethical path. As Bill George at *U.S. News & World Report* points out: "[Buffett's] commitment to sound ethics and principles, his self-discipline and consistency, his transparency in disclosing mistakes, his criticism of Wall Street fees and compensation of underperforming CEOs, and his pleas for improving corporate governance—all have had a salutary influence on the corporate community."[107] In reality, no other businessman has applied such pressure to ensure that ethical practices are woven into the regulatory fabric that governs Wall Street. And in the end, Buffett's donation of his fortune to philanthropy will be, at nearly $40 billion, the largest in history. Fittingly, rather than set up his own ostentatious foundation, he is donating his money to the philanthropic organization of his good friends Bill and Melinda Gates, where he knows it will be put to wise use.

No, one need not be Machiavellian to be successful. But being able to recognize that shades of gray in others helps.

Healthy Cynicism

Over the years, I've found that nice people (that is, the majority of people) generally fall into two categories—those who *have* dealt with and have been wounded by the successfully sinister, and those who *haven't*. Those who haven't—which naturally includes many younger people—often simply don't believe that the successfully sinister exist. After all, since elementary school they've been told that virtually anyone can somehow be reasoned with. Even if a problem does arise, the naif thinks, surely the seemingly sinister person can be taught how to act more reasonably, perhaps through the proper modeling of patience, understanding, and compassion. Explaining the true nature surrounding the cognitive dysfunction and emotional imbalance of the successfully sinister to a naif is a little like trying to explain color to a blind person—it is no wonder that such naivete continues even when someone is warned point blank to be wary.* People simply aren't generally raised and educated to understand that small percentages of the population—some of whom are outwardly very successful—are quite capable of masking deeply disturbed personalities. Sometimes, sadly, the devastating reality of these "unfixable" personalities becomes clear only after marriage and children. (As relationship expert Russell Friedman once quipped: "You can't love someone into mental health.")[108]

On the other hand, those who *have* dealt with the successfully sinister usually know instantly what I'm talking about. When I describe the concept behind this book, within seconds and without a further word from me, people I barely know will unwrap and describe psychic wounds that they've carried privately for years—the ex-wife who left the kids and a trail of credit card debt; the supervisor who made life a living hell; the friend who wormed close, mimicking hair, dress, even a

---

*One example, which I've witnessed and heard about in various universities around the country, is when a more Machiavellian professor seduces a student into doing an independent study project under his or her direction. When the student is particularly hardworking and good-hearted, sometimes one can't resist attempting to give warning. Such warnings are almost invariably shrugged off—"It's only a two credit project," the student might say, "What could Professor X really do to me?" The student always remarks on how nice Professor X seems, and one can practically hear them thinking: *Why am I being warned about somebody who really likes me and is obviously such a great professor? Maybe* you're *the one with the problem.*

A semester or two later, the tale of woe generally runs along the lines of "I'm just an undergrad, but Professor X wants me to read twenty-five journal articles and use all this information to create an advanced new theory. It's crazy! And he never shows up for meetings, never can help me with anything, and really—*I don't think he understands what he's doing!*" The other faculty all know what's going on, but in the topsy-turvy world of tenure, there is virtually no ability to discipline. Besides, X is so Machiavellian that even the administration is afraid to discipline him (or her). And of course, the administration has its own sinister cast of characters with their own Machiavellian mackerel to—well, you get the idea.

The students one really feels for are those from overseas, who arrive looking like lost lambs and sign on with the first friendly-looking professor they talk to. That professor is often X, in happy, charming, chameleon mode. Students don't realize that once they've signed, it can be virtually impossible to escape. However many experiments they run, papers they publish, or patents they obtain, at many universities, X is not required to allow them to graduate. Some students work *full time* on their doctorates for seven or more years, indentured servants who plump their masters' credentials through their thankless, endless, low-paying work. In these cases in particular, nice guys really do finish last.

way of talking—and stole a boyfriend; the uncle who took his grandmother's life savings and left her to die unattended in a filthy nursing home. "I can't believe there might be some kind of scientific explanation for this," the *have-dealt-withs* tell me time after time, "I never even talk about it because no one would believe me." Without knowledge of recent studies, people have little way of figuring out that their seemingly isolated experience was far more common than they'd realized—and that extraordinarily enlightening explanations are becoming available.

In an ironic twist of justice, it appears that the worst of all human crimes—genocide—often occurs simply because people can't believe that heretofore noncriminal humans can perpetrate horrendous acts such as mass murder or gratuitous torture. "I don't believe you," said Supreme Court Justice Felix Frankfurter, when told by an eyewitness of the "naked corpses in the Warsaw ghetto, yellow stars, starving children, Jew hunts, and the smell of burning flesh." Frankfurter interrupted to add: "I do not mean you are lying. I simply said I cannot believe you."[109] The justice literally could not conceive of the atrocities being described. Samantha Power describes one of the key causes of genocide in her Pulitzer Prize–winning *"A Problem from Hell"*: "Despite graphic media coverage, American policymakers, journalists, and citizens are extremely slow to muster the imagination needed to reckon with evil. Ahead of the killings, they assume rational actors will not inflict seemingly gratuitous violence. They trust in good-faith negotiations and traditional diplomacy."[110] Sadly, ordinary people often have little exposure to the research regarding Machiavellians that could do much to help prevent genocide. Only by recognizing Machiavellians for what they are and how they operate can we begin to stop them.

But if the dark shades of the successfully sinister are sometimes evil—and I believe they are—it is important to understand that that evil is complex. The dictionary definition of evil, after all, is "Morally bad or wrong; wicked: *an evil tyrant*."[111] That definition implies a gestalt sense of evil—not evil in every particular. Shades of gray lurk, sometimes darker, sometimes lighter. And the occasional genuinely decent act from a successfully sinister person can't help but confuse our gut feeling that someone like Hitler, or on a more mundane level, the man who throws his screaming children from a fifteenth-floor hotel balcony, must be *totally* evil. With popular emphasis on "the sociopath next door,"[112] people often don't understand that deeply dysfunctional, even unquestionably evil individuals can have genuinely decent aspects to their personality. "You could not imagine what a good heart Adi has,"[113] one man exclaimed, after witnessing Adolf Hitler, as a penniless young man, protesting against the unjust treatment of an employee of a coffeehouse. And Hitler gratefully recognized the Jewish doctor who treated his mother's terminal cancer, protecting him until he emigrated in late 1941.[114]

In the ultimate world with its shades of gray, where "bad" traits can be used for good purposes, and "good" traits can be seen in bad people, some things, it seems, are relative. But not everything.*[115] It seems that normal people worldwide use the same neural mechanisms to process moral questions. Thus the basic features of morality appear to be hardwired, and not a product of culture.[116] Harvard biologist Marc Hauser, who has done extraordinary work in this area, has found that although there are some differences in morality between cultures, there are limits to those variations. There are some things, in other words, that religion or education can't easily instill in us, because it goes against our natural intuition—our built in moral compass.

Ultimately then, religion, education, and even family may have less of an impact on our innate sense of morality than we may think.[117] Ethics classes, in other words, really may just preach to the choir. Those few who are wired differently—and we are beginning to learn how the wiring's awry—march to their own moral tune, no matter what they are taught.

---

*Even Einstein's Theory of Relativity was nearly instead named Invariance Theory, after the fact that, although space and time each vary individually, space-time itself, along with the speed of light, is invariant.

CHAPTER 12

THE SUN ALSO SHINES ON THE WICKED

"Obsolete power corrupts obsoletely."
—Jon Stewart

Not long after Carolyn took the trip to Europe with the emphysemic Ted, my mother passed away. A second aneurysm while she was sitting alone on a couch one evening erased her mind—her body failed a few hours later. I traveled from Michigan back to the Olympic Peninsula, where my brother had arranged the funeral. The Sequim Valley Chapel was small and plain; it filled quickly with people my mother had known, many of whom I'd never met. I stood teary-eyed as strangers reminded me of my mother's kindness, cheery nature, and ready willingness to help. My brother stood morosely beside me, steadily shaking hands and acknowledging condolences.

Carolyn sat in the front pew, her crutches beside her, eerily serene.

With my mother's and Ted's deaths, my sister was left largely to her own resources. Except, of course, for my father, who would stop by her tiny apartment to say hello, or to help by examining her cat with his practiced veterinarian's eye, or to take her garbage out, clinking his way toward the curb with plastic bags full of empty gin bottles. Her alcoholism left him loath to give her money; he had stopped bringing her by his cabin because of her incessant pilfering. But sometimes he would take her for outings. One rare photograph captures the two together around 1990, at the top of Hurricane Ridge in the Olympic Mountains, near Sequim. She smiles knowingly, hiding her thinness under bulky clothing. But the contrast of his rugged outdoorsman build with her wasted frame stands out despite her careful pose.

In the years not long before my father's death, as the tentacles of Alzheimer's began to slither more deeply through his mind, his thoughts turned more frequently toward Carolyn—worrying what would become of her. He made an appointment with an estate lawyer, his low-key good humor serving to make his final sentient preparations for death seem as simple as picking out a shirt at Wal-Mart. He carefully set aside part of whatever monies might be left after his illness in a separate trust fund for Carolyn that I was to administer. "Don't give her any money directly, ever," he warned me. "Just use it to buy things she needs." Later, I was to discover just how deeply ingrained my father's caution was. Toward the end, when he would have psychotic episodes at the nursing home a mile from my house in Michigan, the nurses would try to soothe him: "It's okay—your daughter is coming." He would roll over, pausing his frantic, senseless kicking. "Which one?" he would ask suspiciously.

Fig. 12.1.

But Carolyn was also my father's daughter—his oldest daughter. Perhaps holding on to the image of an adorable, rambunctious three-year-old, suddenly rag doll–limp and terror-stricken through no fault of her own, he never stopped loving her, fretting about her, wondering how in God's name she had become what she had become.

As I wondered in turn.

***

Until January 2007. That's when an electronic prepublication version of a study popped up on one of my noodling, doodling, double-checking searches.[1] My eyes widened as I read the abstract. This study—a massive one involving 4,660 Danish polio patients—was intended to discover whether polio could be associated with subsequent risk of hospitalization for psychiatric disorders.

It turns out that medical data in Denmark is tracked in a particularly useful fashion, making it relatively straightforward to compare polio survival with demographically similar individuals. It should be remembered that many polio survivors, unlike Carolyn, develop compulsively goal-oriented, high-achieving personalities—and they do not lack for kindness.[2] Even so, this study found that polio survivors appeared to have a 40 percent increased risk of being hospitalized for various psychiatric disorders. This sounds high, and *is* high, but still, in all, equates to only a small bump in the incidence of psychiatric disorders in polio survivors. For those who contracted polio before age seven, as Carolyn did, the risk of psychiatric hospitalization was even higher.

It was no wonder I hadn't seen a pattern, despite my careful searches. Such a slight bump upward in the number of psychiatric disorders in polio survivors versus healthy people is tough to discern, except using the large numbers of patients that Danish researcher Nete Munk Nielsen and her colleagues were able to study. It's easy to suppose that if there was a bump upward in *clinically* diagnosed psychiatric disorders, there would have been a similar, or perhaps even larger increase in *subclinical* disorders. (Carolyn, remember, was never diagnosed with any psychiatric disorder.) Perhaps without polio, Carolyn would have had the creative, emotionally dynamic personality of so many in my family. But the polio left a tragic overlay on her character.

Was this due to neural damage from the poliovirus itself? Or trauma related to the horrific experience of having polio? Or genetic predisposition?

All of the above, it seems.

Part of Carolyn's problem was that she skated ahead of the wave of research findings. When the full force of her disordered personality began to flower in her teens, it was still over a decade before borderline personality would become a diagnostic entity. No one knew the real significance of the fact that poliovirus could invade not only motor neurons, but the reticular activating system—that crucial section of our brain that maintains our attention and alertness.[3]

Carolyn's attentional system was dysfunctional.

This would have meant that neurotransmitter systems throughout Carolyn's brain were also dysfunctional, perhaps causing her to have some of the same sweeping neurological differences we've seen in the brains of those who are clinically diagnosed with a personality disorder.

The clues were all there, even from the beginning of my research. It's just that until prompted by this final, capstone study, I wasn't able to put those clues together. I knew that polio's invasion and partial destruction of the reticular activating system also affects the survivor's attention.[4] But I hadn't associated this finding with those of Michael Posner's group, which indicated that minor differences in the attentional network appear to be vitally influential in the development of full-blown borderline personality disorder—for those with the genetic predisposition. And I also hadn't connected polio's damage to the attentional network with Joseph Newman's work. Newman, if you'll remember, had shown the importance of a dysfunctional attentional network in the development of psychopathy—a disorder he has also shown to not necessarily be related to violence.

Despite her sometimes psychopathic-like behavior, Carolyn seemed so intelligent because, indeed, she was. Polio never invades the non-motor neurons of the cortical areas; Carolyn's natural intelligence was left intact, floating on a surreal, dysfunctional emotional foundation. Through decade after decade of manipulation and deceit, no one could know that Carolyn's strange, uncaring attitude was not a conscious choice but was almost certainly due to shaky neural underpinnings, in all probability caused by a perfect storm of neural damage due to the poliovirus infection, extraordinary stress from the consequent social isolation and ostracization, and underneath it all, a genetic predisposition.

Was the genetic predisposition related to the genetics that helped form her personality? Probably. Our family certainly seems to have had more than its share of idiosyncratic personalities. But perhaps even more importantly, the predisposition related to the genetics of the receptors on Carolyn's neurons—receptors she shared with both of her similarly paralyzed cousins, each of whom was also a descendant of my mother's father. Paralysis from polio, remember, often runs in families.[5]

But there is another little oddity. The gene that makes the key neural receptor that the poliovirus uses to slip into a cell is found on chromosome 19. The chromosome 19, if you'll remember, is also where APOE4—the allele that predisposes people to Alzheimer's—is found. It turns out that people who have been paralyzed by polio rarely get Alzheimer's.

Why?

The same APOE4 allele that increases the risk of Alzheimer's reduces the risk of getting polio.[6]

Evil genes indeed.

Evil and Free Will

Did Carolyn have free will in how she led her life? In some sense, the question is meaningless. Does a cat have a choice when she affectionately licks her kittens? Does a killer whale have a choice when it toys with a terrified seal pup? If I've learned anything through these many years of research, it's that Carolyn's choices were a bit like the choices a tree on a windy shoreline has in deciding how tall and how bent to grow. Sure, others, as for example, George Washington and Mahatma Gandhi, were probably able to produce real changes in their neurological makeup through their conscious choices—strengthening their top-down control even if they were unable to adjust their bottom-up passions. Research is in fact showing that extraordinary neural shifts can take place through long-term conscious efforts.[7]

But what of those, like Carolyn, who don't seem to have the requisite neural apparatus to understand that there is a problem, not with drinking, or with others, but rather, with themselves? What motivation could such a person have to even attempt a change? What if the ability to exert focused mental effort is itself dysfunctional as a result of some varying combination of genetic predisposition and environmental factors, as was probably the case with Carolyn?[8] In point of fact, how many people have Washington's or Gandhi's strength of character—a trait probably intimately connected with a genetically based ability to focus—to put forth the prodigious effort needed to overcome an innate predisposition?

Perhaps neuroscientist Adrian Raine put it best when he wrote:

> If some individuals have damaged brains, can they be said to be fully in control of their actions and cognitions? Do they have complete freedom of will, or does the brain damage place constraints on such freedom? At one extreme, many theologians, philosophers, and scientists would argue that, barring exceptional circumstances such as severe physical and mental illness, each and every one of us has full control over our actions. We choose whether to commit sin or not, and thus our criminal actions (sins) are a product of a will that is under our full control. At the other extreme, some scientists take a more reductionist approach and eschew the idea of a disembodied soul that has its own free will. Francis Crick, for example, believes that free will is nothing more than a large assembly of neurons (probably involving the anterior cingulate cortex), and that under a certain set of assumptions it would be possible to build a machine that would believe it has free will . . .

> I would instead argue for a middle ground between these two extremes. I suspect that freedom of will lies on a continuum, with some people having almost complete freedom in their actions, while others have relatively little freedom of will. Rather than viewing intent in black and white, all-or-nothing terms as the law (with a few exceptions) does, it is likely that there are shades of gray, with most of us lying between the extremes. I would argue that early social, biological, and genetic mechanisms play substantial roles in shaping freedom of will . . . and that for some, freedom of will is constrained early in life due to brain dysfunction beyond their control. Brain dysfunction would be a primary process in constraining free will.[9]

Carolyn was one of the unlucky ones, someone whose genes *and* environment colluded to give her little freedom of will to leap the narcissistic, self-serving, self-destructive bonds that guided her thoughts and actions. If my sister *was* lucky, it was only in the protective shield her dysfunction provided—she remained oblivious to the hurt she caused others and retained an intermittent sense she could control herself and her destiny.

There is nothing romantic about the sufferings of those with personality disorders.[10] Perhaps the future holds real possibilities for altering the unlucky fates of those, like Carolyn, who are doomed by dysfunction to sometimes horrific behavior. Once the genetics and neural mechanisms underlying these multifaceted dysfunctions are more plainly understood, new cognitive therapies and drugs might be able to provide early intervention for those with unusual emotional deficits or cognitive disturbances. Already, for example, imaging techniques are being used to prove that emotional arousal to negative stimuli can be reduced in those suffering from borderline personality disorder by using dialectic-behavioral therapy.[11] Someday—perhaps sooner than we think—the genes involved may even be reengineered by inserting a sly nucleotide here and a tandem sequence there, simultaneously repairing the coding that might have been altered by drugs such as alcohol. Perhaps even new growth can be encouraged in areas where neurons have been destroyed. It will be the brilliant researchers at the National Institutes of Health and at laboratories and universities worldwide—perhaps led by an ambitious, prickly narcissist or two—who will pioneer these new approaches.

But always lurking in the background is the haunting question of where pathology truly begins. Could we end up drugging ourselves into some Stepford baseline? Or, as Cambridge neuropsychologist Barbara Sahakian asks, "Do we want to become a *Minority Report* society where we're preventing crimes that might not happen?"[12]

And what will be the impact of these truly remarkable neuroscientific breakthroughs on the legal system? At this point, few believe that neuroscience will overturn the concept of free will or personal responsibility in the context of the law. Many of the nation's top neuroscientists and lawyers, believe the influence of neuroscience will be felt most strongly in mitigation ("he's not fully responsible, because his brain pathology made him unable to think rationally") and in perception of risk ("this guy has brain factors that predict future violence").[13] We've already seen, however, that the concept of mitigation has hampered research in critical areas such as sadism. And, in a related issue, it is becoming apparent that America's high prison population compared to other countries may simply be a consequence of the fact that Americans have fewer involuntary patients in mental institutions.[14] (Deinstitutionalization—a result of the political left's push for patients' rights and the right's push for cost-savings—has had far-reaching, unanticipated consequences.) Certainly the debate surrounding free will and responsibility, which has occupied philosophers for centuries, is not likely to end soon.*[15]

But let's return to Carolyn.

Was Carolyn herself "successfully" sinister? Yes and no. In some

---

*Stephen Morse, the Ferdinand Wakeman Hubbell Professor of Law & Professor of Psychology and Law in Psychiatry at the University of Pennsylvania Law School, notes that "free will or lack of it is not a criterion for criminal responsibility or non-responsibility." He also adroitly points out the corollary that free will is also not used as a basis for any psychological diagnosis. Fundamentally, if a psychiatric "syndrome does not sufficiently impair the defendant's capacity for rationality, it will have no excusing force [from a legal perspective] whatsoever, no matter how much of a causal role it played. . . . To say that a syndrome caused a crime tells us nothing about whether the defendant deserves excuse or mitigation (except in New Hampshire . . .)."

sense, she wasn't a complete failure—whatever her flaws and deceits, at the very least she stayed out of jail. In a perverse way, her handicap was enlightening. Pinned to the ground, so to speak, both physically and mentally, Carolyn settled for dating drunks, among others, and keeping letters and diaries for amusement's sake. And it is only in reading her diaries that it becomes clear that Carolyn was an impenetrable person because, in some ways, there was little there to penetrate. Her empathy was evanescent because it was only superficial. She showed little guilt because there was little to show.

Both high-tech neuroscience and Carolyn's old-fashioned journal entries have helped me to realize that Carolyn, and people like her, often don't consciously intend to be evil and certainly don't see themselves as evil—despite the blindingly obvious and sometimes terrible consequences of their actions. Instead, these are people who are constrained by the quirks of their neural machinery—often carved by both genes and environment—to act in self-serving, manipulative, and deceitful ways. Evil though the consequences of their actions may be, such Machiavellians are still real people, not caricatures—they can become heartbreakingly lonely, monumentally sad, and their eyes can become filled with tears of pity—even if it is only self-pity.

Whatever their interior feelings or potential for change, however, a dispassionate look at the evidence points to extreme caution in dealing with the successfully sinister. At the personal level, the Carolyns of the world, whether created through nature, or nurture, or both, can turn families into minefields and friendships into feuds. At the professional level, they can beguile and mislead, savaging their companies with their distorted, self-serving cognitions even as they set subordinates, colleagues, and superiors at each other's throats. At the spiritual level, they can twist good intentions into ill, and set entire populations aflame with hatred. At a political level, they can play master puppeteer in the lives of millions, snuffing out entire populations with a wave of the hand and without a second thought.

WHO *ARE* THE SUCCESSFULLY SINISTER?

Before Hitler's seizure of power, psychiatrist Ernst Kretschmer remarked: "In normal times we diagnose them; in disturbed times they govern us."[16] In my reading, however, Kretschmer's quip misses the mark in a number of crucially important ways.

Rather than being diagnosed "in normal times," it appears that most people who interact with the successfully sinister, even trained psychologists and psychiatrists, have no idea with whom they're dealing—not unless these analysts are given twenty-twenty hindsight clues such as a dead body or unexplained missing millions from a company's accounts. A charming, highly successful lawyer, for example, who beats and abuses his wife and children can almost literally get away with murder without being caught.[17] A major company like Enron can run a flagrant Ponzi scheme where dozens of insiders are in a position to know something seriously strange is going on—and *still* no one says a word publicly.[18] Pedophile priests in the Catholic Church can be responsible for the rape of tens of thousands of children, and the church hierarchy not only manages to keep the offenses hidden but knowingly moves the priests to new parishes, where fresh prey await.[19] Key members of the United Nations can literally be in *"Complicity with Evil,"* as described in Adam LeBor's meticulously researched book of that name, in the commission of genocide after genocide. And yet those who allowed these disgracefully corrupt and malign episodes to proceed are granted a golden retirement with plaudits.[20] And individuals like Mao not only kill tens of millions but are worshiped in godlike fashion and touted as countercultural icons. Incidental death totals equivalent to a dozen or more Nazi Holocausts are minimized or tucked away from public discussion.

No, rather than being diagnosed, per Kretschmer's quip, highly successful Machiavellians appear to lurk in every human population. With their extraordinary ability to stack any deck in their favor, their relentless need for control, and their self-serving ruthlessness, those with at least a modicum of talent, looks, and assertiveness are more likely to be found in positions of power. This means the closer you climb toward the nexus of power in any given social structure, the more likely you'll be able to find a person with Machiavellian tendencies. It really doesn't matter what the underlying political system is—democratic, fascist, communist, or religious—or whether the social structure involves a company, university, schoolboard, religious group, city council, state government, federal government, or UN-style supragovernment; the larger the social structure and the bigger the payoff, the more Machiavellians eventually seem to find a way to creep to the top in numbers all out of proportion to their underlying percentage in society. Don't forget the growing body of research literature that reveals how people selectively sort themselves into positions congenial to their personalities.[21]

Machiavellians can have an incalculably restrictive, demoralizing, and corrupt effect on those in their sphere of influence. But what is worse is that Machiavellian behavior in a family, company, religious institution, school, union, or governmental unit—in fact, in virtually any social group—often seems to reach awe-inspiring proportions before anyone feels compelled to take solid action.*[22] Many people simply prefer to go about their everyday lives rather than take up a righteous cause; it is often much easier to simply ignore, evade, justify, or silence the speech of anyone who does speak out than to constructively act against unsavory activities. Ordinary people's emote control also means that sinister behavior can be seen as less important or—because of calcified beliefs about an ideology, institution, or person—even justifiable. Moreover, the utter ruthlessness of some Machiavellians can mean that even the most sincere and altruistic keep quiet because of realistic concerns for themselves and their loved ones. Taking action against a Machiavellian is often a dangerous proposition, and no one takes on such a task lightly.[23] (Friends in the know are often

*This behavior can't help but evoke shades of psychologist Stanley Milgram's work. In a classic set of experiments, Milgram revealed that many ordinary people will go to absurd lengths—even giving electric shocks to shocking screaming victims—in their blind tendency to obey authority. Perhaps this relates to Posner's research involving people's varying ability to resolve conflicting information and Wilson's studies related to decision making and synchronized neural rhythms.

just being reasonable when they recommend cautious silence.) All of these factors serve to keep a stable sinister system intact, despite the fact that such a system is often less effective than other, more open systems that make more effective use of the "wisdom of crowds." (Machiavellians, in fact, often work behind the scenes to ensure their system is not put in a position of competing with other systems.)

Opaque organizations, systems, and ideologies that easily allow for underhanded interactions play to Machiavellians' strong suit, allowing them to conceal their deceitful practices more easily. Idealistic systems such as communism and some religious or quasi-religious creeds are perfect for Machiavellians because they often lack checks and balances, or don't use them.

When kindhearted people are unaware that a few leading individuals in "their group" are likely to be sinister, they are ripe for victimization. Their own kindness can be turned against them and others. Hitler's greatest strength, for example, was his ability to appeal not only to the worst characteristic—hatred—but also to people's best qualities—faith, hope, love, and sacrifice. As with most Machiavellians, he was a master at turning people's best traits against them. "He confided the secret of his approach to an intimate: 'When I appeal . . . for sacrifice, the first spark is struck. The humbler the people are, the greater the craving to identify themselves with a cause bigger than themselves.'"[24]

Such factors as political instability with no end in sight, worsening economic disaster, and rapid social changes have been pointed out as critical to the rise of successfully sinister dictators such as Hitler.[25] In reality, what these factors appear to do is merely allow the successfully sinister—always loitering near the top of every significant social structure—to not only gain ascendancy but also to rewrite the rules. As power is consolidated, the sycophantic cocoon that a leading Machiavellian is able to encase himself in can, it seems, reinforce his own narcissistic thought patterns. (As Ovid is said to have observed over two thousand years ago: "All things may corrupt when minds are prone to evil.")[26] In light of all this, it becomes clear that Kretschmer's comment "in disturbed times they govern us" is true but misleading.

Machiavellians are *always* present in every system that relates to power. It's just that in times of troubles and in nontransparent systems, it's easier for them to reach the pinnacle.

This is not to say that everyone at higher levels is Machiavellian. (One British study, for example, found that only one in six supervisors is thought by their subordinates to be a psychopath.)[27] But certainly there appear to be high enough percentages of deeply Machiavellian individuals at powerful social levels to make for very different social interactions in that milieu. In such a high-powered setting, even if one is not deeply Machiavellian by nature, it is difficult to survive without using some Machiavellian strategies oneself.

The devious methods for success used by the sinister help explain why systems of ethics can at times be so surprisingly ineffectual and sometimes even counterproductive. Altruists who draw up rules and legislation to deter Machiavellian behavior are often surprised to find their policy turned on its head and used by Machiavellians for nefarious purposes. "Bad whistle-blowers," for example, can make frivolous allegations that trigger costly, mandatory, and ultimately fruitless investigations. "Moral entrepreneurs" can find law firm Web sites extolling the money to be made from turning in minor, easily resolvable transgressions.[28] Politicians, litigators, scientists—almost anyone with a grudge—can become a Javert of their chosen Jean Valjean, raising their own profile even as they destroy careers. As Peter Morgan and Glenn Reynolds point out in *The Appearance of Impropriety*:

> Over the last twenty-odd years, this nation has engaged in a far-reaching effort to increase public confidence in institutions through the use of ethics rules that stress appearances and procedures. Governmental ethics rules have expanded exponentially, to the point where an entire bureaucracy exists just to interpret and explain them. In the corporate world, ethics codes have proliferated wildly, and business ethics consulting is itself a major industry, worth over one billion dollars a year by some estimates. Yet judged on its own terms, that experiment has been a failure. . . . In fact, faith in government and corporate America has probably never been lower.[29]

Many, if not most, new enterprises appear to be started by talented individuals who enlist the help of their friends and others known to be competent and visionary. In evoking shades of gray, it can be helpful if some of these individuals have a few Machiavellian characteristics, such as a ruthlessly competitive streak or an ability to intimidate opponents. But, if the enterprise is successful, and as the group of people participating grows larger and gains more power, the structure begins to act like a magnet, attracting more hard-core, brutally self-serving Machiavellians who gradually find ways to insert themselves into positions of power. This is particularly easy to do if the top positions are held by leaders who are oblivious to the machinations of upwardly aspiring Machiavellians, or who want others to do their dirty work. Indeed, sometimes these upstart Machiavellians are competent—or even, like robber baron Andrew Carnegie, brilliant—and their tactics can prove helpful for the enterprise. The Machiavellian may ruthlessly eliminate business units, for example, that need the pruning others have found too painful to undertake, or they may seize opportunities others, perhaps with more compassionate concerns, have shied away from. More often, however, it seems Machiavellians use their unsavory tactics to climb above their real level of talent. At these undeserved power levels, the Machiavellian can play a cutthroat game of backbiting, back scratching, building a personal power base, reporting falsely rosy pictures of their work, demonizing adversaries, and siphoning assets—activities that strengthen the Machiavellian even as they weaken the enterprise. Such phenomena are obvious in business and academia. But they are also apparent in government, where gerrymandering and blocking of transparency rules for earmarking, for example, are the tip of the iceberg for protecting the more Machiavellian incumbents from public scrutiny and truly democratic processes.

Eventually then, after decades or perhaps even centuries, a corrupt, rickety enterprise, having been kept afloat by well-meaning cadres, can be overtaken or overrun by other, newer social structures —freshly formed businesses, religious offshoots, or political parties— that have had less time for relatively untalented Machiavellians to insert themselves into the highest echelons of the system. This might also explain why the monolithic, protected, noncompetitive nature of the American educational system—which allows Machiavellians to find their secure place in faculties, unions, committees, teaching-related societies, school boards, and school districts—sometimes performs so poorly.[30]

How Can You Tell?

The very rarity of Machiavellians at most social levels can make them difficult to pick out. Shades of gray involving quasi-Machiavellians, or the very real "good side" of a Machiavellian, can make detection even more difficult. (Even *I* am fooled on occasion. To never be fooled, though, I'd have to be completely paranoid.) But protection can be had by simply being aware of the existence of these deeply deceitful chameleons who, it should be remembered, are often propelled by very different neurological processes. On a personal level, such awareness can cause subtle investigation of relationships that look to become significant. Gossip can be surprisingly helpful here. While a Machiavellian's hoodwinked supervisor, for example, may rave about the Machiavellian's sincerity and talent, coworkers, underlings, janitors, roommates, teammates, cellmates, or simple acquaintances may have a very different story—if you happen to gain their confidence. (That's why books on hiring often recommend, after all the high-level interviews have taken place, seeing what the secretary thinks of a candidate.) Likewise, a boyfriend's mother may warble on about his giving nature, but his many former girlfriends may tell a *very* different story. The more powerful or influential a person's position or potential position, the more fraught his personal relations may become. Power, in fact, is a magnet for Machiavellians and brings out the very best of their beguiling charm—as when "little Kathy," Paris Hilton's mother, targeted her mogul, megamillionaire heir Rick Hilton.[31] Sometimes it can be almost impossible to believe that someone could be as chameleon-like as others say, and the tendency for those in power might be to ignore the message.* And it's true that the source for negative gossip herself may be Machiavellian—as with the legendary Roxalena and her duplicitous tales about the sultan's beloved son.

At slightly more distant social levels, the camera goes out of focus—it is difficult to detect whether a mayor, school board member, or union leader, for example, is Machiavellian, unless one has access to people in the know. But at still higher social levels—say, senatorial or presidential—information about a person eventually becomes available, at least in open democratic societies. A strong mismatch between public and private lifestyles is a telling, though not surefire, mark of the chameleon-like behavior of a Machiavellian. Loveless marrying for money or a series of low-key scandals may also provide indicators—indicted Hollingsworth CEO Conrad Black, for example, was early on expelled from an elite private school for selling purloined solutions to examination papers, while Russian president Vladimir Putin—whose critics so often suffer unusual and horrific deaths—appears to have plagiarized large chunks of his doctoral dissertation.[32] Intellectuals may snicker at journalism's bad boy Matt Drudge or at publications like the *National Enquirer*, but there is a reason that totalitarian regimes such as the People's Republic of China ban similar reporting within their borders. (Perhaps surprisingly, the *Drudge Report* is *the* must-see Web site for top-ranked journalists, while the *National Enquirer* carries an excellent reputation among those same journalists for investigative reporting.)[33]

At high social levels, the game of "find the Machiavellian" can become a house of mirrors, because disinformation is always rampant. Each political party, which naturally includes its own Machiavellians,

---

*When I served as a communications officer in the army in the late 1970s, I still remember the enlisted men complaining to me about one of my sergeants. "He's a completely different person around us than around you," warned one of the cable apes, as they called themselves. One day I happened to overhear my creepily servile sergeant abusing the company clerk, and I suddenly understood what the men were talking about. Becoming aware of this type of brown-nosing was good preparation for becoming a professor. It's amazing how often I'll hear a professor proudly describe how his firm chewing out set a cheating student back on the right path. Yet I hear through the grapevine how that same student continues to brag about his cheating prowess.

has a vested interest in characterizing the other candidates and parties as Machiavellian. At the same time, each party's followers can't help but reassure themselves that their candidate and party couldn't possibly be Machiavellian, aside, perhaps, from a cursory jot and tiddle. Or perhaps they suspect "their guy" has some Machiavellian traits, but they believe the end justifies the means—at the very least, their candidate is achieving the group's desired goal. Besides, a Machiavellian—with his frequent finger-pointing and rabble-rousing—can often be a lot more exciting than a more ho-hum reasonable sort. (*Sure he's a loaded cannon, but he stirs things up around here!*) Machiavellian leaders might also actually attract emotionally off-kilter followers who identify with the label of victim and enjoy the heightened stimulation they receive from the Machiavellian's rants.

Perhaps at these high levels, the best an ordinary person can do is to try to lay aside his or her own ideological blinkers and look honestly at public figures. If a given individual seems most interested in vilifying others, proceeds to characterize his own in-group as having been unduly victimized, is ruthlessly vindictive, and finally, is discovered to have cozy, self-serving financial deals, there are reasonable grounds to assume that a person is more than a little Machiavellian and that his or her leadership may be aimed more toward self than public service. Unfortunately, our own tendency, at least regarding leaders who purport to share our ideology, is often to avoid looking too closely.

Both at home and in other countries, the choice is sometimes more difficult, with no right answer at hand. As psychiatrist and political commentator Charles Krauthammer has written: "The essence of foreign policy is deciding which son of a bitch to support and which to oppose—in 1941, Hitler or Stalin; in 1972, Brezhnev or Mao; in 1979, Somoza or Ortega. One has to choose. A blanket anti-son of a bitch policy . . . is soothing, satisfying and empty. It is not a policy at all but righteous self-delusion."[34]

Sometimes, in other words, there are no satisfying choices to be had.

CAROLYN

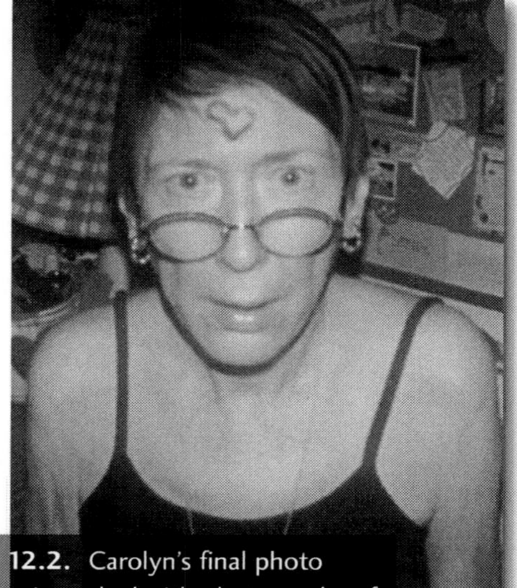

Fig. 12.2. Carolyn's final photo album is packed with photographs of feline knickknacks and her cat. But this haunting picture—different from all the others—somehow meant enough for her to keep.

So many years chasing Carolyn. I understand better now how my sister's life was spent tracing the skewed patterns of her neurons. Did she have a deep-seated genetic predisposition for her sinister behavior? Quite possibly. Did the polio exacerbate her instability? Almost certainly. Did she drink to fill the emptiness of her ephemeral, underperforming neurotransmitters, and did that contribute to her behavioral problems and her anorexia? Absolutely.

Carolyn's ultimate psychiatric diagnosis? Perhaps it doesn't matter. After all, borderline, psychopath, and any number of other diagnoses are all just words we clumsily apply to the complex processes that can be associated with both nature and nurture as a dysfunctional brain develops.

Do I forgive Carolyn?

Certainly writing this book has given me a compassion I never felt before. I can understand now some of the neurological and genetic quirks that spun Carolyn in directions she had little control over, whatever her conscious feelings to the contrary. But I also know that if, through some cosmic trick, Carolyn were to suddenly be alive again, smiling in my living room as she played with one of my children, something visceral would still rise in me, as if I'd just found my children toying with a snake. Perhaps surprisingly, I think this means I'm "at a good place" now. Reason has given me an understanding and peace that might be called forgiveness. But emotions remain, serving as a shield that helps protect both my loved ones and myself from Carolyn and people like Carolyn. Researching this book hasn't saddened me, or at least not too much, about the things that can't be changed in people like Carolyn. It has just given me a gentle reminder that in the end, we should perhaps not worry so much about changing others. Rather, we might think instead about changing ourselves.

Still, when all is said and done, I can't but wonder—did Carolyn have the capacity for love? Not cheaply expressed "love" for an easily manipulated paramour or for an obedient and at least temporarily useful caretaker, but real love that included the idea of sacrifice and sorrow and joy for another? Did she love, for example, our father—the person who, through all the years, loved her most unwaveringly, most loyally, most honestly?

A journal entry tells me:

> May 6th: Penny watered the artichokes and fed the birds; also, cleaned up the dried parsley I accidentally spilled. Barb called—Dad died. My request for help with periodontal care seemed self-serving; but apparently this will be handled through a trust fund.

I page onward through the entries. The diary continues on routinely through May, June, and July: she changed sheets, read cookbooks, pulled a muscle while inebriated, went shopping, watched television. One Thursday she presciently notes: "There is absolutely no reason to keep a journal. Actually self-indulgent! My hubris will catch up with me."

Actually, it wasn't hubris. It was her boyfriend, Jack.

Jack himself had apparently called 911 to report Carolyn's death, but when the police arrived at the apartment, Jack had proved curiously evasive. When asked if Carolyn drank, after much prodding, Jack confessed she drank a third of a bottle of alcohol a day. As Officer

Crispin related in his subsequent phone call to me, this amount of booze was unimpressive: "Most alcoholics drink about a half gallon a day," he told me. "A third of a bottle shouldn't have killed her." Crispin had ordered toxicology tests; he suspected by Jack's jittery demeanor that he had somehow drugged her.

And we did our own investigating. According to her caretaker, Carolyn had apparently been edgy over the previous month about pains in her chest. Jack had moved in to ease her fears and be around in the case of emergency. Instead, the visit had somehow turned into a two-week mutual bender.

Three days after Carolyn's death, Jack himself called, voice slurred though it was only mid-afternoon, and admitted that Carolyn was in no pain when she died. "She was just setting there and she got kind of quiet." By then we'd talked with the forensic pathologist, who'd penciled in coronary artery atherosclerosis as the cause of death. (Actually, the pathologist added offhandedly, the cirrhosis of the liver was so bad that it could really have been a toss-up.)

Fig. 12.3. Carolyn circa 1948.

"What time of day did Carolyn die?" my husband asked Jack inanely. It's a little hard to know what to say to someone who's just intimated he was too drunk to call for help when

Carolyn had had, right in front of him, the heart attack she'd been fearing.

"By the time I was available again," Jack answered elliptically, "it was history."

Carolyn's last written words reverberate: *"Back to the real world after panic attack. Must ease Jack out. Can't tolerate the smoke or the late night 'sloppies.' He is still a good friend to have."*

I rifle through her diaries. Nothing more. Nothing of interest.

Except one day. August 9th—three months after my father died. Two years before her own death.

> Nancy did nice lady visitation to thank me for goodies. I fell very much in like with her. I was too talky—I had lost last night's supper and had swollen eyes from tears shed over that actor Charlton Heston's announcement of Alzheimer's diagnosis in lieu of Dad's recent demise. Great buy on ripe brie at Sunny Farms Grocery with sour grapes and mediocre plum and nectarine.

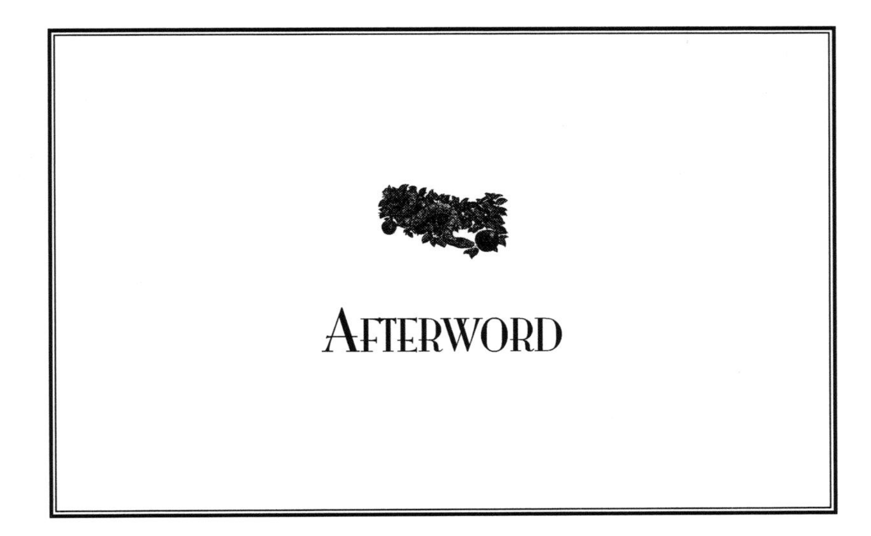

Afterword

Professor Barbara Oakley's *Evil Genes* provides a terrific "big picture" overview of those who carry traits similar to borderline personality disorder. What Barb didn't tell you, however, was that when she first began to look for information related to her sister, she stumbled across my *Stop Walking on Eggshells*. "That's what started the boulder rolling," she told me.

Many who have read *Evil Genes* have asked Dr. Oakley what they should do if they are dealing with a troubled person in their life who they suspect has some—or many—of the symptoms of borderline personality disorder. As it turns out, my most recent book, *The Essential Family Guide to BPD: New Tools and Techniques to Stop Walking on Eggshells*, deals with precisely that topic. That's why Barb asked me to write this afterword.

As you know, many of those who have symptoms of borderline personality disorder remain undiagnosed. Some of these people can

---

Portions of this afterword are from *The Essential Family Guide to BPD: New Tools and Techniques to Stop Walking on Eggshells*, by Randi Kreger (Hazelden Publishing, 2008).

function at prominent levels—as doctors, ministers, business leaders, or other highly respected individuals. If you suspect someone you care about or work with has some or many of the traits of borderline personality disorder, you may be left feeling helpless and hopeless. However, you *do* have the power to improve your situation whether or not the borderline-like person ("BP") you know chooses to change.

Your feelings are normal, and you are not alone.

The following are common emotions among those who affiliate with BPs (called non-BPs here for short):

- **Self-doubt**: People question their own sanity because BPs forcefully and consistently maintain that the non-BP's perceptions are completely wrong.
- **Guilt/shame**: Over time, the accusations have a brainwashing effect. The non-BP comes to believe that he or she is, indeed, the source of all the problems.
- **Depression/grief/exhaustion**: Non-BPs being devalued by a BP in once loving relationships cherish powerful memories of the times when the BP loved them unconditionally and thought they could do no wrong. Non-BPs can ultimately feel as if the person who loved them has died and been replaced by someone unknown. The "love/hate" cycle is completely unpredictable—it can take minutes, hours, days, months, or years.
- **Isolation**: If you are in a relationship with a BP, you may find that the BP's unpredictable behavior and moodiness can make friendships difficult. Some non-BPs find that friends and family do not believe them when they describe the BP's behavior. Other non-BPs say friends suggest solutions they consider simplistic or unacceptable. The non-BP's self-esteem suffers, and making friends eventually seems impossible. In addition, the BP may insist that the non-BP cut off ties with friends or family. The non-BP often feels there is little choice but to comply.
- **Feeling trapped and helpless**: Staying in a relationship with a BP may seem hopeless, but leaving may not seem like a realistic

option either. Getting angry does not work—neither does reasoning or anything else. Who *wouldn't* feel helpless under these circumstances?

If "your" BP is a family member, don't tell him you think he has BPD or try to force therapy upon him.

If you do, as Dr. Oakley pointed out earlier in *Evil Genes*, you can be *sure* your family member will become furious and say that *you're* the one who has borderline personality disorder—especially if you and your BP have been playing the "blame game." Even if the diagnosis was coming from a doctor, it would be meaningless unless your BP wanted help. Making major changes is tough even when people *want* to make them.

The question people ask most often is, "How do I make my family member go to a therapist?" The answer is, "you can't." You may be able to threaten the person into going, but all you get is a warm body, not an engaged or open mind. Rather than focus on what your family member *thinks*, which you can't control, focus on his *behavior* toward you, and think about what you will and will not accept.

Recognize the common triggers of shame and fear of abandonment.

Whether your BP is in your family, or simply someone you must work with, you can create more predictability in your life by identifying triggers (sensitive areas) that lead to borderline defense mechanisms. The most common ones are shame and fear of abandonment.

Everyone fears abandonment to some extent. However, BPs react to real or *imagined* threats. And they have a good imagination. In a family situation, normal things such as spending time with your own friends or working long hours can trigger an over-the-top expression of fear that cannot be easily assuaged because it originates from within the BP.

People with borderline personality disorder feel innately bad, worthless, defective, or toxic at the very core of their being—not for something they have done, but simply for being. This makes them sensitive to rejection. They anxiously await it, see it when it isn't there, and overreact to it whether it's there or not. This is why small slights—or *perceived* small slights—can cause major messes.

Leave if your family member rages or becomes violent.

Don't listen to family members berate and call you names. At that point in time, they can't see your point of view or think through the effects of their interactions with others. It's not that they *won't*—it's that they *can't*, not without having had the proper treatment. Remember—their emotional toolkit is missing some key components.

Verbal abuse harms you: ongoing, repeated verbal assaults can be every bit as emotionally devastating as physical battering—especially when it is meted out by an intimate or by someone in a position of authority.

If your family member loses control, you can:

- Retreat to a room that is off-limits to everyone else
- Call a friend and go over to her house
- Call a friend and have the friend come to your house
- Go to a movie
- Put on some headphones and listen to music
- Take a taxi home
- Turn off your answering machine

Think through your options and make concrete plans ahead of time. Then discuss your plan with the BP before these situations occur so that you have a shared understanding of what will happen. Assure your family member that if you leave, you will be back. Let your BP make her own choices, and give the gift of allowing her to take responsibility for the outcome of her decisions.

Use phrases like:

- "I want to hear about your feelings, but it's hard for me when things get too emotional" (instead of "you get too emotional").
- "We'll talk later, when things calm down. I want to give you my full attention, and that's too hard for me to do right now."
- "I can't listen right now. Not until things are calmer."
- "Let me have a little while to calm down and then we can talk."

Become an advocate for your borderline child.

If you believe your minor child has BPD, run to your nearest bookstore and purchase a copy of *Borderline Personality Disorder in Adolescents* (Fair Winds Press, 2007) by Blaise Aguirre, MD.

You will hear over and over again that a person can only receive a diagnosis of BPD after the age of eighteen. This is an old and entrenched way of thinking. Today, experienced clinicians believe it's crucial to address borderline-like behavior *as soon as possible.* The longer the disorder goes untreated, the more entrenched it can become.

Learn and practice the five core skills that will empower you to improve your own life.

Borderline personality disorder is multifaceted and borderline-like behaviors can be wildly unpredictable. Those with BP family members, or who are dealing with a BP in their work life, need more than information: they need skills training, emotional support, and often their own therapist to help them determine what is and isn't "normal."

My book *The Essential Family Guide to Borderline Personality Disorder* (Hazelden, 2008) identifies five major tools family members need to organize their thinking, learn specific skills, and focus on what they need to do instead of becoming overwhelmed. If you are dealing with a BP in your work life, you will also find these tools to be useful.

Here, I'll briefly go over the first three: take care of yourself, uncover what keeps you feeling stuck, and communicate to be heard. (The fourth and fifth have to do with setting limits and reinforcing the right behavior. These are too complicated to discuss given the amount of space available.)

Tool 1: Take Care of Yourself

Non-BPs tend to become isolated. Don't let this happen to you. Reach out to others. Don't let yourself be embarrassed into isolation or pushed into it by threats, implied or outright.

Depend on friends to give you reality checks. Non-BPs almost always lose sight of what is normal. Join a support group, whether online or in real life (see links at BPDCentral.com). Get enough sleep and live a healthy lifestyle.

Tool 2: Uncover What Keeps You Feeling Stuck

Most often, you don't recognize situations that have the ability to make you feel trapped and stuck until you're deeply involved and unable to tell what is "normal" anymore. These feelings of helplessness and lack of control can cause just as much suffering as the presence of the personality disorder itself.

Feelings of fear—for example, fear of conflict, fear of being alone, fear of financial problems, and fear of losing the relationship—make some people feel trapped. Susan Jeffers, the author of *Feel the Fear and Do It Anyway*, says that all we need to do to diminish fear is to develop more trust in our ability to handle whatever comes our way.

Sometimes survival is dependent upon giving up the myths of the ideal parent, sister, or other family member, and accepting reality, no matter how much we wish it were otherwise. We may cherish our feelings of loyalty and carry a strong sense of obligation, but sometimes those feelings can mislead us into doing things that are ultimately harmful for both ourselves and our loved one.

Take a close look at your beliefs and decide which ones are based on myths and which ones are based on reality. This consists of asking questions such as: What do I do because of a sense of obligation? What feelings rise up when I ask myself that question? Which of my obligations feel good to me? Which ones do not?

Guilt drives parents to lose their sense of judgment and go to ridiculous lengths to assuage their guilt. Society says that parents who have a child who isn't functioning well are not up-to-snuff. Acknowledge your guilt, but formulate a plan to overcome it so it doesn't act as a barrier.

Rescuers love their family member and want to help. So they "help," often by doing things they don't want to do or by giving up things they don't want to give up. When things don't improve (or get even worse), they "help" some more—even though they resent it and think it's unfair. Rescuers do things for people that the people could do for themselves, which encourages their dependence.

Tool 3: Communicate to Be Heard

The first rule of communication is knowing when *not* to attempt it. When someone with symptoms of borderline personality disorder has intense feelings, the emotional centers of his brain "hijack" the logical centers. Asking your family member to process factual information is like asking a baby to drive a car. It's not that he doesn't *want* to, it's that he *can't*.

If you're feeling criticized or blamed—especially when it's quite undeserved—the natural response is to defend yourself. But this only makes things worse because the message the borderline receives is, "Your feelings are wrong." While no one wants to be told their emotions are baseless, borderline individuals have an especially intense, negative reaction to having their feelings "invalidated" (meaning rejected, denied, ignored, mocked, judged, or diminished).

To avoid doubling his anger at and pain of being told he's incorrect about his own feelings, you need to separate your BP's distorted *thinking* from the intense, overwhelming *feelings*, and then empathically acknowledge those emotions to your family member *without necessarily agreeing with the thoughts that link the two.*

So how do you do that? Through a technique called "empathic acknowledging"—a blend of empathy, active listening skills, and acknowledging.

EMPATHY

Empathy is emotionally putting yourself in someone else's place to the point when you can vicariously experience her thoughts and feelings.

Metaphorically, people who express sympathy are like people who drive by the scene of an accident, slow down, give an encouraging expression to the driver of the banged-up car, then speed up and go along their merry way. On the other hand, people who express empathy pull over, get out of their car, clasp the shoulder of the driver, and say, "Oh wow, I bet this is the *last* thing you needed to happen right now."

ACTIVE LISTENING

Active listening means suspending your judgments and opinions, and pushing everything out of your mind except your family member—you're about to enter her world. Focus on what she's feeling as well as saying, with her words, tone of voice, facial expressions, and body language.

Listen with 100 percent of your attention without interrupting, asking questions, offering solutions, or thinking about what you're going to say next. This says, "You and what you say are so important that I'm giving you my undivided time and attention. I am willing to listen to you with an open mind."

Acknowledging (Verbal and Nonverbal)

Verbally, use encouragers such as "oh," "hmm," and "really?" Reflect their feelings ("That sounds frustrating"). Show involvement ("I'm happy for you"). Punctuate intense emotions ("Oh no!").

Your most powerful communication tool is your face and body, not your mouth. Research shows that people convey just 7 percent of their attitudes and beliefs through their actual words. The other 93 percent comes from our tone of voice (38 percent) and our facial expressions (55 percent).

Make your eyes soft and steady, showing interest. Have a relaxed facial expression (no tightening or scrunching up) with a neutral expression or genuine half-smile. Relax your body (whether seated or standing) and have your arms loose by your sides. Don't stare, glare, look away, grimace, frown, or scowl.

Asking validating questions is another form of acknowledgment. Ask specific, clarifying questions in a way that shows genuine interest and is not provoking. BPs frequently make general, black-and-white comments like "You're selfish." Ask your BP just what she means by these vague words. What exactly did you do that showed selfishness? How often did you do it? What makes your family member think you don't care about her?

You might say, "I really want to understand you, but I'm having trouble appreciating the depth of your feelings about this. Can you try explaining this in another way? I care, I just need to understand better."

Defuse Aggression

Use noncombative statements that help you to reach your goal and inject some reality into the situation. Examples include:

- "You are the expert on yourself."
- "I appreciate what you said, but what I mean is . . ."
- "At the time my motivation was . . ."

Manage the conversation:

- "Could we get back on the subject?"
- "Let's discuss that at a later time. I'd like to keep the focus on . . ."
- "I was hoping we could talk about this. I don't see how we can resolve the situation if you won't talk with me."

Create a climate of cooperation:

- "Maybe we can find a way to . . ."
- What we *do* agree on is that . . ."

Respond to unwarranted criticism or abusive statements:

- "I won't stand here and listen to you abuse/yell/attack me, so I'm going to leave if this continues."
- "No, I won't tolerate that kind of language, so I'm going to leave if it continues."

DON'T MAKE YOUR HAPPINESS CONTINGENT ON YOUR FAMILY MEMBER'S DEGREE OF RECOVERY.

Recovered BP A. J. Mahari advises family members to let go of their desire to control their borderline family member's recovery. "This is *their* journey, not yours," she says. "You can support them, but it can't be your life plan."

Some non-BPs have embraced detaching with love, a concept promoted by Al-Anon, an organization for people whose lives are affected by someone who abuses alcohol. "Detachment" is neither kind nor unkind. It doesn't imply judgment or condemnation. It is simply a way that encourages each person in a relationship to make his or her own decisions and live with the consequences of those decisions.

Strive to make sure you're neither used nor abused. Avoid manipulating situations to help the BP avoid looking at his own behavior. Neither create a crisis nor prevent one if it is in the natural course of events.

Evil Genes may have opened your eyes to a whole 'nother side of human behavior. Although it may seem that a BP you know is consciously choosing his or her troubling behavior, try to remember that if your brain were similarly wired, you would in all probability be acting the same way. Join us at www.BPDCentral.com if you need friends and support as you retool your life in light of your new knowledge.

Randi Kreger
www.BPDCentral.com
Coauthor of *Stop Walking on Eggshells* and
author of *Stop Walking on Eggshells Workbook* and
The Essential Family Guide to BPD

For Pondering

1. Have you ever known a successfully sinister person such as those described in *Evil Genes*? How did his or her actions duplicate or differ from those of the typical Machiavellian described in this book? Did "your" Machiavellian have any redeeming traits?
2. Are you dealing with a successfully sinister person now in your life? How will you deal differently with him or her based on the information you've gleaned from *Evil Genes*?
3. Do you think that you interact with people differently because of your own past experiences with the successfully sinister?
4. Do you see Machiavellian traits in yourself? Are they healthy? How would you know?
5. Have you ever found that one of your kindest traits was used to manipulate you into doing something wrong or hurtful to others or yourself? If so, did this cause changes in your subsequent behavior?
6. Oakley struggled initially to write *Evil Genes* as a conventional—and impersonal—nonfiction book. Do you think that her ultimate decision to include memories of her sister made the book more

powerful? Do you think it was difficult for her to share intimate information about her family? Would you have been willing to do the same?

7. Is there a danger that ascribing Machiavellian traits to gene and brain dysfunction might result in persecution or discrimination ("bad seed," "born bad," "no hope")? Or might it simply make treatment more probable?
8. Oakley pointed out how oblivious people can be to the influence of their emotions, which forms a cognitive blind spot they are *certain* they don't have. What do you think is one of your worst cognitive blind spots? Can you gain some inkling of why a well-meaning, intelligent person might disagree with you on this matter? (Watch for the little emotional uplift you get when you struggle and then conclude how fundamentally *wrong* the other opinion is.) Do you think that examining your most profound cognitive blind spots gives you insight into the cognitive disturbances of borderline-like thinking?
9. Some dictators, like Hitler, Milosevic, or Saddam Hussein, were killed or ousted while still in relative middle age. Others, like Stalin and Mao, lived until their midseventies, exerting virtually complete control over their countries until their deaths. What is the difference between the long- and the relatively short-lived dictators?
10. Someday dictators will have access to technology to have themselves cloned, allowing for an endless procession of "mini-me's." What effect might this have on evil dictatorships of the future?
11. Over time, whatever political party is in power nationwide often appears to lose focus and become mired in corruption and schemes that enrich their supporters but do not benefit the nation or their constituents. Oakley suggests that this is because over time more and more Machiavellians work their way into positions of leadership in the winning party. Do you agree? Or is there some other set of social mechanisms at work? Do you see similar corrupt processes occurring at a local political level, in your religious group, or at your place of work?

12. Would it be effective to teach high school and college students about manipulative strategies such as gaslighting and projection and how to best react to such manipulation? Might this type of training have long-term benefits for society, such as reducing the percentage of marriages that end in divorce and shortening the period of time where a person fruitlessly tries to "fix" an "unfixable" Machiavellian? Or would it have the contrary effect of teaching Machiavellians to improve their strategy?
13. Should high schools ever implement programs that teach students to be wary of possible Machiavellian motives behind emotional appeals for "fairness" or "justice"? Why might high school teachers oppose teaching about Machiavellian motives? Would opposition to such teaching be a form of censorship or manipulation?
14. Did Carolyn and Jack love each other?
15. Did Carolyn love her father?

ACKNOWLEDGMENTS

A number of important people have been responsible for getting this literary car on the road and keeping it on track. The first is my agent, Ben Salmon of Rights Unlimited, who plucked the early, rough manuscript from obscurity and took it to the top. The second is my Prometheus editor, Linda Greenspan Regan, who *is* the top—her support, guidance, and superb pointed criticisms have been of enormous benefit to this work. I am also deeply indebted to my editor, Audrey Perkins, faculty at Linn-Benton Community College in Albany, Oregon. Audrey has waded through every word of the developing manuscript numerous times (and since the manuscript was originally half again as long, that's saying a lot). Her careful and incisive editorial suggestions have been invaluable in improving the manuscript's quality and lifting it to a higher level. Audrey's colleague, Richard Liebaert, formerly of the Linn-Benton Biology Department, has also waded twice through the manuscript and made terrific contributions with his constructive criticism, which extended from a wide range of scientific insights to "stuff" in general. Professor Robin

Hemley, director of the Nonfiction Writing Program at the University of Iowa, helped provide guidance during a crucial period in the book's formation, as did Dr. Richard Bruno, a leading authority on post-polio syndrome. My dear friend Dr. Gabrielle Stryker has given profoundly appreciated guidance and assistance related to both immunology and the book's illustrations.

A number of experts have reviewed various portions of the manuscript as well as the manuscript as a whole. However, any book that covers so many different disciplines is bound to be deemed problematic at times. No one can possibly have read all the relevant literature—virtually every book I cite, for example, has tantalized me with ten or more alluring references. The more discerning specialists will undoubtedly be able to think of exceptions, objections, counterexamples, and contradictions to the possibilities, generalizations, and conclusions that have been presented here. I apologize in advance to those who might take offense in this regard and note that any errors that might be found are mine and mine alone. Additionally, I would like to point out that those who have helped me on certain aspects of this work should not be taken to share all of my views. My intention with this book was to provoke thought, since we have just reached the edge of knowledge where science has become extraordinarily thought provoking.

I could not have written this book without the "wisdom of crowds." I am deeply indebted to the following individuals for their many contributions: Elizabeth Abbott, Craig Becker, Mary Tracy Bee (thanks for making cutting up corpses fun!), Giuseppe Biamonti, Warren Buffett, Rand Careaga, Rita Carter, Cindy Collins, Richard Felder, Paul Frick, Marc Haeringer, Jon Halliday, Sabine Herpertz, Linda Jack, Quinn Tyler Jackson, Kent Kiehl, Grace Kwok, Shailesh Lal, Eric LaRock, Doreen Lawrence, Cameron Leith, Guruprasad Madhavan, Peter McConville, Sam McFarland, Ken McLeod, Mark Milstein, Stephen J. Morse, Dmitri Nabokov, Andrew Nathan, Randolph Nesse, Joseph Newman, Nete Munk Nielsen, Jerry Oppenheimer, Jim Phelps, Cliff Pickover, Robert Plomin, Lucian Pye, Xianggui (Harvey) Qu, Adrian Raine, James M. Royer, Pat Santy,

Kwai Sim Shek, Ken Silk, Daniel J. Simons, Helen Smith, Margaret Soltan, Richard Stamps, Bella Stander, David and Laura Stiles, Glenn Storey, Eugene Subbotsky, Ron Summers, Essi Viding, Daniel Weinberger, Margaret Willard-Traub, David Sloan Wilson, Matthew A. Wilson, and Ke Xu.

Judith Rich Harris, author of *The Nurture Assumption*, gave me gentle impetus at the beginning of this project—for her graciousness I am utterly grateful. I could not have written this book if she had not "broken trail."

At Prometheus Books, Julia DeGraf's copy-editing skills are remarkable, and deeply appreciated. Grace Zilsberger's cover design is inspired (those who look closely might spot the DNA double-helix on the snake's back). Chris Kramer, Jill Maxick, Rich Snyder, Gretchen Kurtz, Mark Hall, Marcia Rogers, Bruce Carle, and Lynn Pasquale round out the crew of consummate Prometheus professionals who I've been fortunate indeed to work with. No question—Prometheus president Jonathan Kurtz runs a taut ship.

At Oakland University, my dean, Pieter Frick, and department chair, Chris Wagner, have provided congenial support and encouragement, as have my "part-time" department chairs, Manohar Das and Gary Barber. I am very grateful. I also take this opportunity to thank the indefatigable staff at the Oakland University Interlibrary Loan Department, especially Patricia Clark, Diane Boving, and Dante Manes Rance, who I'm sure have long since concluded that I have unhealthy interests for an engineering professor.

Some of my friends at Oakland University have contributed in ways too numerous to count to this project: Kris Allen, Farid Badar, Gary Barber, Len Brown, Dan Chang, Todd Estes, Christine Hansen, Richard Haskell, Peggy Lin, Kathy Pfeiffer, Pat Piskulich, Andrew Rusek, Brian Sangeorzan, Donna, John and Carina Searight, Meir "Fiki" Shillor, Anna Spagnuolo, and most especially, Cathy Starnes. Thank you all for making every day a fun day to come to work.

Our younger daughter's line drawings are deeply appreciated. (Artistic talent not only skipped my generation—it left a vacuum.) Our older daughter's eagle eyes and insightful comments were invaluable. Our adopted sons Bafti and Irfan have each contributed in many ways and made my life and this book richer and more complete. I also thank my in-laws, Jane Raley and Daniel Oakley, whose patient encouragement and interest over the many years of this project has been greatly appreciated. My brother, as always, has provided faith in the fundamental decency of this branch of the family.

My husband, Philip, swept me off my feet with a three-week engagement at the South Pole in Antarctica nearly twenty-five years ago. I knew it was a risk, but I thought—*this man is worth the risk.* How right I was! Surprisingly for me, a writer, I can't find the words to describe my luck. But I can tell you this: I could have done none of this without Philip.

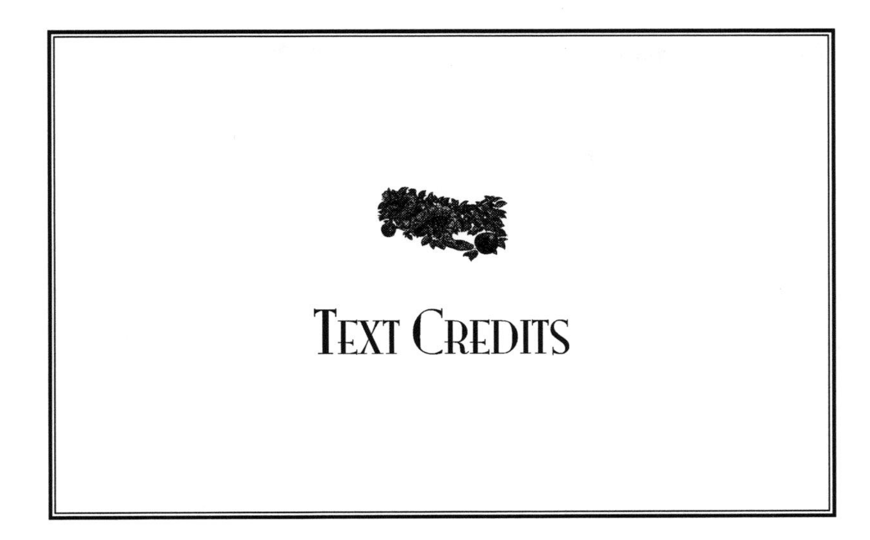

Text Credits

1. Excerpt from Regina Pally, "The Neurobiology of Borderline Personality Disorder: The Synergy of 'Nature and Nurture,'" *Journal of Psychiatric Practice* 8, no. 3 (2002): 133–42. With kind permission of Lippincott Williams & Wilkins.

2. Excerpt from Adrian Raine, "Psychopathy, Violence, and Brain Imaging," in *Violence and Psychopathy*, ed. Adrian Raine and José Sanmartín (New York: Kluwer Academic/Plenum Publishers, 2001), pp. 50–51. With kind permission of Springer Science and Business Media and Adrian Raine.

3. Excerpt from Vladimir Nabokov's "An Evening of Russian Poetry" by arrangement with the Estate of Vladimir Nabokov. All rights reserved.

Illustration Credits

Fig. Intro.1 Permission by author.

Fig. 1.1 Reproduced with the permission of Punch, Ltd., http://www.punch.co.uk/.

Fig. 1.2 Image courtesy Matthew Henry Hall, http://www.matthewhenryhall.com.

Fig. 2.1 Two pie charts created by author from data available in Essi Viding, R. James, R. Blair, Terrie E. Moffitt, and Robert Plomin, "Evidence for Substantial Genetic Risk for Psychopathy in 7-Year-Olds," *Journal of Child Psychology and Psychiatry* 46, no. 6 (2005): 592–97.

Fig. 3.1 Genetics Home Reference (Internet), Bethesda, MD: National Library of Medicine (US); 2003 (updated December 15, 2006; accessed December 22, 2006), "How Many Chromosomes Do People Have?" Adapted from http://ghr.nlm.nih.gov/handbook/basics/howmanychromosomes.

Fig. 3.2 Image courtesy Darryl Leja, National Human Genome Research Institute, National Institutes of Health.

Fig. 3.3 Adapted from National Center for Biotechnology Information (Internet), Bethesda, MD: National Library of Medicine (US), 2003 (accessed December 22, 2006), Homo sapiens build 36.2.

Fig. 3.4 Illustration by Gabrielle Stryker.

Fig. 3.5 Illustration after image from National Institute of Mental Health, "Depression Gene May Weaken Mood-Regulating Circuit," Bethesda, MD: National Institute of Mental Health, Clinical Brain Disorders Branch, National Institutes of Health, US Department of Health and Human Services, 2005, http://www.nimh.nih.gov/press/short circuit .cfm (accessed November 1, 2006). The clever additional captions are from Dr. Jim Phelp's Web site at http://www.psycheducation.org.

Fig. 3.6 Illustrations after image from National Institute of Mental Health, "Depression Gene May Weaken Mood-Regulating Circuit," Bethesda, MD: National Institute of Mental Health, Clinical Brain Disorders Branch, National Institutes of Health, US Department of Health and Human Services, 2005, http://www.nimh.nih.gov/press/shortcircuit.cfm (accessed November 1, 2006).

Fig. 4.1 Reprinted with slight modifications from Kent A. Kiehl, Andra M. Smith, Robert D. Hare, Adrianna Mendrek, Bruce B. Forster, Johann Brink, and Peter F. Liddle, "Limbic Abnormalities in Affective Processing by Criminal Psychopaths As Revealed by Functional Magnetic Resonance Imaging," *Biological Psychiatry* 50: 677–84, Copyright 2001, with permission from the Society of Biological Psychiatry.

Fig. 4.2 Modified from image available courtesy National Institute on Drug Abuse, National Institutes of Health.

Fig. 4.3 Illustration by R. Oakley.

Fig. 4.4 Illustration by R. Oakley.

Fig. 4.5 Illustration by Gabrielle Stryker.

Fig. 4.6 Slightly modified for viewing in black and white from Adrian Raine and Yaling Yang, "Neural Foundations to Moral Reasoning and Antisocial Behavior," *Social Cognitive Affective Neuroscience* 1, no. 3 (2006): 203–13. By permission of Oxford University Press.

Fig. 5.1 Illustration courtesy R. L. Bruno of midbrain damage after poliovirus infection in humans based on 158 autopsies performed by Dr. David Bodian.

Fig. 5.2 Permission by author.

Fig. 5.3 Permission by author.

Fig. 5.4 Permission by author.

Fig. 5.5 Permission by author.

Fig. 5.6 Public domain.

Fig. 6.1 Permission by author.

Fig. 7.1 Mark H. Milstein/Northfoto.

Fig. 7.2 AP/Wide World Photos.

Fig. 8.1 "Diebold Variations" © 2004–06 Rand Careaga, http://homepage.mac.com/rcareaga/diebold/adworks.htm#the_weaselese.

Fig. 8.2 Illustrations loosely adapted from those of Henry Gray, *Anatomy of the Human Body* (Lea & Febiger, Philadelphia, 1918).

Fig. 8.3 Modified from image available courtesy National Institute on Drug Abuse, National Institutes of Health.

Fig. 8.4 Illustration by Gabrielle Stryker.

Fig. 8.5 From Robert O. Friedel, *Borderline Personality Disorder Demystified*, Copyright © 2004 Robert O. Friedel. Appears by permission of the publisher, Marlowe & Company, A Division of Avalon Publishing Group, Inc.

Fig. 8.6 Reprinted from S. C. Herpertz and others, "Evidence of Abnormal Amygdala Functioning in Borderline Personality Disorder: A Functional MRI Study," *Biological Psychiatry* 50: 295, Copyright (2001), with permission from the Society for Biological Psychiatry.

Fig. 8.7 From Robert O. Friedel, *Borderline Personality Disorder Demystified*, Copyright © 2004 Robert O. Friedel. Appears by permission of the publisher, Marlowe & Company, A Division of Avalon Publishing Group, Inc.

Fig. 8.8 Modified from Peter A. Johnson and others, "Understanding Emotion Regulation in Borderline Personality Disorder: Contributions of Neuroimaging," *Journal of Neuropsychiatry and Clinical Neurosciences* 15, no. 4 (2003). By permission of American Psychiatric Publishing, Inc.

Fig. 8.9 Slightly modified from F. D. Juengling, C. Schmahl, B. Heßlinger, D. Ebert, J. D. Bremner, J. Gostomzyk, M. Bohus, and K. Lieb, "Positron Emission Tomography in Female Patients with Borderline Personality Disorder," *Journal of Psychiatric Research* 37: 113, Copyright (2003), with permission from Elsevier.

Fig. 8.10 Slightly modified to clarify point of interest from F. D. Juengling, C. Schmahl, B. Heßlinger, D. Ebert, J. D. Bremner, J. Gostomzyk, M. Bohus, and K. Lieb, "Positron Emission Tomography in Female Patients with Borderline Personality Disorder," *Journal of Psychiatric Research* 37: 112, Copyright (2003), with permission from Elsevier.

Fig. 8.11 From Robert O. Friedel, *Borderline Personality Disorder Demystified*, Copyright © 2004 Robert O. Friedel. Appears by permission of the publisher, Marlowe & Company, a division of Avalon Publishing Group, Inc.

Fig. 9.1 Permission by author.

Fig. 9.2 Photograph by Harrison Forman, from the American Geographical Society Library, University of Wisconsin-Milwaukee Libraries.

Fig. 9.3 Photograph ca. 1915, provided courtesy of the Hoover Archives at Stanford (Joshua B. Powers Collection).

Fig. 9.4 AP/World Wide Photos.

Fig. 10.1 Public domain.

Fig. 11.1 AP/Wide World Photos.

Fig. 11.2 AP/Wide World Photos.

Fig. 12.1 Permission by author.

Fig. 12.2 Permission by author.

Fig. 12.3 Permission by author.

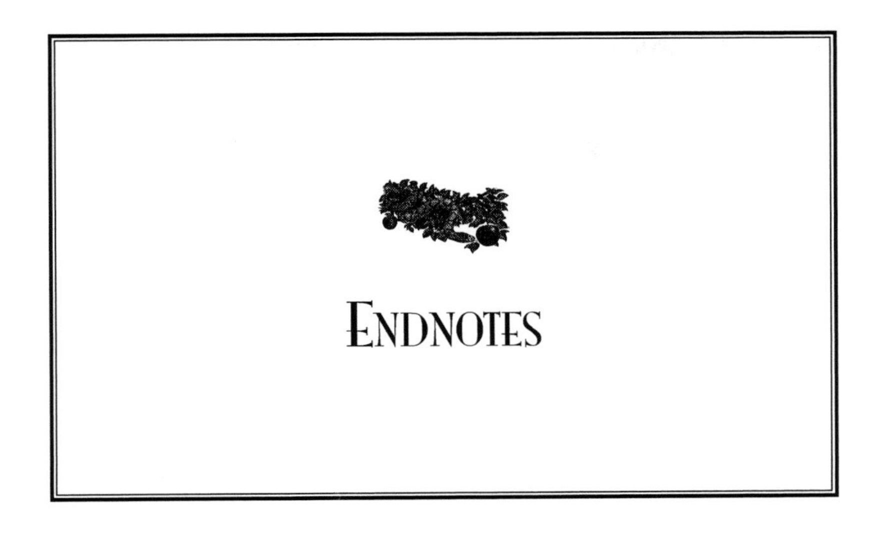

ENDNOTES

1. Paul Johnson, *A History of the American People* (New York: Harper-Perennial, 1999), p. 445. Via Hugh Hewitt, "The War of the World," www.hughhewitt.townhall.com, December 26, 2006 (accessed December 26, 2006).

INTRODUCTION

1. Sterling Seagrave and Peggy Seagrave, *Dragon Lady: The Life and Legend of the Last Empress of China*, 1st ed. (New York: Knopf, 1992), pp. 261–62. The book, which relies on largely English sources, lays the blame for the Empress's wicked reputation at the feet of ruthless hucksters whose largely fictional accounts of the Empress's life sold well.

2. Robert Conquest, *The Great Terror: A Reassessment* (New York: Oxford University Press, 1990); John Earl Haynes and Harvey Klehr, *In Denial: Historians, Communism & Espionage* (San Francisco: Encounter Books, 2003), p. 23. Martin Amis, *Koba the Dread: Laughter and the Twenty Million* (New York: Talk Miramax Books, 2002), p. 10.

3. Robert D. Hare, "Psychopaths: New Trends in Research," *Harvard Mental Health Letter* 12 (1995): 4–5.

4. Simon Sebag Montefiore, *Stalin: The Court of the Red Tsar* (New York: Knopf, 2004), pp. 422–23.

5. Ibid., p. 48.

6. Christopher Byron, *Testosterone, Inc.: Tales of CEOs Gone Wild* (Hoboken, NJ: John Wiley & Sons, 2004), pp. 25–26.

7. Ibid., pp. 18–19.

8. Pamela Fayerman, "Hitler's Defeat after Allied Invasion Attributed to Parkinson's Disease," *Vancouver Sun*, July 27, 1999; Betty Glad, "Why Tyrants Go Too Far: Malignant Narcissism and Absolute Power," *Political Psychology* 23, no. 1 (2002): 1–36; Colin Martindale, Nancy Hasenfus, and Dwight Hines, "Hitler: A Neurohistorical Formulation," *Confinia Psychiatrica* 19, no. 2 (1976): 106–16; Hyman Muslin, "Adolf Hitler: The Evil Self," *Psychohistory Review* (Sangamon State University) 20, no. 3 (1992): 251–70; Jerrold Post, *When Illness Strikes the Leader: The Dilemma of the Captive King* (New Haven, CT: Yale University Press, 1993), p. 51; Fritz Redlich, *Diagnosis of a Destructive Prophet* (New York: Oxford University Press, 1998), pp. 18, 231–35; David Ronfeldt, *Beware the Hubris-Nemisis Complex* (Santa Monica, CA: RAND, 1994); Ron Rosenbaum, *Explaining Hitler: The Search for the Origins of His Evil* (New York: Random House, 1999), p. 87; John C. Sonne, "On Tyrants as Abortion Survivors," *Journal of Prenatal & Perinatal Psychology & Health* 19, no. 2 (2004): 149–67.

9. As of February 21, 2007.

10. Glad, "Why Tyrants Go Too Far."

11. Tony Becher and Paul Trowler, *Academic Tribes and Territories*, 2nd ed. (Philadelphia, PA: Open University Press, 2001); Thomas Kuhn, *The Structure of Scientific Revolutions*, 3rd ed. (Chicago: University of Chicago Press, 1996).

12. Judith Rich Harris, *The Nurture Assumption* (New York: Free Press, 1998); Steven Pinker, *The Blank Slate: The Modern Denial of Human Nature* (New York: Viking, 2002).

Chapter 1: In Search of Machiavelli

1. Albert Szent-Gyorgyi, in L. J. Peter, *Peter's Quotations: Ideas for Our Time* (New York: Bantam Books, 1979), p. 123.

2. Richard Christie, "Why Machiavelli?" in *Studies in Machiavellianism*, ed. Richard Christie and Florence Geis (New York: Academic Press, 1970), p. 2.

3. David M. Buss, *Evolutionary Psychology* (Boston: Allyn & Bacon, 1999), pp. 24–30.

4. Annie M. Paul, "One Mean Renaissance Man," *Salon.com*, September 13, 1999, http://www.salon.com/books/it/1999/09/13/machiavelli/print.html (accessed April 1, 2005).

5. Christie and Geis, *Studies in Machiavellianism*.

6. Christie, "Why Machiavelli?" p. 4.

7. Theodor W. Adorno et al., *The Authoritarian Personality* (New York: Harper, 1950); Hans J. Eysenck, *The Psychology of Politics* (London: Routledge and K. Paul, 1954); *Kung-san Yang, The Book of Lord Shang*, trans. J. J. L. Duyvendak (Chicago: University of Chicago Press, 1928).

8. Niccolo Machiavelli, *The Prince. The Discourses* (New York: Modern Library, 1940).

9. Christie, "Why Machiavelli?" p. 8.

10. Ibid.

11. Paul, "One Mean Renaissance Man."

Chapter 2: Psychopathy

1. J. W. McHoskey, W. Worzel, and C. Szyarto, "Machiavellianism and Psychopathy," *Journal of Personality and Social Psychology* 74, no. 1 (1998): 192–210; Paul, "One Mean Renaissance Man."

2. Claire Valier, *Theories of Crime and Punishment*, Longman Criminology Series (Longman Publishing Group, 2002), pp. 14–18, 22.

3. S. D. Hart and R. D. Hare, "Psychopathy and Antisocial Personality Disorder," *Current Opinion in Psychiatry* 9 (1996): 129–32; Jorge Moll, Ricardo de Oliveira-Sousa, and Paul J. Eslinger, "Morals and the Human Brain: A Working Model," *NeuroReport* 14, no. 3 (2003): 299–305.

4. Aristotle, *Nicomachean Ethics*, II, I. Patricia Smith Churchland, "A Case Study in Neuroethics: The Nature of Moral Judgment," in *Neuroethics*, ed. Judy Illes (New York: Oxford University Press, 2006); K. Kafetsios and Eric LaRock, "Cognition and Emotion: Aristotelian Affinities with Contempo-

rary Emotion Research," *Theory and Psychology* 15 (2005): 639–57. S. Morse, "Brain Overclaim Syndrome and Criminal Responsibility: A Diagnostic Note," *Ohio State Journal of Criminal Law* 3 (2006): 397–412; Adrian Raine and Yaling Yang, "Neural Foundations to Moral Reasoning and Antisocial Behavior," *Social Cognitive Affective Neuroscience* 1, no. 3 (2006): 203–13.

5. J. Reid Meloy, *Violence Risk and Threat Assessment* (San Diego, CA: Specialized Training Services, 2000), p. 142.

6. Dan J. Stein, "The Neurobiology of Evil: Psychiatric Perspectives on Perpetrators," *Ethnicity & Health* 5, no. 3/4 (2000): 303–15.

7. David J. Cooke, "Psychopathy, Sadism and Serial Killing," in *Violence and Psychopathy*, ed. Adrian Raine and José Sanmartín (New York: Kluwer Academic/Plenum Publishers, 2001), p. 128; Robert D. Hare, David J. Cooke, and Stephen D. Hart, "Psychopathy and Sadistic Personality Disorder," in *Oxford Textbook of Psychopathy*, ed. Theodore Millon, Paul H. Blaney, and Roger D. Davis (New York: Oxford University Press, 1999), pp. 555–84.

8. Meloy, *Violence*, p. 113.

9. W. John Livesley, "Behavioral and Molecular Genetic Contributions to a Dimensional Classification of Personality Disorder," *Journal of Personality Disorders* 19, no. 2 (2005): 131–55.

10. Stein, "Neurobiology of Evil."

11. Hare, Cooke, and Hart, "Psychopathy."

12. Essi Viding et al., "Evidence for Substantial Genetic Risk for Psychopathy in 7-Year-Olds," *Journal of Child Psychology and Psychiatry* 46, no. 6 (2005): 592–97.

13. Judith Rich Harris, *No Two Alike* (New York: Norton, 2006).

14. Essi Viding, "Annotation: Understanding the Development of Psychopathy," *Journal of Child Psychology and Psychiatry* 45, no. 8 (2004): 1329–37.

Chapter 3: Evil Genes

1. Robert Plomin, *Nature and Nurture: An Introduction to Human Behavioral Genetics* (Belmont, CA: Wadworth/Thomson Learning, 2004), p. 67. See also Lisabeth F. DiLalla and Irving I. Gottesman, *Behavior*

Genetics Principles, 1st ed. (Washington, DC: American Psychological Association, 2004).

2. Plomin, *Nature and Nurture*, p. 112. L. B. Koenig et al., "Genetic and Environmental Influences on Religiousness: Findings for Retrospective and Current Religiousness Ratings," *Journal of Personality* 73, no. 2 (2005): 471–88.

3. K. P. Harden et al., "Marital Conflict and Conduct Disorder in Children-of-Twins," *Child Development* 78, no. 1 (2007): in press.

4. Joshua Roffman et al., "Neuroimaging-Genetic Paradigms: A New Approach to Investigate the Pathophysiology and Treatment of Cognitive Deficits in Schizophrenia," *Harvard Review of Psychiatry* 14, no. 2 (2006): 78–91.

5. Richard Redon et al., "Global Variation in Copy Number in the Human Genome," *Nature* 444, no. 7118 (2006): 444–54.

6. David J. Buller, *Adapting Minds: Evolutionary Psychology and the Persistent Quest for Human Nature* (Cambridge, MA: MIT Press, 2005), p. 437.

7. E. Meshorer et al., "SC35 Promotes Sustainable Stress-Induced Alternative Splicing of Neuronal Acetylcholinesterase mRNA," *Molecular Psychiatry* 10, no. 11 (2005): 985–97; E. Meshorer and H. Soreq, "Virtues and Woes of AChE Alternative Splicing in Stress-Related Neuropathologies," *Trends in Neurosciences* 29, no. 4 (2006): 216–24; R. Valgardsdottir et al., "Structural and Functional Characterization of Noncoding Repetitive RNAs Transcribed in Stressed Human Cells," *Molecular Biology of the Cell* 16, no. 6 (2005): 2597–2604.

8. John Rose, *Human Stress and the Environment: Health Aspects* (Taylor & Francis, 1994), pp. 1, 8, 133.

9. Sridhar Prathikanti and Daniel R. Weinberger, "Psychiatric Genetic —the New Era: Genetic Research and Some Clinical Implications," *British Medical Bulletin* 73 and 74 (2005): 107–22. But see Robert Plomin, "Finding Genes in Child Psychology and Psychiatry: When Are We Going to Be There?" *Journal of Child Psychology and Psychiatry* 46, no. 10 (2005): 1030–38, for skepticism regarding this approach.

10. Daniel Weinberger, in discussion with the author, June 1, 2006.

11. C. H. Chen et al., "Brain Imaging Correlates of Depressive Symptom Severity and Predictors of Symptom Improvement after Antidepressant Treatment," *Biological Psychiatry* [Epub ahead of print] (2007).

12. J. L. Roffman et al., "Neuroimaging and the Functional Neuroanatomy of Psychotherapy," *Psychological Medicine* 35, no. 10 (2005): 1385–98.

13. Weinberger, in discussion with the author, June 1, 2006.

14. D. L. Murphy et al., "Brain Serotonin Neurotransmission," *Journal of Clinical Psychiatry* 59, Suppl 15 (1998): 4–12.

15. Robert Plomin, M. J. Owen, and P. McGuffin, "The Genetic Basis of Complex Human Behaviors," *Science* 264, no. 1733–39 (1994); A. Reif and K. P. Lesch, "Toward a Molecular Architecture of Personality," *Behavioural Brain Research* 139, no. 1–2 (2003): 1–20.

16. Berend Olivier, "Serotonin and Aggression," *Annals of the New York Academy of Sciences* 1036 (2004): 382–92.

17. A. S. New et al., "Impulsive Aggression Associated with HTR1B Genotype in Personality Disorder," in *Annual Meeting of the American Psychiatric Association* NR 388 (1999); A. S. New et al., "Serotonin Related Genotype and Impulsive Aggression," in *Annual Meeting of the Society of Biological Psychiatry* 45, Abstract #387 (1999); F. Rybakowski et al., "The 5-HT2A-1438 A/G and 5-HTTLPR Polymorphisms and Personality Dimensions in Adolescent Anorexia Nervosa: Association Study," *Neuropsychobiology* 53, no. 1 (2006): 33–39.

18. T. Iidaka et al., "A Variant C178T in the Regulatory Region of the Serotonin Receptor Gene HTR3A Modulates Neural Activation in the Human Amygdala," *Journal of Neuroscience* 25, no. 27 (2005): 6460–66.

19. X. Ni et al., "Association between Serotonin Transporter Gene and Borderline Personality Disorder," *Journal of Psychiatric Research* 40, no. 5 (2006): 448–53; L. Pezawas et al., "5-HTTLPR Polymorphism Impacts Human Cingulate-Amygdala Interactions: A Genetic Susceptibility Mechanism for Depression," *Nature Neuroscience* 8, no. 6 (2005): 828–34; H. Steiger et al., "The 5HTTLPR Polymorphism, Psychopathologic Symptoms, and Platelet [3H-] Paroxetine Binding in Bulimic Syndromes," *International Journal of Eating Disorders* 37, no. 1 (2005): 57–60.

20. Ahmad R. Hariri et al., "A Susceptibility Gene for Affective Disorders and the Response of the Human Amygdala," *Archives of General Psychiatry* 62, no. 2 (2005): 146–54; Pezawas et al., "5-HTTLPR."

21. S. Eddahibi et al., "Serotonin Transporter Overexpression Is Responsible for Pulmonary Artery Smooth Muscle Hyperplasia in Primary Pulmonary Hypertension," *Journal of Clinical Investigation* 108, no. 8 (2001): 1141–50; S. Eddahibi et al., "Hyperplasia of Pulmonary Artery Smooth Muscle Cells Is Causally Related to Overexpression of the Serotonin Transporter in Primary Pulmonary Hypertension," *Chest* 121, no. 3 (2002): 97S–98S.

22. R. Cacabelos et al., "Molecular Genetics of Alzheimer's Disease and Aging," *Methods and Findings in Experimental Clinical Pharmacology* 27,

no. Suppl A (2005): 1–573; D. K. Lahiri, C. Ghosh, and Y. W. Ge, "A Proximal Gene Promoter Region for the Beta-Amyloid Precursor Protein Provides a Link between Development, Apoptosis, and Alzheimer's Disease," *Annals of the New York Academy of Sciences* 1010 (2003): 643–47.

23. Reinaldo B. Oriá et al., "Role of Apolipoprotein E4 in Protecting Children against Early Childhood Diarrhea Outcomes and Implications for Later Development," *Medical Hypotheses* 68, no. 5 (2007): 1099–1107.

24. Christian R. A. Mondadori et al., "Better Memory and Neural Efficiency in Young Apolipoprotein Eε4 Carriers," *Cerebral Cortex*, epub ahead of print (2006): bhl103.

25. Nancy Touchette, "Gene Variation Affects Memory," *Genome News Network*, 2003, http://www.genomenewsnetwork.org/articles/08_03/memory.shtml (accessed May 11, 2006).

26. S. Sen et al., "A BDNF Coding Variant Is Associated with the NEO Personality Inventory Domain Neuroticism, a Risk Factor for Depression," *Neuropsychopharmacology* 28, no. 2 (2003): 397–401.

27. Jim Phelps, "Connecting Anxiety and Depression via the Serotonin Transporter Gene," *PsychEducation.org: Extensive Mental Health Information on Specific Topics*, 2006, http://www.psycheducation.org/mechanism/4WhyShortsLongs.htm (accessed July 10, 2006).

28. M. Ribases et al., "Contribution of NTRK2 to the Genetic Susceptibility to Anorexia Nervosa, Harm Avoidance and Minimum Body Mass Index," *Molecular Psychiatry* 10, no. 9 (2005): 851–60.

29. J. Strauss et al., "BDNF and COMT Polymorphisms: Relation to Memory Phenotypes in Young Adults with Childhood-Onset Mood Disorder," *Neuromolecular Medicine* 5, no. 3 (2004): 181–92.

30. A. Thapar et al., "Catechol O-methyltransferase Gene Variant and Birth Weight Predict Early-Onset Antisocial Behavior in Children with Attention-Deficit/Hyperactivity Disorder," *Archives of General Psychiatry* 62, no. 11 (2005): 1275–78.

31. Ke Xu and D. Goldman, "Catechol-O-methyltransferase Genotype, Intermediate Phenotype, and Psychiatric Disorders," in *Cell Biology of Addiction*, ed. Bertha Madras et al. (Cold Spring Harbor, NY: Cold Spring Harbor Laboratory Press, 2005), pp. 29–44.

32. K. Xu, M. Ernst, and D. Goldman, "Imaging Genomics Applied to Anxiety, Stress Response, and Resiliency," *Neuroinformatics* 4, no. 1 (2006): 51–64.

33. Xu and Goldman, "Catechol-O-methyltransferase Genotype."

34. Michael N. Smolka et al., "Catechol-O-methyltransferase val158met Genotype Affects Processing of Emotional Stimuli in the Amygdala and Prefrontal Cortex," *Journal of Neuroscience* 25, no. 4 (2005): 836–43.

35. Ibid.; Xu, Ernst, and Goldman, "Imaging Genomics."

36. Xu, Ernst, and Goldman, "Imaging Genomics."

37. I. W. Craig, "The Role of Monoamine Oxidase A, MAOA, in the Aetiology of Antisocial Behaviour: The Importance of Gene-Environment Interactions," *Novartis Foundation Symposium* 268 (2005): 227–37; discussion 237–41, 242–53; A. Serretti et al., "Temperament and Character in Mood Disorders: Influence of DRD4, SERTPR, TPH and MAO-A Polymorphisms," *Neuropsychobiology* 53, no. 1 (2006): 9–16.

38. Christian P. Jacob et al., "Cluster B Personality Disorders Are Associated with Allelic Variation of Monoamine Oxidase A Activity," *Neuropsychopharmacology* 30, no. 9 (2005): 1711–18.

39. Andreas Meyer-Lindenberg et al., "Neural Mechanisms of Genetic Risk for Impulsivity and Violence in Humans," *Proceedings of the National Academy of Sciences* 103, no. 16 (2006): 6269–74.

40. Ibid.

41. H. G. Brunner et al., "Abnormal Behavior Associated with a Point Mutation in the Structural Gene for Monoamine Oxidase A," *Science* 262, no. 5133 (1993): 578–80.

42. Avshalom Caspi et al., "Role of Genotype in the Cycle of Violence in Maltreated Children," *Science* 297 (2002): 851–54.

43. Meshorer et al., "SC35"; Meshorer and Soreq, "Virtues and Woes."

44. T. E. Moffitt, "The New Look of Behavioral Genetics in Developmental Psychopathology: Gene-Environment Interplay in Antisocial Behaviors," *Psychological Bulletin* 131, no. 4 (2005): 533–54.

45. S. M. Brown et al., "A Regulatory Variant of the Human Tryptophan Hydroxylase-2 Gene Biases Amygdala Reactivity," *Molecular Psychiatry* 10, no. 9 (2005): 884–88, 805; Serretti et al., "Temperament"; G. Zaboli et al., "Tryptophan Hydroxylase-1 Gene Variants Associate with a Group of Suicidal Borderline Women," *Neuropsychopharmacology* 31, no. 9 (2006): 1982–90.

46. J. Auerbach et al., "Dopamine D4 Receptor (D4DR) and Serotonin Transporter (5-HTTLPR) Polymorphisms in the Determination of Temperament in 2-Month-Old Infants," *Molecular Biology* 4 (1999): 369–73.

47. "Common Gene Version Optimizes Thinking—but with a Possible

Downside," *NIH News*, February 8, 2007, http://www.nih.gov/news/pr/feb 2007/nimh-08.htm (accessed February 25, 2007).

48. A. Knafo et al., "Individual Differences in Allocation of Funds in the Dictator Game Associated with Length of the Arginine Vasopressin 1a Receptor RS3 Promoter Region and Correlation between RS3 Length and Hippocampal mRNA," *Genes, Brain, and Behavior* 7, no. 3 (2008): 266–75.

49. D. T. Lykken et al., "Emergenesis: Genetic Traits That May Not Run in Families," *American Psychologist* 47, no. 12 (1992): 1565–77.

50. R. Bachner-Melman et al., "AVPR1a and SLC6A4 Gene Polymorphisms Are Associated with Creative Dance Performance," *PLoS Genetics* 1, no. 3 (2005): e42; R. Oerter, "Biological and Psychological Correlates of Exceptional Performance in Development," *Annals of the New York Academy of Sciences* 999 (2003): 451–60.

51. E. F. Torrey and R. H. Yolken, "*Toxoplasma gondii* and Schizophrenia," *Emerging Infectious Diseases* 9, no. 11 (2003): 1375–80.

52. Ni et al., "Association between Serotonin"; A. E. Skodol et al., "The Borderline Diagnosis II: Biology, Genetics, and Clinical Course," *Biological Psychiatry* 51, no. 12 (2002): 951–63.

53. Norbert Wiener, *Ex-Prodigy: My Childhood and Youth* (Cambridge, MA: MIT Press, 1953), p. 158.

54. Larry J. Siever, Harold W. Koenigsberg, and Deidre Reynolds, "Neurobiology of Personality Disorders: Implications for a Neurodevelopmental Model," in *Neurodevelopmental Mechanisms in Psychopathology*, ed. Dante Cicchetti and Elaine Walker (New York: Cambridge University Press, 2003), pp. 416–17; A. E. Skodol and Donna S. Bender, "Why Are Women Diagnosed Borderline More Than Men?" *Psychiatric Quarterly* 74, no. 4 (2003): 349–60; S. Torgersen, "Genetics in Borderline Conditions," *Acta Psychiatrica Scandinavica Supplementum* 379 (1994): 19–25.

55. Ibid., pp. 11–12.

56. Nicholas Wade, "Researchers Say Intelligence and Diseases May Be Linked in Ashkenazic Genes," *New York Times*, June 3, 2005.

Chapter 4: Using Medical Imaging to Understand Psychopaths

1. "Functional Families, Dysfunctional Brains," *Science Daily*, April 10, 1998, http://www.sciencedaily.com/releases/1998/04/980410101830.htm (accessed November 24, 2005).

2. An excellent article about both the strengths and weaknesses of imaging approaches can be found in Malcolm Gladwell, "The Picture Problem: Mammography, Air Power, and the Limits of Looking," *New Yorker*, December 13, 2004.

3. Kent A. Kiehl et al., "Limbic Abnormalities in Affective Processing by Criminal Psychopaths as Revealed by Functional Magnetic Resonance Imaging," *Biological Psychiatry* 50, no. 9 (2001): 677–84.

4. S. W. Anderson et al., "Impairment of Social and Moral Behavior Related to Early Damage in Human Prefrontal Cortex," *Nature Neuroscience* 2 (1999): 1032–37; A. Bechara et al., "Insensitivity to Future Consequences Following Damage to Human Prefrontal Cortex," *Cognition* 50 (1994): 7–15; A. R. Damasio, D. Tranel, and H. Damasio, "Individuals with Sociopathic Behavior Caused by Frontal Damage Fail to Respond Autonomically to Social Stimuli," *Behavioural Brain Research* 41 (1990): 81–94.

5. Kent Kiehl in communication with the author, July 5, 2005.

6. Adrian Raine et al., "Corpus Callosum Abnormalities in Psychopathic Antisocial Individuals," *Archives of General Psychiatry* 60 (2003): 1134–42.

7. Adrian Raine and Yaling Yang, "The Neuroanatomical Bases of Psychopathy: A Review of Brain Imaging Findings," in *Handbook of Psychopathy*, ed. Christopher J. Patrick (New York: Guilford Press, 2006), pp. 278–95.

8. G. P. Shumyatsky et al., "*Stathmin*, a Gene Enriched in the Amygdala, Controls Both Learned and Innate Fear," *Cell* 123, no. 4 (2005): 697–709.

9. James Blair, Derek Mitchell, and Karina Blair, *The Psychopath: Emotion and the Brain* (Malden, MA: Blackwell Publishing, 2005), p. 139; R. J. Blair, "Neurobiological Basis of Psychopathy," *British Journal of Psychiatry* 182 (2003): 5–7; F. Schneider et al., "Functional Imaging of Conditioned Aversive Emotional Responses in Antisocial Personality Disorder," *Neuropsychobiology* 42 (2000): 192–201.

10. Sharon Ishikawa and Adrian Raine, "Prefrontal Deficits and Antisocial Behavior: A Causal Model," in *Causes of Conduct Disorder and Juvenile Delinquency*, ed. Avshalom Caspi, Benjamin B. Lahey, and Terrie Moffitt (New York: Guilford Press, 2003), pp. 277–304.

11. "Functional Families, Dysfunctional Brains."

12. Adrian Raine et al., "Reduced Prefrontal and Increased Subcortical Brain Functioning Assessed Using Positron Emission Tomography in Predatory and Affective Murderers," *Behavioral Sciences and the Law* 16 (1998): 319–32.

13. Bechara et al., "Insensitivity"; A. Bechara et al., "Different Contributions of the Human Amygdala and Ventromedial Prefrontal Cortex to Decision-Making," *Journal of Neuroscience* 19, no. 13 (1999): 5473–81; L. K. Fellows and M. J. Farah, "Different Underlying Impairments in Decision-Making Following Ventromedial and Dorsolateral Frontal Lobe Damage in Humans," *Cerebral Cortex* 15, no. 1 (2005): 58–63.

14. This example is after that of Meloy, *Violence*.

15. Adrian Raine, "Psychopathy, Violence, and Brain Imaging," in *Violence and Psychopathy*, ed. Adrian Raine and José Sanmartín (New York: Kluwer Academic/Plenum Publishers, 2001), pp. 35–56.

16. Adrian Raine, "Murderous Minds: Can We See the Mark of Cain?" *Cerebrum* 1, no. 1 (1999): 15–30.

17. Kent A. Kiehl et al., "Temporal Lobe Abnormalities in Semantic Processing by Criminal Psychopaths as Revealed by Functional Magnetic Resonance Imaging," *Psychiatry Research: Neuroimaging* 130, no. 1 (2004): 27–42.

18. Terrence W. Deacon, *The Symbolic Species* (New York: W. W. Norton & Company, 1997), pp. 427, 459; Andreas Meyer-Lindenberg et al., "Neural Correlates of Genetically Abnormal Social Cognition in Williams Syndrome," *Nature Neuroscience* 8, no. 8 (2005): 991–95.

19. David Dobbs, "The Gregarious Brain," *New York Times Magazine*, July 8, 2007.

20. Jef Allbright, "Scientists Watch the Brain Wrestle with Moral Dilemmas," *Jef's Web Files*, 2004, http://www.jefallbright.net/node/2691 (accessed December 29, 2005); Jorge Moll et al., "The Neural Correlates of Moral Sensitivity: A Functional Magnetic Resonance Imaging Investigation of Basic and Moral Emotions," *Journal of Neuroscience* 22, no. 7 (2002): 2730–36.

21. Dharol Tankersley, C. Jill Stowe, and Scott A. Huettel, "Altruism Is

Associated with an Increased Neural Response to Agency," *Nature Neuroscience* 10 (2007): 150–51.

22. E. J. Mundell, "Why Do Good? Brain Study Offers Clues," *HealthDay*, January 22, 2007, http://www.healthday.com/Article.asp?AID=601147 (accessed January 23, 2007).

23. H. Takahashi et al., "Brain Activation Associated with Evaluative Processes of Guilt and Embarrassment: An fMRI Study," *Neuroimage* 23, no. 3 (2004): 967–74. See also S. Berthoz et al., "Affective Response to One's Own Moral Violations," *Neuroimage* 31, no. 2 (2006): 945–50; E. C. Finger et al., "Caught in the Act: The Impact of Audience on the Neural Response to Morally and Socially Inappropriate Behavior," *Neuroimage* 33, no. 1 (2006): 414–21; C. L. Harenski and S. Hamann, "Neural Correlates of Regulating Negative Emotions Related to Moral Violations," *Neuroimage* 30, no. 1 (2006): 313–24; Moll, Oliveira-Sousa, and Eslinger, "Morals and the Human Brain."

24. D. J. Stein and D. Kaminer, "Forgiveness and Psychopathology: Psychobiological and Evolutionary Underpinnings," *CNS Spectrums* 11, no. 2 (2006): 87–89.

25. Raine and Yang, "Neural Foundations." This paper contains a comprehensive review of virtually all neurological imaging findings related to antisocial behavior, which is much more extensive than the basic findings presented here.

26. Paul J. Frick and Monica A. Marsee, "Psychopathy and Developmental Pathways to Antisocial Behavior in Youth," in *Handbook of Psychopathy*, ed. Christopher J. Patrick (New York: Guilford Press, 2005), pp. 353–74.

27. Meloy, *Violence*, pp. 121–22.

28. "Psychologist Adds Scientific Insight to Loaded Label of 'Psychopath,'" Physorg.com, June 28, 2006, http://www.physorg.com/news70728146.html (accessed July 1, 2006).

29. Joseph Newman, in correspondence with the author, February 1, 2007.

30. Sandra Blakeslee, "Cells That Read Minds," *New York Times*, January 10, 2006.

31. R. James Blair and Karina S. Perschardt, "Empathy: A Unitary Circuit or a Set of Dissociable Neuro-Cognitive Systems?" *Behavioral and Brain Sciences* 25 (2002): 27–28.

32. Deacon, *Symbolic Species*, p. 402; Linda Mealey and Stuart Kinner, "The Perception-Action Model of Empathy and Psychopathic 'Cold-Heartedness,'" *Behavioral and Brain Sciences* 25 (2002): 42–43.

33. Raine and Yang, "Neuroanatomical Bases"; Yaling Yang et al., "Volume Reduction in Prefrontal Gray Matter in Unsuccessful Criminal Psychopaths," *Biological Psychiatry* 57, no. 10 (2005): 1103–1108.

34. Jamie Talan, "Lying Liars," *Scientific American Mind* (April 2006): 8; Y. Yang et al., "Prefrontal White Matter in Pathological Liars," *British Journal of Psychiatry* 187 (2005): 320–25.

35. Lisa Desai, "'Corporate Psychopaths' at Large," *CNN*, August 26, 2004, http://edition.cnn.com/2004/BUSINESS/08/26/corporate.psychopaths/ (accessed September 28, 2005). See also Paul Babiak and Robert Hare, *Snakes in Suits: When Psychopaths Go to Work* (New York: HarperCollins, 2006).

36. Desai, "'Corporate Psychopaths' at Large."

Chapter 5: Insights from My Sister's Love Letters

1. Richard L. Bruno, *The Polio Paradox: Uncovering the Hidden History of Polio to Understand and Treat "Post-Polio Syndrome" and Chronic Fatigue* (New York: Warner Books, 2002), pp. 47–54.

2. Richard Bruno in correspondence with the author, February 16, 2007.

3. R. L. Bruno et al., "The Neuroanatomy of Post-Polio Fatigue," *Archives of Physical Medicine and Rehabilitation* 75, no. 5 (1994): 498– 504; R. L. Bruno, N. M. Frick, and J. Cohen, "Polioencephalitis, Stress, and the Etiology of Post-Polio Sequelae," *Orthopedics* 14, no. 11 (1991): 1269–76.

4. Richard Bruno in correspondence with the author, February 18, 2007.

5. Bruno, Frick, and Cohen, "Polioencephalitis."

6. E. Meyer, "Psychological Considerations in a Group of Children with Poliomyelitis," *Journal of Pediatrics* 31 (1947): 34–48; John R. Paul, *A History of Poliomyelitis* (New Haven: Yale University Press, 1971). Meyer's study also examined older and younger children—52 in total.

7. Bruno, *Polio Paradox*, pp. 105–106.

8. S. J. Creange and R. L. Bruno, "Compliance with Treatment for Postpolio Sequelae: Effect of Type A Behavior, Self-Concept, and Loneliness," *American Journal of Physical Medicine & Rehabilitation* 76, no. 5 (1997): 378–82.

9. Meyer, "Psychological Considerations."

10. R. L. Bruno and N. M. Frick, "The Psychology of Polio as Prelude to Post-Polio Sequelae: Behavior Modification and Psychotherapy," *Orthopedics* 14, no. 11 (1991): 1185–93.

11. Ibid.

Chapter 6: The Connection between Machiavellianism and Personality Disorders

1. I. I. Gottesman, "Defining Genetically Informed Phenotypes for the DSM-V," in *Descriptions and Prescriptions: Values, Mental Disorders and the DSMs*, ed. J. Z. Sadler (Baltimore, MD: Johns Hopkins University Press, 2002).

2. A. C. Ruocco, "Re-evaluating the Distinction between Axis I and Axis II Disorders: The Case of Borderline Personality Disorder," *Journal of Clinical Psychology* 61, no. 12 (2005): 1509–23.

3. Roger K. Blashfield and Vincent Intoccia, "Growth of the Literature on the Topic of Personality Disorders," *American Journal of Psychiatry* 157 (2000): 472–73.

4. John W. McHoskey, "Machiavellianism and Personality Dysfunction," *Personality and Individual Differences* 31 (2001): 791–98.

5. Robert O. Friedel, *Borderline Personality Disorder Demystified* (New York: Marlowe & Company, 2004), p. 24.

6. M. Bohus, C. Schmahl, and K. Lieb, "New Developments in the Neurobiology of Borderline Personality Disorder," *Current Psychiatry Reports* 6 (2004): 43–50; M. F. Lenzenweger et al., "Detecting Personality Disorders in a Nonclinical Population," *Archives of General Psychiatry* 54 (1997): 345–51.

7. Jerold J. Kreisman and Hal Straus, *Sometimes I Act Crazy* (New York: John Wiley & Sons, Inc., 2004), p. 4; Paul T. Mason and Randi Kreger, *Stop Walking on Eggshells: Taking Your Life Back When Someone You Care about Has Borderline Personality Disorder* (Oakland, CA: New Harbinger Publications, 1998).

8. Peter A. Johnson et al., "Understanding Emotion Regulation in Borderline Personality Disorder: Contributions of Neuroimaging," *Journal of Neuropsychiatry and Clinical Neurosciences* 15, no. 4 (2003): 397–402.

9. Joel Paris, "Antisocial and Borderline Personality Disorders: Two Separate Diagnoses or Two Aspects of the Same Psychopathology?" *Comprehensive Psychiatry* 38, no. 4 (1997): 237–42.

10. Information in the box is adapted from Steven Leichter and Elizabeth Dreelin, "Borderline Personality Disorder and Diabetes: A Potentially Ominous Mix," *Clinical Diabetes* 23 (2005): 101–103; Mason and Kreger, *Eggshells*; Timothy J. Trull and Christine A. Durrett, "Categorical and Dimensional Models of Personality Disorder," *Annual Review of Clinical Psychology* 1 (2005): 355–80.

11. Barry Kiehn and Michaela Swales, "An Overview of Dialectical Behaviour Therapy in the Treatment of Borderline Personality Disorder," *Psychiatry On-Line*, http://www.priory.com/dbt.htm (accessed January 25, 2007).

12. Joel Paris, "Borderline Personality Disorder," *James Wood, PhD*, http://www.jwoodphd.com/borderline_personality_disorder.htm (accessed December 3, 2006).

13. Kreisman and Straus, *Crazy*.

14. Ibid., p. v.

15. Mason and Kreger, *Eggshells*, p. 46.

16. Ibid., p. 47.

17. Alan Butterfield, "Pin Thin: 105 lbs . . . and Getting Thinner!" *National Enquirer*, October 17, 2005, pp. 4–5; C. A. Magill, "The Boundary between Borderline Personality Disorder and Bipolar Disorder: Current Concepts and Challenges," *Canadian Journal of Psychiatry* 49, no. 8 (2004): 551–56; Jim Phelps, "What's the Difference between Bipolar Disorder and 'Borderline Personality Disorder'?" *PsychEducation.org*, 2006, http://www.psych education.org/depression/borderline.htm (accessed February 2, 2007).

18. Kreisman and Straus, *Crazy*, p. 50.

19. Leichter and Dreelin, "Borderline Personality Disorder."

20. H. W. Koenigsberg et al., "Are the Interpersonal and Identity Disturbances in the Borderline Personality Disorder Criteria Linked to the Traits of Affective Instability and Impulsivity?" *Journal of Personality Disorders* 15, no. 4 (2001): 358–70.

21. Mason and Kreger, *Eggshells*, p. 36.

22. Koenigsberg et al., "Interpersonal."

23. Mason and Kreger, *Eggshells*, p. 43.

24. Ibid., pp. 44, 45.

25. Ibid., p. 45.

26. From an Amazon.com review of *Stop Walking on Eggshells* entitled "THIS BOOK SAVED MY VERY SANITY!!! A revelation . . . ," October 8, 2005, by Christopher Francis.

27. Theodore L. Dorpat, *Gaslighting, the Double Whammy, Interrogation, and Other Methods of Covert Control in Psychotherapy and Analysis* (Northvale, NJ: Jason Aronson, 1996), p. 31.

28. Victor Santoro, *Gaslighting: How to Drive Your Enemies Crazy* (Port Townsend, WA: Loompanics, 1994), p. 5.

29. "Brain Structures 'Tune In' to Rhythms to Coordinate Activity," *MIT*, 2005, http://www.scienceblog.com/cms/brain_structures_tune_in_to_rhythms_to_coordinate_activity_9405 (accessed December 3, 2005); Matthew W. Jones and Matthew A. Wilson, "Theta Rhythms Coordinate Hippocampal-Prefrontal Interactions in a Spatial Memory Task," *PLoS Biology* 3, no. 12 (2005).

30. R. A. Moore et al., "Theta Phase Locking across the Neocortex Reflects Cortico-Hippocampal Recursive Communication during Goal Conflict Resolution," *International Journal of Psychophysiology* 60, no. 3 (2006): 260–73.

31. T. G. Guthiel, "Suicide, Suicide Litigation, and Borderline Personality Disorder," *Journal of Personality Disorders* 18 (2004): 248–56.

32. Roberto Ceniceros, "Personality Disorder Presents Disability Challenge," *Crain's Detroit Business*, 2005, http://www.crainsdetroit.com/cgi-bin/article.pl?articleId=27824 (accessed October 8, 2005).

Chapter 7: Slobodan Milosevic: The Butcher of the Balkans

1. Jeffrey Fleishman, "Slobodan Milosevic 1941–2006," *Los Angeles Times*, March 12, 2006.

2. Adam LeBor, *Milosevic: A Biography* (New Haven: Yale University Press, 2002), p. 111.

3. Ibid., p. 249.

4. Ibid., p. 43.

5. Ibid., p. 41.

6. Dusko Doder and Louise Branson, *Milosevic: Portrait of a Tyrant* (New York: The Free Press, 1999), pp. 141, 142.

7. Ibid., p. 26.

8. Ibid., pp. 26, 42, 137, 138.

9. Fleishman, "Slobodan Milosevic."

10. Doder and Branson, *Milosevic*, p. 20.

11. Friedel, *Demystified*, p. 15.

12. Guy Lesser, "War Crime and Punishment: What the United States Could Learn from the Milosevic Trial," *Harper's Magazine*, January 2004, pp. 37–52.

13. T. Wilkinson-Ryan and D. Westen, "Identity Disturbance in Borderline Personality Disorder: An Empirical Investigation," *American Journal of Psychiatry* 157 (2000): 528–41; Rebekah Bradley and Drew Westen, "The Psychodynamics of Borderline Personality Disorder: A View from Developmental Psychopathology," *Development and Psychopathology* 17, no. 4 (2005): 927–57.

14. S. Akhtar, "The Syndrome of Identity Diffusion," *American Journal of Psychiatry* 141 (1984): 1381–85.

15. Slavoljub Djukic, *Milosevic and Marković*, trans. Alex Dubinsky (London: McGill-Queen's University Press, 2001), p. 160.

16. After K. Lieb et al., "Borderline Personality Disorder," *Lancet* 364, no. 9432 (2004): 453–61. American Psychiatric Association, *Diagnostic and Statistical Manual of Mental Disorders, Fourth Edition, Text Revision* (Washington, DC: American Psychiatric Association, 2000), p. 654.

17. Djukic, *Milosevic and Marković*, p. 163.

18. Doder and Branson, *Milosevic*, p. 103.

19. Djukic, *Milosevic and Marković*, p. 161.

20. Doder and Branson, *Milosevic*, p. 272.

21. Ibid., p. 52.

22. Djukic, *Milosevic and Marković*, p. 162.

23. V. L. Willour et al., "Attempted Suicide in Bipolar Disorder Pedigrees: Evidence for Linkage to 2p12," *Biological Psychiatry* 61, no. 5 (2007): 725–27.

24. Doder and Branson, *Milosevic*, p. 9.

25. Djukic, *Milosevic and Marković*, p. 5.

26. Doder and Branson, *Milosevic*, p. 154.

27. Ibid., pp. 137, 138.

28. Djukic, *Milosevic and Marković*, p. 69.

29. LeBor, *Milosevic*, p. 219.

30. Doder and Branson, *Milosevic*, pp. 137, 138.

31. "The Lesson of Slobodan Milosevic's Trial and Tribulation," *Economist*, February 15, 2003, pp. 45–46.

32. Robert L. Spitzer and Jerome C. Wakefield, "DSM-IV Diagnostic Criterion for Clinical Significance: Does It Help Solve the False Positives Problem?" *American Journal of Psychiatry* 156, no. 12 (1999): 1856–65. The phrase "clinically significant" was also included "to exclude from the DSM listing those disorders that clinicians were unlikely to see or treat"—a reason exposed by Spitzer and Wakefield as being inherently flawed. The DSM-IV itself notes the difficulty of deducing whether a diagnostic criterion is met, writing that it "is inherently a difficult clinical judgment." Drew Westen and Laura Arkowitz-Westen, "Limitations of Axis II in Diagnosing Personality Pathology in Clinical Practice," *American Journal of Psychiatry* 155, no. 12 (1998): 1767–72.

33. B. J. Board and Katarina F. Fritzon, "Disordered Personalities at Work," *Psychology, Crime and Law* 11 (2005): 17–32; Joshua Wolf Shenk, *Lincoln's Melancholy: How Depression Challenged a President and Fueled His Greatness* (New York: Houghton Mifflin Company, 2005), p. 24.

34. Kreisman and Straus, *Crazy*, p. 9. For an interesting proposed solution to this problem, see Drew Westen, Jonathan Shedler, and Rebekah Bradley, "A Prototype Approach to Personality Disorder Diagnosis," *American Journal of Psychiatry* 163, no. 5 (2006): 846–56.

35. Friedel, *Demystified*, p. 2.

36. Michael S. McCloskey, K. Luan Phan, and Emil F. Coccaro, "Neuroimaging and Personality Disorders," *Current Psychiatry Reports* 7 (2005): 65–72.

37. LeBor, *Milosevic*, p. 198.

38. "That's What They Want You to Believe: Why Are Conspiracy Theories So Popular?" *Economist*, December 21, 2002, pp. 45–46.

39. Doder and Branson, *Milosevic*, p. 253.

40. LeBor, *Milosevic*, p. 184.

41. Ibid., p. 267.

42. Doder and Branson, *Milosevic*, pp. 265, 261.

43. M. C. Anderson et al., "Neural Systems Underlying the Suppression of Unwanted Memories," *Science* 303, no. 5655 (2004): 232–35.

44. Lenard Cohen, "Unravelling the Milosevic Mystery," *Simon Fraser University News*, 2001, http://www.sfu.ca/mediapr/sfnews/2001/May17/cohen.html (accessed June 12, 2005).

45. LeBor, *Milosevic*, p. 197.

46. Doder and Branson, *Milosevic*, p. 24.

47. N. H. Donegan et al., "Amygdala Hyperreactivity in Borderline Personality Disorder: Implications for Emotional Dysregulation," *Biological Psychiatry* 54, no. 11 (2003): 1284–93.

48. Bohus, Schmahl, and Lieb, "New Developments."

49. Cohen, "Unravelling the Milosevic Mystery."

50. Doder and Branson, *Milosevic*, p. 19.

51. LeBor, *Milosevic*, p. 316.

52. Ibid., p. 147.

53. Ibid., pp. 270–71.

54. David J. Cooke and Christine Michie, "Refining the Construct of Psychopathy: Towards a Hierarchical Model," *Psychological Assessment* 13, no. 2 (2001): 171–88.

55. Nicholas P. Swift and Harpal S. Nandhra, "'Borderpath' for Cluster B Personality Disorder?" *Psychiatric Services* 55 (2004): 193–94.

56. H. Chabrol and F. Leichsenring, "Borderline Personality Organization and Psychopathic Traits in Nonclinical Adolescents," *Bulletin of the Menninger Clinic* 70, no. 2 (2006): 160–70; Falk Leichsenring, Heike Kunst, and Jürgen Hoyer, "Borderline Personality Organization in Violent Offenders," *Bulletin of the Menninger Clinic* 67, no. 4 (2003): 314–27; Jennifer L. Skeem et al., "Psychopathic Personality or Personalities?" *Aggression & Violent Behavior* 8, no. 5 (2003): 513–46.

57. Skeem et al., "Psychopathic Personality or Personalities?"

58. Hare, Cooke, and Hart, "Psychopathy."

59. "Interview: Ambassador William Walker," *PBS Frontline*, http://www.pbs.org/wgbh/pages/frontline/shows/kosovo/interviews/walker.html (accessed February 17, 2007).

60. Doder and Branson, *Milosevic*, p. 250.

61. Michael Leverson Meyer, "Dead Where They Lay," *Newsweek*, June 24, 2002, p. 29.

62. LeBor, *Milosevic*, p. 208.

63. Ibid., p. 220.

Chapter 8: Lenses, Frames, and How Broken Brains Work

1. Barbara Oakley, *Hair of the Dog: Tales from Aboard a Russian Trawler* (Pullman, WA: WSU Press, 1996), p. 45.

2. Vladimir Vladimirovich Nabokov, *Poems and Problems* (New York: McGraw-Hill, 1970), pp. 158–59.

3. Associated Press, "Chinese, English Speakers Use Brains Differently to Tackle Math," *Fox News*, June 27, 2006, http://www.foxnews.com/story/0,2933,201048,00.html (accessed January 25, 2007).

4. P. Kochunov et al., "Localized Morphological Brain Differences between English-Speaking Caucasians and Chinese-Speaking Asians: New Evidence of Anatomical Plasticity," *NeuroReport* 14, no. 7 (2003): 961–64.

5. Hannah Faye Chua, Julie E. Boland, and Richard E. Nisbett, "Cultural Variation in Eye Movements during Scene Perception," *Proceedings of the National Academy of Sciences* 102, no. 35 (2005): 12629–33.

6. Charles J. Fillmore, "Linguistics in the Morning Calm," in *Linguistics in the Morning Calm* (Seoul: Hanshin Publishing Company, 1982). See also G. Lakoff and M. Johnson, *Philosophy in the Flesh: The Embodied Mind and Its Challenge to Western Thought* (New York: Basic Books, 1999); C. Siewert, *The Significance of Consciousness* (Princeton, NJ: Princeton University Press, 1998).

7. K. Amunts et al., "Hand Skills Covary with the Size of Motor Cortex: A Macrostructural Adaptation," *Human Brain Mapping* 5 (1997): 206–15; E. A. Maguire et al., "Navigation-Related Structural Change in the Hippocampi of Taxi Drivers," *Proceedings of the National Academy of Sciences* 97, no. 8 (2000): 4398–4403; A. H. Watson, "What Can Studying Musicians Tell Us about Motor Control of the Hand?" *Journal of Anatomy* 4 (2006): 527–42.

8. Oakley, *Hair of the Dog*, p. 139. Believing that other people and cultures share our perspectives is such a common misperception that it has earned its own name: "projection bias." This bias is thought by the Central Intelligence Agency to be among our most fundamentally dangerous false beliefs. Andrew Newberg and Mark Robert Waldman, *Why We Believe What We Believe: Uncovering Our Biological Need for Meaning, Spirituality, and Truth* (New York: Free Press, 2006), p. 255.

9. E. Bazanis et al., "Neurocognitive Deficits in Decision-Making and Planning of Patients with DSM-III-R Borderline Personality Disorder," *Psychological Medicine* 32, no. 8 (2002): 1395–1405.

10. Jules Lobel and George Loewenstein, "Emote Control: The Substitution of Symbol for Substance in Foreign Policy and International Law," *Chicago Kent Law Review* 80 (2004): 1045–90.

11. Douglas S. Massey, "A Brief History of Human Society: The Origin and Role of Emotion in Social Life," *American Sociological Review* 67, no. 1 (2002): 1–29.

12. Lobel and Loewenstein, "Emote Control."

13. "The Roger Coleman Case: Did Virginia Execute an Innocent Man?" *wbur.org Boston's NPR® News Source*, http://www.insideout.org/documentaries/dna/thestories2.asp (accessed January 14, 2006).

14. Maria Glod and Michael Shear, "DNA Tests Confirm Guilt of Executed Man," *Washington Post*, January 13, 2006.

15. Drew Westen et al., "The Neural Basis of Motivated Reasoning: An fMRI Study of Emotional Constraints on Political Judgment during the U.S. Presidential Election of 2004," *Journal of Cognitive Neuroscience* 18, no. 11 (2006): 1947–58.

16. Bradley and Westen, "Psychodynamics."

17. Emory University Health Sciences Center, "Emory Study Lights Up the Political Brain," *EurekAlert!* January 24, 2006, http://www.eurekalert.org/pub_releases/2006-01/euhs-esl012406.php (accessed July 2, 2006).

18. Ibid.

19. Stanley A. Renshon, *The Psychological Assessment of Presidential Candidates* (New York: New York University Press, 1996), p. 36.

20. This might be equated to the "dark side" of blink. Malcolm Gladwell, *Blink: The Power of Thinking without Thinking* (Boston: Little, Brown and Company, 2005), p. 75.

21. P. Brambilla et al., "Anatomical MRI Study of Borderline Personality Disorder Patients," *Psychiatry Research* 131, no. 2 (2004): 125–33.

22. K. M. Putnam and K. R. Silk, "Emotion Dysregulation and the Development of Borderline Personality Disorder," *Developmental Psychopathology* 17, no. 4 (2005): 899–925.

23. Donegan et al., "Amygdala."

24. Heather A. Berlin, Edmund T. Rolls, and Susan D. Iversen, "Borderline Personality Disorder, Impulsivity, and the Orbitofrontal Cortex," *American Journal of Psychiatry* 162 (2005): 2360–73.

25. Friedel, *Demystified*, p. 103.

26. Johnson et al., "Understanding"; M. Leyton et al., "Brain Regional

Alpha-[11C]methyl-L-tryptophan Trapping in Impulsive Subjects with Borderline Personality Disorder," *American Journal of Psychiatry* 158 (2001): 775–82.

27. Robert O. Friedel, "Dopamine Dysfunction in Borderline Personality Disorder: A Hypothesis," *Neuropsychopharmacology* 29 (2004): 1029–39; Joel Paris et al., "Neurobiological Correlates of Diagnosis and Underlying Traits in Patients with Borderline Personality Disorder Compared with Normal Controls," *Psychiatry Research* 121 (2004): 239–52; Skodol et al., "Borderline Diagnosis II"; Andrew E. Skodol et al., "The Borderline Diagnosis I: Psychopathology, Comorbidity, and Personality Structure," *Biological Psychiatry* 51, no. 12 (2002): 933–35.

28. Koenigsberg et al., "Interpersonal."

29. F. D. Juengling et al., "Positron Emission Tomography in Female Patients with Borderline Personality Disorder," *Journal of Psychiatric Research* 37 (2003): 109–15.

30. A. S. New et al., "Blunted Prefrontal Cortical ^{18}fluorodeoxyglucose Positron Emission Tomography Response to Meta-chlorophenylpiperazine in Impulsive Aggression," *Archives of General Psychiatry* 59, no. 7 (2002): 621–29; P. H. Soloff et al., "A Fenfluramine-Activated FDG-PET Study of Borderline Personality Disorder," *Biological Psychiatry* 47 (2000): 540–47.

31. Eva Irle, Claudia Lange, and Ulrich Sachsse, "Reduced Size and Abnormal Asymmetry of Parietal Cortex in Women with Borderline Personality Disorder," *Biological Psychiatry* 57 (2005): 173–82.

32. Leanne M. Williams et al., "'Missing Links' in Borderline Personality Disorder: Loss of Neural Synchrony Relates to Lack of Emotion Regulation and Impulse Control," *Journal of Psychiatry and Neuroscience* 31, no. 3 (2006): 181–88.

33. Irle, Lange, and Sachsse, "Reduced Size."

34. M. I. Posner et al., "An Approach to the Psychobiology of Personality Disorders," *Development and Psychopathology* 15, no. 4 (2003): 1093–1106. M. I. Posner et al., "Attentional Mechanisms of Borderline Personality Disorder," *Proceedings of the National Academy of Sciences* 99, no. 25 (2002): 16366–70.

35. Posner et al., "Psychobiology," citing Jin Fan et al., "Assessing the Heritability of Attentional Networks," *BioMed Central Neuroscience* 2, no. 14 (2001): Online (open access) at http://www.biomedcentral.com/1471-2202/1472/1414; H. H. Goldsmith et al., "Genetic Analyses of Focal Aspects of Infant Temperament," *Developmental Psychology* 35, no. 4 (1999): 972–85.

36. Posner et al., "Psychobiology."

37. Bechara et al., "Insensitivity"; Bechara et al., "Different Contributions"; A. Bechara et al., "Deciding Advantageously before Knowing the Advantageous Strategy," *Science* 275 (1997): 1293–95; Fellows and Farah, "Different Underlying Impairments."

38. Marc D. Hauser, *Moral Minds: How Nature Designed Our Universal Sense of Right and Wrong* (New York: Ecco, 2006), pp. 231–32. M. Koenigs et al., "Damage to the Prefrontal Cortex Increases Utilitarian Moral Judgements," *Nature* 446, no. 7138 (2007): 908–11.

39. Gary R. Brendel, Emily Stern, and David A. Silbersweig, "Defining the Neurocircuitry of Borderline Personality Disorder: Functional Neuroimaging Approaches," *Development and Psychopathology* 17 (2005): 1197–1206.

40. Norbert Bromberg and Verna Volz Small, *Hitler's Psychopathology* (New York: International Universities Press, 1983), p. 181.

41. Bradley and Westen, "Psychodynamics."

42. E. D. London et al., "Orbitofrontal Cortex and Human Drug Abuse: Functional Imaging," *Cerebral Cortex* 10 (2000): 334–42.

43. Bazanis et al., "Neurocognitive Deficits."

44. Bohus, Schmahl, and Lieb, "New Developments"; M. Driessen et al., "Magnetic Resonance Imaging Volumes of the Hippocampus and the Amygdala in Women with Borderline Personality Disorder and Early Traumatization," *Archives of General Psychiatry* 57 (2000): 1115–22; N. Rüsch et al., "A Voxel-Based Morphometric MRI Study in Female Patients with Borderline Personality Disorder," *NeuroImage* 20, no. 1 (2003): 385–92; C. G. Schmahl et al., "Magnetic Resonance Imaging of Hippocampal and Amygdala Volume in Women with Childhood Abuse and Borderline Personality Disorder," *Psychiatry Research* 122 (2003): 193–98; L. Tebartz van Elst et al., "Frontolimbic Brain Abnormalities in Patients with Borderline Personality Disorder: A Volumetric Magnetic Resonance Imaging Study," *Biological Psychiatry* 54 (2003): 163–71.

45. Paul M. Thompson et al., "Mapping Adolescent Brain Change Reveals Dynamic Wave of Accelerated Gray Matter Loss in Very Early-Onset Schizophrenia," *Proceedings of the National Academy of Sciences* 98, no. 20 (2001): 11650–55.

46. E. A. Hazlett et al., "Reduced Anterior and Posterior Cingulate Gray Matter in Borderline Personality Disorder," *Biological Psychiatry* 58, no. 8 (2005): 614–23.

47. L. Tebartz van Elst et al., "Subtle Prefrontal Neuropathology in a Pilot Magnetic Resonance Spectroscopy Study in Patients with Borderline Personality Disorder," *Journal of Neuropsychiatry and Clinical Neurosciences* 13 (2001): 511–14. Alessandro Bertolino et al., "Common Pattern of Cortical Pathology in Childhood-Onset and Adult-Onset Schizophrenia as Identified by Proton Magnetic Resonance Spectroscopic Imaging," *American Journal of Psychiatry* 155, no. 10 (1998): 1376–84.

48. John F. Clarkin and Michael I. Posner, "Defining Mechanisms of Borderline Personality Disorder," *Psychopathology* 38 (2005): 56–63; A. C. Ruocco, "The Neuropsychology of Borderline Personality Disorder: A Meta-Analysis and Review," *Psychiatry Research* 137, no. 3 (2005): 191–202.

49. Wilkinson-Ryan and Westen, "Identity Disturbance," citing S. Akhtar, *Broken Structures: Severe Personality Disorders and Their Treatment* (Northvale, NJ: Jason Aronson, Inc., 1992).

50. Putnam and Silk, "Emotion Dysregulation."

51. Douglas A. Granger, Nancy A. Dreschel, and Elizabeth A. Shirtcliff, "Developmental Psychoneuroimmunology: The Role of Cytokine Network Activation in the Epigenesis of Developmental Psychopathology," in *Neurodevelopmental Mechanisms in Psychopathology*, ed. Dante Cicchetti and Elaine Walker (New York: Cambridge University Press, 2003), pp. 293–323; M. Marcenaro et al., "Rheumatoid Arthritis, Personality, Stress Response Style, and Coping with Illness. A Preliminary Survey," *Annals of the New York Academy of Sciences* 876 (1999): 419–25.

52. N. Kopeloff, L. M. Kopeloff, and M. E. Raney, "The Nervous System and Antibody Production," *Psychiatry Quarterly* 7, no. 1 (1933): 84–106.

53. M. Zimmerman and J. I. Mattia, "Axis I Diagnostic Comorbidity and Borderline Personality Disorder," *Comprehensive Psychiatry* 40 (1999): 245–52.

54. I. Goethals et al., "Brain Perfusion SPECT in Impulsivity-Related Personality Disorders," *Behavioural Brain Research* 157, no. 1 (2005): 187–92.

55. McCloskey, Phan, and Coccaro, "Neuroimaging."

Chapter 9: The Perfect "Borderpath": Chairman Mao

1. K. S. Lo, "Introduction," in *Selected Works of Contemporary Yixing Potters*, ed. Hong Kong Museum of Art (Hong Kong: Urban Council, Hong

Kong, 1994), p. 9. In one of those twists that makes collecting interesting, revolution teapots are still expensive, despite their pedestrian artistry, because the zisha being mined during that period allows for a roughness that results in an exceptionally beautiful patina upon use.

2. Nien Cheng, *Life and Death in Shanghai* (London: HarperCollins, 1995), p. 96.

3. Jung Chang and Jon Halliday, *Mao: The Unknown Story*, 1st ed. (London: Jonathan Cape, 2005), pp. 537, 566.

4. R. J. Rummel, "Getting My Reestimate of Mao's Democide Out," *Democratic Peace*, 2005, http://freedomspeace.blogspot.com/2005/11/getting-my-reestimate-of-maos-democide.html (accessed December 16, 2005).

5. R. J. Rummel, *Death by Government* (New Brunswick, NJ: Transaction Publishers, 1994), p. 9.

6. Lucian W. Pye, "Rethinking the Man in the Leader," *China Journal* 35 (1996): 107–12. See also Ross Terrill, "Mao in History," *National Interest* 52 (1998): 54–63.

7. Ross Terrill, *Mao: A Biography* (Stanford, CA: Stanford University Press, 1999), p. 16.

8. Chang and Halliday, *Mao*, pp. 6, 7.

9. Shenk, *Lincoln's Melancholy*, pp. 14, 29.

10. Francis Fukuyama, *Trust: The Social Virtues and the Creation of Prosperity* (New York: Free Press, 1995), pp. 85, 86.

11. Philip Short, *Mao: A Life* (New York: Henry Holt and Company, 1999), p. 69.

12. Ibid.

13. Ibid., pp. 144, 145.

14. Chang and Halliday, *Mao*, pp. 51–65.

15. Short, *Mao*, p. 226; Jonathan Spence, *Mao Zedong* (New York: Lipper/Viking, 1999), p. 80.

16. William Cardasis, Jamie A. Hochman, and Kenneth R. Silk, "Transitional Objects and Borderline Personality Disorder," *American Journal of Psychiatry* 154, no. 2 (1997): 250–55; Mason and Kreger, *Eggshells*, p. 44; Lawrence A. Labbate and David M. Benedek, "Bedside Stuffed Animals and Borderline Personality," *Psychological Reports* 79 (1996); T. A. Stern and R. L. Glick, "Significance of Stuffed Animals at the Bedside and What They Can Reveal about Patients," *Psychosomatics* 34 (1993). Borderline patients will frequently bring teddy bears and other such "transitional objects" with

them for hospital stays. This is sometimes used by doctors to intuit problematic patients. For example, one study found that 61% of psychiatric inpatients with stuffed animals in their rooms had a diagnosis of borderline personality disorder, as opposed to only 17% having borderline personality disorder in the overall psychiatric unit during the same period of time. Be careful if you decide to bring Binky to the psychiatric hospital with you.

17. Chang and Halliday, *Mao*, pp. 89, 90.

18. Ibid., pp. 631–32; Zhisui Li, *The Private Life of Chairman Mao* (New York: Random House, 1994), pp. 382–84.

19. Li, *Private Life*, p. 63.

20. Spence, *Mao Zedong*, p. 98.

21. Chang and Halliday, *Mao*, p. 69.

22. Ibid., p. 98.

23. Li, *Private Life*, p. 81.

24. Ibid., p. 505.

25. Ibid., p. 363.

26. Ibid., p. 121.

27. Ibid.

28. Ibid., p. 125.

29. Ibid., p. 338.

30. Li, *Private Life*, pp. 382–84.

31. Hazlett et al., "Reduced Anterior."

32. Spence, *Mao Zedong*, p. 144.

33. Ibid., p. 124.

34. Kreisman and Straus, *Crazy*, p. 50.

35. Li, *Private Life*, p. 180.

36. Andrew J. Nathan, foreword to Li, *Private Life*, p. viii.

37. Short, *Mao*, pp. 69, 70.

38. Ibid., p. 71.

39. Leland M. Heller, *Life at the Border: Understanding and Recovering from the Borderline Personality Disorder* (Okeechobee, FL: Dyslimbia Press, 2000), p. 17.

40. Koenigsberg et al., "Interpersonal."

41. Stuart R. Schram, "Mao Tse-tung and Theory of Permanent Revolution," *China Quarterly* 46 (1971): 221–44.

42. Short, *Mao*, p. 384.

43. R. J. Waldinger, "The Role of Psychodynamic Concepts in the Diag-

nosis of Borderline Personality Disorder," *Harvard Review of Psychiatry* 1 (1993): 158–67.

44. Chang and Halliday, *Mao*, p. 35.

45. Bohus, Schmahl, and Lieb, "New Developments." See also H. W. Koenigsberg et al., "Characterizing Affective Instability in Borderline Personality Disorder," *American Journal of Psychiatry* 159, no. 5 (2002): 784–88.

46. Li, *Private Life*, p. 107.

47. T. Ebisawa, "Circadian Rhythms in the CNS and Peripheral Clock Disorders: Human Sleep Disorders and Clock Genes," *Journal of Pharmacological Sciences* 103, no. 2 (2007): 150–54; R. Grant Steen, *The Evolving Brain: The Known and the Unknown* (Amherst, NY: Prometheus Books, 2007), p. 66.

48. Li, *Private Life*, p. 364.

49. Ibid., p. 359.

50. William A. Henkin and Patrick J. Carnes, "Is Sex Addiction a Myth?" in *Clashing Views on Abnormal Psychology*, ed. Susan Nolen-Hoeksema (New York: McGraw-Hill, 1997), pp. 196–215.

51. J. Bancroft and Z. Vukadinovic, "Sexual Addiction, Sexual Compulsivity, Sexual Impulsivity, or What? Toward a Theoretical Model," *Journal of Sex Research* 41, no. 3 (2004): 225–34.

52. Chang and Halliday, *Mao*, p. 346.

53. Short, *Mao*, pp. 475, 586.

54. Ibid., p. 505.

55. Li, *Private Life*, pp. 339, 340.

56. Ibid., p. 387.

57. Robert Jay Lifton, *Revolutionary Immortality: Mao Tse-tung and the Chinese Cultural Revolution* (New York: Vintage Books, 1968), p. 32.

58. C. C. Dickey et al., "A MRI Study of Fusiform Gyrus in Schizotypal Personality Disorder," *Schizophrenia Research* 64, no. 1 (2003): 35–39.

59. Spence, *Mao Zedong*, p. 51.

60. Short, *Mao*, p. 103.

61. Terrill, *Mao*, p. 18.

62. Li, *Private Life*, p. 84.

63. Ibid., p. xix.

64. Terrill, *Mao*, p. 18.

65. Li, *Private Life*, p. 351.

66. Ibid., p. 120.

67. Chang and Halliday, *Mao*, p. 95.

68. Short, *Mao*, p. 467.

69. Chang and Halliday, *Mao*, pp. 437, 438.

70. Li, *Private Life*, p. 106.

71. Chang and Halliday, *Mao*, p. 569.

72. Ibid., p. 546.

73. Meloy, *Violence*, p. 20.

74. Jay R. Kaplan et al., "Central Nervous System Monoamine Correlates of Social Dominance in Cynomolgus Monkeys (*Macaca fascicularis*)," *Neuropsychopharmacology* 26, no. 4 (2002): 431–43; Jeffrey Rogers et al., "Genetics of Monoamine Metabolites in Baboons," *Biological Psychiatry* 55, no. 7 (2004): 739–45.

75. Gary Marcus, *Birth of the Mind: How a Tiny Number of Genes Creates the Complexities of Human Thought* (New York: Basic Books, 2004), p. 170.

76. Short, *Mao*, p. 471.

77. Chang and Halliday, p. 544.

78. Ibid., p. 42.

79. Ibid.

80. Spence, *Mao Zedong*, p. 100.

81. Li, *Private Life*, p. 125.

82. Chang and Halliday, *Mao*, pp. 144, 199.

83. Albert Mohler, "Chairman Mao's Reign of Terror—Finally the Truth Comes Out," *www.AlbertMohler.com*, October 20, 2005, http://www.albertmohler.com/commentary_print.php?cdate=2005-10-20 (accessed November 1, 2005).

84. Chang and Halliday, *Mao*, p. 332.

85. Li, *Private Life*, pp. 115, 124.

86. Spence, *Mao Zedong*, p. 101.

87. Roderick MacFarquhar and Michael Schoenhals, *Mao's Last Revolution* (Cambridge, MA: Harvard University Press, 2006), p. 262.

88. Short, *Mao*, p. 550.

89. Chang and Halliday, *Mao*, p. 13.

90. See also Delroy L. Paulhus and Kevin M. Williams, "The Dark Triad of Personality: Narcissism, Machiavellianism, and Psychopathy," *Journal of Research in Personality* 36, no. 6 (2002): 556–63.

91. W. John Livesley et al., "Genetic and Environmental Contributions

of Dimensions of Personality Disorder," *American Journal of Psychiatry* 150, no. 12 (1993): 1826–31.

92. Carl Vogel, "A Field Guide to Narcissism," *Psychology Today*, January/February 2006, pp. 68–74.

93. A. Benvenuti et al., "Psychotic Features in Borderline Patients: Is There a Connection to Mood Dysregulation?" *Bipolar Disorders* 7 (2005): 338–43; D. Pizzagalli et al., "Brain Electric Correlates of Strong Belief in Paranormal Phenomena: Intracerebral EEG Source and Regional Omega Complexity Analyses," *Psychiatry Research* 100, no. 3 (2000): 139–54; A. Sbrana et al., "The Psychotic Spectrum: Validity and Reliability of the Structured Clinical Interview for the Psychotic Spectrum," *Schizophrenia Research* 75, no. 2 (2005): 375–89.

94. Li, *Private Life*, p. 233.

95. Ibid., p. 443.

96. Meloy, *Violence*, pp. 30–31.

97. Short, *Mao*, p. 590.

98. Ibid., p. 594.

99. Ibid., p. 600.

100. Jerrold Post, "Current Concepts of the Narcissistic Personality: Implications for Political Psychology," *Political Psychology* 14, no. 1 (1993): 99–121; Vamik Volkan, "Narcissistic Personality Organization and Reparative Leadership," *International Journal of Group Psychotherapy* 30 (1980): 131–52.

101. Post, "Current Concepts of the Narcissistic Personality."

102. Li, *Private Life*, p. 9.

103. Stephen Sherrill, "Acquired Situational Narcissism," *New York Times*, December 9, 2001.

104. Terrill, *Mao*, p. 19.

105. Ibid.

106. Tim Healy and David Hsieh, "Mao Now," *Asiaweek*, September 6, 1996, http://www.asiaweek.com/asiaweek/96/0906/cs1.html (accessed October 30, 2005).

Chapter 10: Evolution and Machiavellianism

1. Linda Mealey, "The Sociobiology of Sociopathy: An Integrated Evolutionary Model," *Behavioral and Brain Sciences* 18 (1995): 523–99; Skeem et al., "Psychopathic Personality or Personalities?"

2. Mealey, "Sociobiology." See also an earlier reference: Adrian Raine, *The Psychopathology of Crime: Crime Behavior as a Clinical Disorder* (San Diego: Academic Press, 1993). There is also the much more comprehensive recent treatment of antisocial and moral behavior from both a neurological and evolutionary perspective: Raine and Yang, "Neural Foundations."

3. Michelle Caruso, "Laci's Ex-Beau Shot His Girl Friend," *National Enquirer*, July 8, 2003.

4. Paulhus and Williams, "Dark Triad of Personality."

5. Hariri et al., "A Susceptibility Gene."

6. L. Mealey, "Primary Sociopathy (Psychopathy Is a Type, Secondary Is Not)," *Behavioral and Brain Sciences* 18, no. 3 (1995): 579–87.

7. Renshon, *Presidential Candidates*, p. 69.

8. Steven D. Levitt and Stephen J. Dubner, *Freakonomics: A Rogue Economist Explores the Hidden Side of Everything* (New York: HarperCollins, 2005), pp. 39–45.

9. Randolph Nesse, "Evolutionary Explanations of Emotions," *Human Nature* 1, no. 3 (1990): 261–89.

10. Raine and Yang, "Neural Foundations."

11. R. W. Byrne and A. Whiten, *Machiavellian Intelligence: The Evolution of Intellect in Monkeys, Apes, and Humans* (Oxford, England: Clarendon Press, 1988); Andrew Whiten and Richard W. Byrne, *Machiavellian Intelligence II: Extensions and Evaluations* (New York: Cambridge University Press, 1997).

12. L. Cosmides and J. Tooby, "Cognitive Adaptions for Social Exchange," in *The Adapted Mind*, ed. J. Barkow, L. Cosmides, and John Toland (New York: Oxford University Press, 1992), pp. 163–228; L. Cosmides et al., "Detecting Cheaters," *Trends in Cognitive Sciences* 9, no. 11 (2005): 505–506; Hauser, *Moral Minds*, p. 287. But see Buller, *Adapting Minds*; David J. Buller, Jerry A. Fodor, and Tessa L. Crume, "The Emperor Is Still Under-Dressed," *Trends in Cognitive Sciences* 9, no. 11 (2005): 508–10. The battle of the reseach titans wages!

13. Robert Wright, *The Moral Animal: The New Science of Evolutionary Psychology* (New York: Pantheon Books, 1994), p. 278, citing Thomas Schelling, *The Strategy of Conflict* (Cambridge, MA: Harvard University Press, 1960).

14. Nicholas Wade, *Before the Dawn: Recovering the Lost History of Our Ancestors* (New York: Penguin Press, 2006), pp. 117–18.

15. Elin McCoy, *The Emperor of Wine* (New York: HarperCollins, 2005), pp. 7, 123, 275.

16. P. D. Evans et al., "Microcephalin, a Gene Regulating Brain Size, Continues to Evolve Adaptively in Humans," *Science* 309, no. 5741 (2005): 1717–20; N. Mekel-Bobrov et al., "Ongoing Adaptive Evolution of ASPM, a Brain Size Determinant in *Homo sapiens*," *Science* 309, no. 5741 (2005): 1720–22. M. Balter, "Evolution: Are Human Brains Still Evolving?" *Science* 309, no. 5741 (2005): 1662–63.

17. Buller, *Adapting Minds*, p. 108.

18. John S. Mattick and Igor V. Makunin, "Non-coding RNA," *Human Molecular Genetics* 15, no. 1 (2006): R17–R29.

19. G. Bloom and P. W. Sherman, "Dairying Barriers and the Distribution of Lactose Malabsorption," *Evolution and Human Behavior* 26 (2005): 301–12.

20. Deacon, *Symbolic Species*, pp. 321–75, citing Bruce H. Weber and David J. Depew, eds., *Evolution and Learning: The Baldwin Effect Reconsidered* (Cambridge, MA: MIT Press, 2004).

21. Wade, *Before the Dawn*, p. 177.

22. J. M. Murphy, "Psychiatric Labeling in Cross-Cultural Perspective," *Science* 141 (1976): 1019–28.

23. David J. Cooke and Christine Michie, "Psychopathy across Cultures: North America and Scotland Compared," *Journal of Abnormal Psychology* 108, no. 1 (1999): 58–68.

24. Wade, *Before the Dawn*, p. 128.

25. Ibid., p. 129. The quote is from Cambridge archaeologist Colin Renfrew. I'd like to acknowledge a debt here to Nicholas Wade's *Before the Dawn*—if you are interested in the development of humanity in the last 150,000 or so years, you couldn't do better than to read it.

26. Steve Sailer, "Genes of History's Greatest Lover Found?" *iSteve.com*, 2003, http://www.isteve.com/2003_Genes_of_History_Greatest_Lover_Found.htm (accessed June 5, 2006).

27. Ibid.

28. Wade, *Before the Dawn*, pp. 236–37.

29. "The First American to Be Able to Claim Descent from Genghis Khan Has Been Discovered," *History News Network*, June 6, 2006, http://hnn.us/ roundup/entries/26356.html (accessed June 7, 2006).

30. Dorothy Einon, "How Many Children Can One Man Have?" *Ethology and Sociobiology* 19, no. 6 (1998): 413–27.

31. Laura L. Betzig, *Despotism and Differential Reproduction* (New York: Aldine Publishing Company, 1986), pp. 88, 89.

32. Christie and Geis, *Studies in Machiavellianism*; Mealey, "Primary Sociopathy."

33. Betzig, *Despotism*, pp. 1, 2.

34. Ibid., p. 88.

35. Hauser, *Moral Minds*, pp. 283–84.

36. Betzig, *Despotism*, p. 81.

37. Ibid., p. 77.

38. Sigal G. Barsade, "The Ripple Effect: Emotional Contagion in Groups," *Yale SOM Working Paper*, no. OB-01, 2000, http://ssrn.com/abstract=250894 (accessed May 28, 2006); Elaine Hatfield, John T. Cacioppo, and Richard L. Rapson, *Emotional Contagion* (New York: Cambridge University Press, 1994).

39. Harris, *No Two Alike*, p. 195.

40. Edward Behr, *Kiss the Hand You Cannot Bite: The Rise and Fall of the Ceauşescus* (New York: Villard, 1991), p. 277.

41. Alev Lytle Croutier, *Harem: The World behind the Veil* (New York: Abbeville Press, 1989), p. 116.

42. Ibid., p. 115.

43. Galina Yermolenko, "Roxolana: The Greatest Empresse of the East," *Muslim World* 95, no. 2 (2005): 231–48.

44. Jerry Oppenheimer, *House of Hilton: From Conrad to Paris* (New York: Crown, 2006), p. 82.

45. Ibid., p. 96.

46. Ibid., p. 34.

47. Ibid., p. 14.

48. Ibid., pp. 9, 10.

49. Croutier, *Harem*, p. 119.

50. John Freely, *Inside the Seraglio: Private Lives of the Sultans in Istanbul* (New York: Penguin, 2000), p. 249.

51. The many valuable footnotes by Bill Thayer in this excerpt have been eliminated for the sake of clarity. "*Historia Augusta*," Loeb Classical Library, 1921, http://penelope.uchicago.edu/Thayer/E/Roman/Texts/Historia_Augusta/Commodus*.html (accessed February 17, 2007).

52. Stanley Bing, *Rome, Inc. The Rise and Fall of the First Multinational Corporation* (New York: W. W. Norton & Company, 2006), pp. 62–63, 180.

53. Peter J. Heather, *The Fall of the Roman Empire: A New History of*

Rome and the Barbarians (New York: Oxford University Press, 2006), p. 258.

54. Kreisman and Straus, *Crazy*, pp. 2, 49. Even as gifted a biographer as Tina Brown was clearly dazzled by Princess Diana's charisma. In Brown's telling, for example, Diana's mood swings were a result of her Prince Charles–induced bulimia, rather than the bulimia being a symptom, along with the mood swings, of a sub-clinical personality disorder that flowered in the stress of palace life. Brown alludes to heredity in only desultory fashion, casually pointing out, for example, that "[b]ody abuse was second nature to the women in Diana's family—like writing thank-you letters." Tina Brown, *The Diana Chronicles* (New York: Doubleday, 2007), pp. 151–52.

55. For an excellent review of the topic, see T. Carnahan and S. McFarland, "Revisiting the Stanford Prison Experiment: Could Participant Self-Selection Have Led to Cruelty?" *Personality and Social Psychology Bulletin* 33, no. 5 (2007): 603–14.

56. Heather, *Rome*; Ramsay MacMullen, *Corruption and the Decline of Rome* (New Haven: Yale University Press, 1988).

57. Christina Asquith, "Trouble at Texas Southern," *Diverse Issues in Higher Education*, December 14, 2006, http://diverseeducation.com/artman/publish/article_6764.shtml (accessed December 20, 2006).

58. Ibid.

59. Ibid. For recent similar tales in academia alone, as monitored by Margaret Soltan of *University Diaries* (http://margaretsoltan.phenominet.com/), see David W. Chen and Laura Mansnerus, "Overseer Finds Kickback Plan at University," *New York Times*, November 14, 2006; Stefan Milkowski and Amanda Bohman, "Legislator Pursues Impeachment," *Fairbanks Daily News-Miner*, February 24, 2007, http://newsminer.com/2007/02/24/5479/ (accessed February 24, 2007); "More Bad News for Bishop State," *Birmingham News*, February 19, 2007, http://www.al.com/opinion/birminghamnews/index.ssf?/base/opinion/1171880185119050.xml&coll=2 (accessed February 24, 2007).

60. R. Pally, "The Neurobiology of Borderline Personality Disorder: The Synergy of 'Nature and Nurture,'" *Journal of Psychiatric Practice* 8, no. 3 (2002): 133–42.

61. Colin Turnbull, *The Mountain People* (New York: Simon & Schuster, 1987), p. 131.

Chapter 11: Shades of Gray

1. Gretchen Rubin, *Forty Ways to Look at Winston Churchill* (New York: Random House, 2004), p. 125.

2. Christopher Hitchens, *The Missionary Position: Mother Teresa in Theory and Practice* (New York: Verso Books, 1995).

3. Henri Troyat, *Catherine the Great*, trans. Joan Pinkham (New York: Plume, 1994); Jack Weatherford, *Genghis Khan and the Making of the Modern World* (New York: Crown Publishers, 2004).

4. Livesley, "Behavioral and Molecular Genetic Contributions."

5. W. Langer, *The Mind of Adolf Hitler: The Secret Wartime Report* (New York: Basic Books, 1972), p. 32.

6. Behr, *Kiss the Hand*, p. 164.

7. "Turkmenbashi Everywhere," *CBS News*, January 4, 2004, http://www.cbsnews.com/stories/2003/12/31/60minutes/main590913.shtml (accessed August 1, 2006).

8. James Shreeve, *The Genome War* (New York: Alfred A. Knopf, 2004), pp. 27–38.

9. Arnold C. Brackman, *A Delicate Arrangement: The Strange Case of Charles Darwin and Alfred Russel Wallace* (New York: Times Books, 1980), p. 124.

10. Ross A. Slotten, *The Heretic in Darwin's Court: The Life of Alfred Russel Wallace* (New York: Columbia University Press, 2004), p. 159.

11. Brackman, *Delicate Arrangement*, p. 34.

12. Michael Shermer, *In Darwin's Shadow: The Life and Science of Alfred Russel Wallace: A Biographical Study on the Psychology of History* (New York: Oxford University Press, 2002).

13. James D. Watson, *The Double Helix: A Personal Account of the Discovery of the Structure of DNA* (New York: Penguin, 1970). Watson's appalling mischaracterization of Rosalind Franklin is well described in Brenda Maddox, *Rosalind Franklin: The Dark Lady of DNA* (New York: HarperPerennial, 2003).

14. Shreeve, *Genome War*, p. 85.

15. Joel N. Shurkin, *Broken Genius: The Rise and Fall of William Shockley, Creator of the Electronic Age* (New York: Macmillan, 2006), p. 126.

16. Ibid., p. 164.

17. Ibid.

18. Ibid., p. 263.

19. Paul A. Offit, *The Cutter Incident: How America's First Polio Vaccine Led to the Growing Vaccine Crisis* (New Haven: Yale University Press, 2005), pp. 127–28.

20. Jason Socrates Bardi, *The Calculus Wars: Newton, Leibniz, and the Greatest Mathematical Clash of All Time* (New York: Thunder's Mouth Press, 2006).

21. Candace B. Pert, *Molecules of Emotion: Why You Feel the Way You Feel* (New York: Scribner, 1997), p. 115.

22. Tom Lewis, *Empire of the Air: The Men Who Made Radio*, 1st ed. (New York: E. Burlingame Books, 1991).

23. Bob Spitz, *The Beatles: The Biography* (Boston: Little, Brown, 2005), p. 647.

24. Raynoma Gordy Singleton, *Berry, Me, and Motown: The Untold Story* (Chicago: Contemporary Books, 1990), p. 333.

25. J. Randy Taraborrelli, *Madonna: An Intimate Biography* (New York: Simon & Schuster, 2001).

26. Slotten, *Heretic*, pp. 172–73.

27. Ken Auletta, *World War 3.0* (New York: Broadway Books, 2001), p. 122.

28. Rubin, *Churchill*, p. 179.

29. Ibid., p. 53.

30. Ibid., p. 32.

31. Ibid., p. 57.

32. Jerry Oppenheimer, *Martha Stewart—Just Desserts* (New York: William Morrow and Company, Inc., 1997), p. 160.

33. Christopher Byron, *Martha, Inc.* (New York: John Wiley & Sons, Inc., 2002), p. 75.

34. Ibid., pp. 284, 311.

35. Ibid., pp. 311–12, 318.

36. Joan Macleod Heminway, ed., *Martha Stewart's Legal Troubles* (Durham, NC: Carolina Academic Press, 2006).

37. Byron, *Testosterone, Inc.: Tales of CEOs Gone Wild*, p. 268.

38. Ibid., pp. 25, 72, 73.

39. William G. Flanagan, *Dirty Rotten CEOs: How Business Leaders Are Fleecing America* (New York: Citadel Press, 2003), p. 183.

40. Kurt Eichenwald, *Conspiracy of Fools* (New York: Broadway Books, 2005), p. 250.

41. Ibid., p. 523.

42. Ibid., p. 172.

43. Ibid., p. 259.

44. Marie Brenner, "The Enron Wars," *Vanity Fair*, April 2002, pp. 180–210.

45. Ibid.

46. Capital News 9 Web Staff, "Former Enron CEO May Have Lost It," *Capital News 9*, April 9, 2004.

47. Eichenwald, *Conspiracy of Fools*, pp. 131, 250, 620.

48. Joseph J. Ellis, *His Excellency: George Washington* (New York: Random House, 2004), pp. xiv, 270.

49. Ibid., p. 272.

50. Edward G. Lengel, *General George Washington: A Military Life* (New York: Random House, 2005), p. 300.

51. Richard Brookhiser, *Founding Father: Rediscovering George Washington* (New York: Free Press, 1996), p. 117.

52. Ellis, *His Excellency*, pp. 153–54; G. B. Singh, *Gandhi: Behind the Mask of Divinity* (Amherst, NY: Prometheus Books, 2004), pp. 134–35.

53. Louis Fischer, *Mahatma Gandhi: His Life and Message for the World* (New York: New American Library, 1954), p. 29.

54. Ann Ruth Willner, *The Spellbinders: Charismatic Political Leadership* (New Haven: Yale University Press, 1984), p. 136.

55. Langer, *Mind of Adolf Hitler*, p. 84.

56. Albert H. Speer, *Inside the Third Reich* (New York: MacMillan, 1970), p. 97.

57. Ian Kershaw, *Hitler: 1889–1936: Hubris* (New York: W. W. Norton & Company, Inc., 2000), p. 281.

58. Nicholas Gage, *Greek Fire: The Story of Maria Callas and Aristotle Onassis* (New York: Warner Books, 2001), p. 11; Michael Gross, *Genuine Authentic: The Real Life of Ralph Lauren* (New York: HarperCollins, 2003), pp. 264, 269; Stephen Manes and Paul Andrews, *Gates: How Microsoft's Mogul Reinvented an Industry and Made Himself the Richest Man in America* (New York: Touchstone, 2003), p. 250; "Not So Saintly Behavior: Kinder, Gentler Ditka Unleashes Tirade at Practice," *Sports Illustrated*/CNN, 1998, http://sportsillustrated.cnn.com/football/nfl/news/1998/08/10/ditka_saints/ (accessed January 28, 2007). Andrew Burstein, *The Passions of Andrew Jackson*, 1st ed. (New York: Alfred A. Knopf, 2003), pp. 57, 235; John

F. Harris, *The Survivor: Bill Clinton in the White House*, 1st ed. (New York: Random House, 2005), pp. 53, 483; Richard Reeves, *President Nixon: Alone in the White House* (New York: Simon & Schuster, 2001), p. 35; Eugen Weber, "Destiny's General," *New York Times*, April 2, 1995; Charles Williams, *The Last Great Frenchman: A Life of Charles de Gaulle* (Boston: Little, Brown and Company, 1993), pp. 62–64; Arianna Huffington, *Maria Callas: The Woman behind the Legend* (New York: Cooper Square Press, 2002).

59. Brookhiser, *Founding Father*, p. 115.

60. Joseph J. Ellis, *American Sphinx: The Character of Thomas Jefferson* (New York: Random House, 1996), p. 39.

61. Ellis, *His Excellency*, p. 246.

62. Peter Michelmore, *The Swift Years: The Robert Oppenheimer Story* (New York: Dodd, Mead, 1969), p. 4.

63. J. Gilleen and A. S. David, "The Cognitive Neuropsychiatry of Delusions: From Psychopathology to Neuropsychology and Back Again," *Psychological Medicine* 35, no. 1 (2005): 5–12.

64. David Dunbar et al., "Debunking 9/11 Myths," *Glenn & Helen Show (Podcast)*, August 15, 2006, http://instapundit.com/archives/031961.php (accessed December 2, 2006).

65. H. Arendt, *Eichmann in Jerusalem: A Report on the Banality of Evil* (New York: Viking Press, 1965); Paul R. Brass, *Theft of an Idol* (Princeton, NJ: Princeton University Press, 1997), p. 9; Daniel J. Goldhagen, *Hitler's Willing Executioners: Ordinary Germans and the Holocaust* (New York: Knopf, 1996); Samantha Power, *"A Problem from Hell": America and the Age of Genocide* (New York: Basic Books, 2002), p. 111; Raine and Yang, "Neural Foundations."

66. P. G. Zimbardo, C. Maslach, and C. Haney, "Reflections on the Stanford Prison Experiment: Genesis, Transformations, Consequences," in *Obedience to Authority: Current Perspectives on the Milgram Paradigm*, ed. T. Blass, (Mahwah, NJ: Lawrence Erlbaum, 2000), p. 194; Philip G. Zimbardo, *The Lucifer Effect: Understanding How Good People Turn Evil* (New York: Random House, 2007), p. 32; T. Carnahan and S. McFarland, "Revisiting the Stanford Prison Experiment." See also the excellent follow-on article that expands on Carnahan and McFarland's findings: S. A. Haslam and S. Reicher, "Beyond the Banality of Evil: Three Dynamics of an Interactionist Social Psychology of Tyranny," *Personality and Social Psychology Bulletin* 33, no. 5 (2007): 615–22.

67. Langer, *Mind of Adolf Hitler*, pp. 82–83.

68. Cohen, "Unravelling the Milosevic Mystery."

69. Langer, *Mind of Adolf Hitler*, p. 227.

70. Norbert Bromberg and Verna Volz Small, *Hitler's Psychopathology* (New York: International Universities Press, 1983), pp. 167–68.

71. George Victor, *Hitler: The Pathology of Evil* (Washington, DC: Brassey's, 2000), p. 115.

72. Langer, *Mind of Adolf Hitler*, p. 95.

73. Bromberg and Small, *Hitler's Psychopathology*, p. 300.

74. "Barry Marshall Interview: Nobel Prize in Medicine," *Academy of Achievement*, October 22, 2006, http://www.achievement.org/autodoc/printmember/mar1int-1 (accessed February 24, 2007); P. H. Duesberg, *Inventing the AIDS Virus* (Lanham, MD: Regnery Publishing, 1997); Kuhn, *Structure of Scientific Revolutions*; B. Martin, "Dissent and Heresy in Medicine: Models, Methods, and Strategies," *Social Science and Medicine* 58, no. 4 (2004): 713–25; J. Moore, "À Duesberg, Adieu!" *Nature* 380, no. 6572 (1996): 293–94.

75. Ellis, *His Excellency*, pp. 180, 158.

76. Ibid., p. 271.

77. Ibid.

78. Martin Meredith, *Our Votes, Our Guns: Robert Mugabe and the Tragedy of Zimbabwe* (New York: Public Affairs, 2003), pp. 154, 228.

79. Robert G. L. Waite, *The Psychopathic God: Adolf Hitler* (New York: Basic Books, 1977), p. 100.

80. Montefiore, *Stalin*, p. 526.

81. Gene N. Landrum, *Profiles of Female Genius: Thirteen Creative Women Who Changed the World* (Amherst, NY: Prometheus Books, 1994).

82. Ibid., p. 339.

83. Ibid.

84. Andrew Mango, *Atatürk: The Biography of the Founder of Modern Turkey* (New York: Peter Meyer Publishers, 2002), p. 466.

85. Conrad Black, *Franklin Delano Roosevelt: Champion of Freedom* (New York: Public Affairs, 2003), p. 1131.

86. Elizabeth Abbott, *Haiti: The Duvaliers and Their Legacy* (New York: Simon & Schuster, 1991).

87. Diana Jean Schemo, "Stroessner, Paraguay's Enduring Dictator, Dies," *New York Times*, August 16, 2006. For a broad overview of brutal right-wing dictators, see: David F. Schmitz, *Thank God They're On Our*

Side: The United States and Right-Wing Dictatorships, 1921–1965 (Chapel Hill, NC: University of North Carolina Press, 1999).

88. Waite, *Psychopathic God*, p. 63.

89. Willner, *Spellbinders*, p. 146.

90. Leycester Coltman, *The Real Fidel Castro* (New Haven: Yale University Press, 2003), p. 9.

91. Montefiore, *Stalin*, p. 233.

92. Ibid., p. 437.

93. Roland Huntford, *Shackleton* (New York: Atheneum, 1986), p. 75.

94. "Thatcher 'Cannot Remember Start of Sentence': Daughter," *Breitbart.com*, December 9, 2005, http://www.breitbart.com/news/2005/12/09/051210030244.lpbfxvhd.html (accessed July 30, 2006).

95. The last fact is pretty obscure—you can find the reference at: Peter Robb, *Midnight in Sicily* (New York: Vintage Books, 1996), pp. 112–13.

96. Landrum, *Profiles*, p. 338.

97. Daniel Chirot, *Modern Tyrants: The Power and Prevalence of Evil in Our Age* (Princeton, NJ: Princeton University Press, 1994).

98. Waite, *Psychopathic God*, p. 435.

99. Matt Drudge with Julia Philips, *Drudge Manifesto* (New York: New American Library, 2000), p. 140.

100. Renshon, *Presidential Candidates*, pp. 304–305.

101. Transcript: Saddam Hussein Interview, CBS News, February 26, 2003, http://www.cbsnews.com/stories/2003/02/26/60II/main542151.shtml (accessed March 5, 2007).

102. Jeff Jacoby, "When Mike Met Mahmoud," *Boston Globe*, August 16, 2006, http://www.boston.com/news/globe/editorial_opinion/oped/articles/2006/08/16/when_mike_met_mahmoud/ (accessed November 25, 2006).

103. Ibid.

104. Roger Lowenstein, *Buffett: The Making of an American Capitalist* (New York: Main Street Books, 1996), p. xiii.

105. Ibid., p. 276, citing L. J. Davis, "Buffett Takes Stock," *New York Times*, April 1, 1990.

106. Warren Buffett, in correspondence with the author, December 15, 2006.

107. Bill George, "The Master Gives It Back," *U.S. News & World Report*, October 30, 2006, pp. 66–68.

108. Russell Friedman, Helen Smith, and Glenn Reynolds, "Russell

Friedman on Moving On," *Glenn & Helen Show (Podcast)*, September 5, 2006, http://instapundit.com/archives/032358.php (accessed February 25, 2007).

109. Power, *Problem from Hell*, pp. 33–34.

110. Ibid., p. xvii.

111. s.v. "evil," Dictionary.com. *The American Heritage® Dictionary of the English Language*, 4th ed. Houghton Mifflin Company, 2004, http://dictionary.reference.com/browse/evil (accessed December 23, 2006).

112. Martha Stout, *The Sociopath Next Door* (New York: Broadway Books, 2005).

113. Redlich, *Destructive Prophet*, p. 27.

114. Ibid., pp. 22, 323.

115. Walter Isaacson, *Einstein: His Life and Universe* (New York: Simon & Schuster, 2007), pp. 131–32.

116. Shankar Vedantam, "If It Feels Good to Be Good, It Might Be Only Natural," *Washington Post*, May 28, 2007.

117. Marc D. Hauser, *Moral Minds: How Nature Designed Our Universal Sense of Right and Wrong* (New York: Ecco, 2006); Greg Ross, "The Bookshelf Talks with Marc Hauser," *American Scientist Online*, 2006, http://www.americanscientist.org/template/InterviewTypeDetail/assetid/52880;jsessionid=baa9 (accessed July 9, 2007).

Chapter 12: The Sun Also Shines on the Wicked

1. Nete Munk Nielsen et al., "Psychiatric Hospitalizations in a Cohort of Danish Polio Patients," *American Journal of Epidemiology* 165, no. 3 (2007): 319–24. See also B. Bandelow et al., "Early Traumatic Life Events, Parental Attitudes, Family History, and Birth Risk Factors in Patients with Borderline Personality Disorder and Healthy Controls," *Psychiatry Research* 134, no. 2 (2005): 169–79; Bruno, Frick, and Cohen, "Polioencephalitis."

2. Bruno and Frick, "Psychology of Polio."

3. Bruno et al., "Neuroanatomy of Post-Polio"; Bruno, Frick, and Cohen, "Polioencephalitis."

4. Nielsen et al., "Psychiatric Hospitalizations," referencing Bruno et al., "Neuroanatomy of Post-Polio"; Shigeo Kinomura et al., "Activation by Attention of the Human Reticular Formation and Thalamic Intralaminar Nuclei," *Science* 271, no. 5248 (1996): 512–15.

5. Bruno, *Polio Paradox*, pp. 47–54.

6. Ibid., pp. 316–17.

7. D. E. Linden, "How Psychotherapy Changes the Brain—the Contribution of Functional Neuroimaging," *Molecular Psychiatry* 11, no. 6 (2006): 528–38; Sharon Begley, *Train Your Mind, Change Your Brain: How a New Science Reveals Our Extraordinary Potential to Transform Ourselves* (New York: Ballantine Books, 2007).

8. M. J. Rietveld et al., "Heritability of Attention Problems in Children: Longitudinal Results from a Study of Twins, Age 3 to 12," *Journal of Child Psychology and Psychiatry* 45, no. 3 (2004): 577–88.

9. Raine, "Psychopathy, Violence, and Brain Imaging," pp. 50–51.

10. Peter D. Kramer, *Against Depression* (New York: Viking, 2005).

11. K. Schnell and S. C. Herpertz, "Effects of Dialectic-Behavioral-Therapy on the Neural Correlates of Affective Hyperarousal in Borderline Personality Disorder," *Journal of Psychiatric Research* 41, no. 10 (2007): 837–47 [Epub ahead of print, October 24, 2006]. Posner et al., "Psychobiology."

12. Ian Sample, "The Brain Scan That Can Read People's Intentions," *Guardian Unlimited*, February 9, 2007, http://www.guardian.co.uk/frontpage/story/0,,2009229,00.html (accessed February 10, 2007).

13. Brent Garland, "Neuroscience and the Law: Brain, Mind and the Scales of Justice," *The American Association for the Advancement of the Science and The Dana Foundation*, 2004, http://www.dana.org/pdf/books/booksummary_neurolaw.pdf (accessed February 25, 2007).

14. Bernard Harcourt, "Institutionalization vs. Imprisonment: Are There Massive Implications for Existing Research?" *The Volokh Conspiracy*, May 2, 2007, http://volokh.com/posts/1178086065.shtml (accessed July 8, 2007).

15. Stephen J. Morse, "The Non-problem of Free Will in Forensic Psychiatry and Psychology," *Behavioral Sciences & the Law* 24 (2006): 1–17.

16. Redlich, *Destructive Prophet*, p. 334.

17. Ann Rule, *Dead by Sunset* (New York: Pocket Books, 1996).

18. Eichenwald, *Conspiracy of Fools*.

19. David France, *Our Fathers: The Secret Life of the Catholic Church in an Age of Scandal* (New York: Broadway Books, 2004).

20. Adam LeBor, *"Complicity with Evil"* (New Haven: Yale University Press, 2006).

21. Carnahan and McFarland, "Revisiting the Stanford Prison Experiment."

22. Stanley Milgram, *Obedience to Authority: An Experimental View* (New York: Harper & Row, 1974).

23. Myron Peretz Glazer and Penina Migdal Glazer, *The Whistleblowers* (New York: Basic Books, 1989).

24. Waite, *Psychopathic God*, p. 396.

25. Victor, *Hitler*, p. 6.

26. Thomas Benfield Harbottle, *Dictionary of Quotations (Classical)* (New York: Macmillan, 1906), p. 198.

27. Rita Carter, *Mapping the Mind* (Los Angeles: University of California Press, 1998), p. 93.

28. Peter W. Morgan and Glenn H. Reynolds, *The Appearance of Impropriety: How the Ethics Wars Have Undermined American Government, Business, and Society* (New York: Free Press, 1997), pp. 113, 115.

29. Ibid., p. 1.

30. Kathryn M. Borman, *Meaningful Urban Education Reform* (Albany: State University of New York Press, 2005); Barbara Oakley et al., "Improvements in Statewide Test Results as a Consequence of Using a Japanese-Based Supplemental Mathematics System, Kumon Mathematics, in an Inner-Urban School District," in *ASEE Annual Conference* (Portland, Oregon, 2005).

31. Oppenheimer, *House of Hilton*, pp. 27, 28, 52, 53.

32. David R. Sands, "Researchers Peg Putin as a Plagiarist over Thesis," *Washington Times*, March 25, 2006, http://www.washingtontimes.com/world/20060324-104106-9971r.htm (accessed November 28, 2006).

33. Ana Marie Cox. "Matt Drudge," *Time*, April 30, 2006, http://www.time.com/time/magazine/article/0,9171,1186874,00.html (accessed February 26 2007); "Drudge Report Sets Tone for National Political Coverage," *ABC News*, October 1, 2006, http://abcnews.go.com/WNT/story?id=2514276&page=1 (accessed February 26 2007); Jeannette Walls, *Dish: How Gossip Became the News and the News Became Just Another Show* (New York: HarperCollins, 2000), pp. 292–304.

34. Charles Krauthammer, "The Clinton Doctrine," *Time*, March 29, 1999, http://www.cnn.com/ALLPOLITICS/time/1999/03/29/doctrine.html (accessed December 16, 2006).

Glossary

affective instability. Moods that shift inappropriately from hour to hour, or even minute to minute, without apparent justification. Often seen in *borderline personality disorder.*

allele. An alternative version of a *gene*. One allele of a *gene* for eye color, for example, might produce brown eyes, while a different allele of that *gene* might help produce blue.

altruism. Unselfish concern for others. See also *posterior superior temporal cortex* and *nucleus accumbens.*

Alzheimer's disease. The most common type of dementia in the elderly; it is characterized by misfolded and tangled proteins in the brain. An *allele* called *APOE4* may increase risk for Alzheimer's, while APOE2 may reduce risk.

amygdala. The part of the brain that determines the emotional significance of sensory information. If necessary, the amygdala reacts with anger or fear, which is why it is sometimes termed the "fight or flight" center of the brain. Disrupted amygdala function may cause the negative emotions a *borderline* feels when first appraising a person or situation.

anterior cingulate cortex. The anterior cingulate cortex, at the front of the *cingulate cortex* deep in the brain, helps us to focus our attention and "tune in" to thoughts. It is a particularly important part of the *executive attentional network* that allows us to quickly resolve conflicting information; it also plays a role in producing feelings of *empathy*. Dysfunction of the anterior cingulated cortex contributes to impulsivity and may inhibit *borderlines*' and *subclinical borderlines*' ability to focus on something they do not wish to hear.

anterior insula. An area that is activated when we feel disgust—either because we've seen something repulsive, or we've seen someone cheating. This area is also associated with food and drug craving. Cigarette smokers who suffer a stroke that damages this area can completely lose their craving for cigarettes.

antisocial personality disorder. A syndrome in which people show a pervasive pattern of disregard for and violation of the rights of others, occurring after the age of fifteen. It is characterized by traits such as deceitfulness, impulsivity, aggressiveness, and irresponsibility. *Psychopaths* form a subset of the worst of those with antisocial personality disorder—they are additionally characterized by a lack of *empathy* or remorse.

APOE4. An *allele* of the APOE *gene* that has been found to increase the risk of *Alzheimer's disease*. The risk is increased when a person has two copies of the *gene*. *Alleles* of other *genes* may amplify or reduce the risk in those with APOE4. The environment also appears to play a role—many neurologists, for example, feel that a concussion can kick Alzheimer's into gear in those with a genetic predisposition.

attentional network. The area of the brain that is able to focus on objects, people, and places. Key areas of the attentional network include the *reticular formation*, the *thalamus*, and the *cingulate cortex*. Both *psychopathy* and *borderline personality disorder* may be related in part to problems with the attentional network. See also *executive attentional network*.

autism. A developmental disorder characterized by severe deficits in social interaction and communication. *Alleles* of certain *genes* increase the risk of autism.

Baldwinian evolution. An indirect evolutionary process where changes in people's behavior cause changes in the environment, and these environmental changes consequently make changes in people's *genes*. Slash-and-burn agriculture, for example, allowed malarial mosquitoes to flourish in the ponds and puddles that replaced forests. This in turn allowed for the rapid spread of genetic mutations that help humans to cope more easily with malaria.

BDNF. "Brain-Derived Neurotrophic Factor" helps support the survival of existing neurons and encourages the growth of new neurons and synapses.

bioengineering. A relatively new discipline that integrates biology and medicine with engineering to solve problems related to living systems. Activities performed by bioengineers include research related to molecules, *genes*, cells, and tissues; design of pacemakers, prosthetic limbs, dialysis machines and medical imaging devices; work as patent attorneys; and creation of hospital and genetic databases.

bipolar personality disorder. A severe disorder that is characterized by cycles of manic "high" episodes, followed by depressive episodes. A shared genetic mechanism is suspected for the mood-shifting aspects of both *borderline* and bipolar personality disorders.

black-and-white thinking. See *splitting.*

borderline. A person with *borderline personality disorder.*

borderline personality disorder. A personality disorder that involves rapid mood swings, emotional instability, very troubled relationships, a chameleon-like ability to change attitudes or behavior, and other problems related to emotion.

borderpath. A person who shows the characteristics of both *psychopathy* and *borderline personality disorder.*

categorical approach. An approach to personality disorders that defines problematic traits for each disorder in yes/no fashion. A person is diagnosed with the disorder if a trained clinician checks enough yes's on those different defining traits. This contrasts with the *dimensional approach.*

caudate nucleus. An area deep inside the brain that is important for learning and memory. It becomes active, for example, when we are falling in love, donating to charity, or punishing cheaters.

cerebral cortex. A thin layer, about a tenth of an inch thick, of gray-colored brain tissue (*gray matter*) that coats the outer surface of the brain. The cerebral cortex is important in complicated brain functions such as thinking, language, consciousness, and memory.

chromosome. A thin, long molecule, like a microscopic thread, that has different *genes* scattered along it. There are 46 chromosomes in most cells in the human body—23 from the mother, and 23 from the father.

cingulate cortex. A vital, emotion-related part of the brain that helps us to focus our attention and "tune in" to thoughts. It is also known to be involved in bonding and social interactions. The cingulate cortex wraps around the *corpus callosum* deep in the brain. See also *anterior cingulate cortex.*

clinically significant. A symptom or trait severe enough that a trained clinician feels that it causes significant distress or impairment. An ordinary person is therefore not qualified to judge whether a person has a clinically significant problem, no matter how flagrant the problem may be.

COMT. A key *gene* underlying general intelligence—it stands for catechol-O-methyltransferase. This *gene* works by serving as the blueprint for an enzyme that breaks down *dopamine* and other *neurotransmitters*. It turns out that the more slowly you metabolize *dopamine*, the smarter you are, so if you have versions of the COMT *gene* that don't process *dopamine* well, chances are you have a higher IQ.

corpus callosum. A bundle of nerve fibers that connect the left and right sides of the brain.

Cultural Revolution. A program launched by Chairman Mao in 1966, supposedly to rid the country of all types of elites. Millions were killed over the subsequent decade. The Cultural Revolution is widely felt both inside and outside China to have been an unmitigated disaster,

dimensional approach. This approach to diagnosing personality disorders contends that dimensions, or traits, are not simply present or absent in any given disorder, but can shade imperceptibly from extremely maladaptive to normal. This contrasts with the *categorical approach* currently used in diagnosing mental disorders, where the traits, and the disorder itself, are coded as simply present or absent. The dimensional approach is useful because it often relates more directly to what is going on neurologically.

DNA. The special molecules that, when strung in a row, form *genes*. When strung in a really long row, *chromosomes* are formed.

dopamine. An important molecule that serves as a messenger to help send signals from one area of the brain to the other. Problems related to the dopamine system have been strongly related to *psychosis* and *schizophrenia*.

dorsolateral prefrontal cortex. The dorsolateral prefrontal cortex is deeply involved in the ability to think logically and rationally about various topics. This is the neural area where a person's plans and concepts are held and manipulated. People with slight problems in their dorsolateral prefrontal cortex appear to act normally; however, they may confidently, even arrogantly, draw bizarre and irrational conclusions. Dysfunction here may also help cause the *gaslighting*, *projection*, and impaired ability to reevaluate negative stimuli seen so frequently in *borderline*-like behavior. Dorsolateral dysfunction also contributes to an inability to learn from punishment.

DSM-IV. An abbreviation for the *Diagnostic and Statistical Manual of Mental Disorders* (the "DSM"—currently on its fourth, text revision edition—the DSM-IV-R). This is the standard reference used by health care professionals to help diagnose and define mental disorders.

emergenesis (emergenic). Emergenesis refers to genetic traits that, surprisingly, *don't* commonly run in families. Emergenic traits can include leadership, many different types of genius, and possibly psychopathological syndromes such as *psychopathy* and *borderline personality disorder*.

emote control. Emotional, as opposed to rational, reasoning. Strong evidence shows that human behavior is the product of both the rational deliberation that takes place in the front areas of the *cerebral cortex*, as well as emote control, which originates in the *limbic system*. Emote control is the default mode of human thought.

empathy. A trait that may be related to many areas of the brain, including the *anterior cingulate cortex* and the *mirror neurons* of the frontal and parietal lobes. People with *Williams syndrome* show great empathy, while people with *borderline* and *schizoid personality disorders*, and *psychopathy*, can show sporadic, reduced, or no empathy. As Dr. Jerold Kreisman and his coauthor Hal Straus note in *Sometimes I Act Crazy*: "The borderline is capable of great sympathy and comforting but often may lack true empathy, the ability to put himself in the other person's shoes, in appreciating how others are impacted by his behavior. Additionally, when they are hurt, their rage at those who have hurt them may be intense and cruel and devoid of concern or understanding for the other party."

evil. *Evil Genes* uses the definition of evil from the *American Heritage Dictionary of the English Language* (Fourth Edition). "Morally bad or wrong; wicked: *an evil tyrant.*"

executive attentional network. A part of the brain rooted in the *anterior cingulate cortex* that is fundamental to the ability to regulate responses to individual objects and people, particularly in conflict situations where several responses are possible.

gaslighting. A form of psychological abuse that involves denying facts. Repeated exposure to gaslighting can cause a person to become anxious, confused, and depressed.

gene. Genes are recipes made up of strings of molecules that tell cells how to form proteins.

genotype. The genetic recipe deep inside the body's cells that provides the instructions to produce, for example, red hair. Also refers to the genetic makeup, as opposed to the physical appearance (*phenotype*), of a person.

gray matter. The parts of neural tissue that are *not* covered with a white insulating substance called myelin. This contrasts with *white matter*, which are the parts of neural tissue that *are* sheathed with myelin. Autistics, who have trouble lying, have more gray matter in their brain, while pathological liars have more *white matter*.

Great Leap Forward. An economic and social plan implemented by Chairman Mao from 1958 to 1968. It was meant to transform China into a modern, industrialized communist society, but instead was effectively a "Great Leap Backwards," responsible for famines that killed millions.

hippocampus. A part of the brain that is involved in the processing of emotions and memory. This organ seems to be associated with a person's ability to "catch" contextual cues. Abnormalities in the hippocampus may explain why those with *borderline personality disorder* don't seem to be able to pay attention to important but placidly unemotional task-relevant information. Instead, their brains seem to key in on emotionally related cues—especially if these cues are negative.

identity disturbance. A poorly understood trait often seen in *borderline personality disorder*—it means "a markedly and persistently unstable self-image or sense of self." People with this trait can show a chameleon-like ability to change attitudes depending on who they are with, which leads others to conclude that they are duplicitous. Such individuals can be so inconsistent in their behavior and attitudes that they appear deeply hypocritical, as for example the person who is a strong proponent of conventional sexual values for others while being personally promiscuous. The identity of a person with this trait often revolves around a cause.

imaging genetics. A medical imaging technique that involves figuring out the size, shape, and function of organs such as the *amygdala* and *cingulate cortex* and then evaluating the same person's *genes* to see how they compare.

intermediate phenotype. *Phenotype* generally refers to a trait you can see on the outside of a body, such as red hair. Intermediate phenotype is related to a trait or part on the inside of the body, such as the liver or the *amygdala*. Many times it's easier to see if a *gene* has an effect on the intermediate phenotype inside the body than on the *phenotype* visible outside the body.

junk DNA. Molecules of *DNA* that seem to serve no useful purpose and lie around generally being bored. Occasionally, however, they turn out to control important activities in the cell.

limbic system. An older part of the brain, in evolutionary terms, that all mammals share; it is an area below our level of awareness. The constant feeding of limbic impulses to the conscious cortical areas means that the limbic system is heavily involved in emotional, rather than logical, thinking. Dysfunction in the limbic system can lead to emotional dysregulation, resulting in moodiness, depression, anxiety, and feelings of emptiness.

long/long. Long versions of certain *genes* produce lots of *serotonin transporter* (*SERT*) molecules—like a copy machine that is turned on for a longer period of time. Long/long means that a person inherited a double dose, from both the mother and father, of the long form of those *serotonin transporter*-related *genes*. This in turn can lead to nice, calm behavior—except that it can also lead to heart problems. See also *short/short*.

Mach-IV. A test pioneered by psychologist Richard Christie that gives an indication of a person's *Machiavellian* tendencies. One can be either a high or low *Mach*, but most people fall in the middle.

Mach (high and low). "High Machs" are those who are found by the *Mach-IV* test to show a substantial number of Machiavellian traits, while "low Machs" have few Machiavellian traits. Most people are in between.

Machiavellian. A person who is charming on the surface, a genius at sucking up to power, but capable of mind-boggling acts of deceit for control or personal gain. Ultimately a Machiavellian, as the term is used throughout *Evil Genes*, is a person whose narcissism combines with subtle cognitive and emotional disturbances in such a fashion as to make him believe that achieving his own desires, and his alone, is a genuinely beneficial—even altruistic—activity. Since the Machiavellian gives more emotional weight to his own importance than to that of anyone or anything else, achieving the growth of his preeminence by any means possible is always justified in his own mind. The subtle cognitive and emotional disturbances of Machiavellians mean they can make judgments that dispassionate observers would regard as unfair or irrational. At the same time, however, the Machiavellian's unusual ability to charm, manipulate, and threaten can coerce others into ignoring their conscience and treading a darker path. A synonym is *successfully sinister.*

magnetic resonance imaging (MRI). A type of imaging where water molecules in the brain are nudged to cause them to burp out electromagnetic waves similar to those that our eyes see as light. Since different types of tissue have different amounts of water, we can use clever technical legerdemain to "see" the different tissue types on photographs that look very similar to x-ray images. Functional magnetic resonance imaging (fMRI) involves sequences of MRIs that are equivalent to a motion picture.

malignant narcissism. A syndrome where features of *narcissistic personality disorder* are thought to be coupled with antisocial and paranoid traits; the term is frequently used as a diagnosis for evil political dictators. In contrast to the tens of thousands of scientific studies found through the Medline search engine in relation to other personality disorders, malignant narcissism has no scientific studies whatsoever.

MAO-A. An abbreviation for an enzyme called *monoamine oxidase A* that helps break down communication molecules like *serotonin* and *dopamine* so they don't continuously build up inside neurons. Versions of the *gene* that don't produce as much MAO-A enzyme may contribute to antisocial behavior in people who have suffered child abuse.

met. Short for *methionine*—an amino acid that is crucial in making proteins. *Genes* are sometimes referred to as being a met version, because the protein made by that *gene* has methionine in it. Met/met means that the *genes* from both the mother and the father are each met versions.

midbrain. An ancient neural area that humans share with reptiles. The midbrain houses the *reticular activating system*, which is responsible for keeping you awake and focusing attention. Overwhelming damage to this area can result in coma or death.

mirror neurons. These neurons are triggered not only when humans perform an action but also when a person witnesses another person performing the same activity—no one knows the mechanism that causes this to happen. Mirror neurons are believed to be in the frontal and parietal lobes in humans. They are thought to be a key element of *empathy*.

narcissistic personality disorder. A disorder characterized by an individual's grandiosity and exhibitionism. People with this disorder lack *empathy*, are hypersensitive to criticism, and possess a constant need for approval and admiration. There appears to be a strong genetic component.

nerve. A bundle of *neurons* that serves as a path to send signals from one part of a body to another. You use nerves to send a signal from your brain to your fingers to tell them to wiggle.

neurasthenia. An old-fashioned term for a psychological malaise that manifests itself physically through psychosomatic symptoms such as insomnia, headaches, chronic pain, dizziness, anxiety attacks, depression, and bad temper.

neuron. An electrically excitable cell that processes and transmits information.

neurotransmitter. A family of molecules that serve as messengers to help send signals from one area of the brain to the other. *Serotonin* and *dopamine* are common neurotransmitters.

nucleotide. Molecules that form *DNA*, and ultimately form *genes*. A single accidental change in a nucleotide—a genetic mutation—can cause problems, as if a cook used a teaspoon of salt instead of yeast in a recipe for bread.

nucleus accumbens. The nucleus accumbens is the common site of action for drugs such as cocaine that produce euphoria; in general, we feel pleasure when this area of the brain is stimulated. Recent research shows that even a simple gift to charity can stimulate this natural reward area.

object constancy. The ability we have to soothe ourselves by remembering the love that others have for us. Those with *borderline personality disorder* often have problems with object constancy—if a loved one is not around, for example, they may more easily fall in love with someone else.

orbitofrontal cortex. One of the three main areas of the *prefrontal cortex*—the orbitofrontal cortex rests right above the eyes. This area appears to have a key role in being able to experience feelings of compassion. Orbitofrontal dysfunction seems to release the normal brakes on aggressive and hostile impulses, can lead to cognitive-perceptual impairment, and may play a role in inflexible attitudes.

orbitomedial prefrontal cortex. One of the three main areas of the *prefrontal cortex*. The orbitomedial area is important in controlling impulses.

paranoia. An irrational distrust of others, often accompanied by delusions of persecution.

paranoid personality disorder. A personality disorder characterized by pervasive suspiciousness, distrust, and resentfulness of others. Those with this disorder can be vindictive, rigid, and good at avoiding blame.

PDQ-4+. The *Personality Diagnostic Questionnaire* is composed of a set of ninety-nine true-false questions that evaluate whether a person shows behavior consistent with any standard personality disorders.

PET. This acronym stands for *positron emission tomography*, an imaging technique that produces three-dimensional images of chemicals in various parts of the brain. Although a purist would object, these images can be thought of as equivalent to color x-rays of tissues.

phenotype. Phenotype generally refers to a trait you can see on the outside of a body, such as red hair. This contrasts with *genotype*, which is the genetic recipe deep inside the body's cells that provides the instructions to produce that red hair.

pleiotropy. The concept that one *gene* can affect many different areas of the body, from the Greek *pleio*, meaning "many," and *trop*, meaning turning to a specific direction because of a stimulus.

polygeny. The concept that a single trait can be influenced by many *genes*. For example, even though *Alzheimer's* is associated with the *APOE4 allele*, other *genes* may ameliorate the errant *allele's* effect.

posterior superior temporal cortex. This neural area is related to perceiving others' intentions and actions. Naturally altruistic people appear to have ramped up activity in this area.

prefrontal cortex. This area lies in the front area of the brain, above and behind the eyes. It allows for social control (helping prevent you from sticking your finger in a delectable pudding) and other types of executive functions, such as the ability to understand the consequences of certain actions. The prefrontal cortex is divided into three areas: lateral, *orbitofrontal* and *orbitomedial prefrontal* areas. Damage to any of these areas can cause antisocial behavior. There are many connections between this area and the *recticular activating system.*

projection. A characteristic where a person is unable to accept responsibility for something, and instead blames ("projects") responsibility for the problem onto someone or something else.

psychopath. Psychologist Robert Hare has perhaps best described psychopaths as "predators who use charm, manipulation, intimidation, and violence to control others and to satisfy their own selfish needs. Lacking in conscience and in feelings for others, they cold-bloodedly take what they want and do as they please, violating social norms and expectations without the slightest sense of guilt or regret." A synonym is *sociopath*, although this later word implies that the person gained his or her nefarious traits because of societal or familial mistreatment.

psychosis. Psychosis is characterized by *paranoia* and a disconnection with reality. It is considered to be a symptom of severe mental illness, but is not considered by mainstream psychiatry to be a diagnosis in itself. Psychosis is not exclusively linked to any particular psychological or physical state, but is often associated with *schizophrenia*, *borderline personality disorder*, *bipolar personality disorder*, severe clinical depression, and drug addiction. Although *psychopathy* and psychosis are both often abbreviated as "psycho," they are two very different terms. Psychosis seems to be affiliated with the neurotransmitter *dopamine*.

psychotic. Characterized by *psychosis*.

quantitative trait loci (QTL) model. This important model centers around the idea that groups of *genes* underlie personality traits. That is, no personality trait is determined by a single *gene* acting alone.

receptor. A receptor is a protein-based molecule on the surface of a cell, on which molecules such as *serotonin* can land; this in turn triggers a cascade of cellular activities. Some receptors serve as unwitting attachment points that viruses such as polio can lock onto and use to creep inside the cell.

reticular activating system. This system is responsible for keeping you awake and focusing attention—it is in the brainstem, a very old part of the brain in evolutionary terms. Overwhelming damage to this area can result in coma or death. See also *attentional network* and *midbrain*.

reticular formation. The pivotal area of the reticular activating system that is responsible for focusing attention, arousal, and vigilance. The poliovirus always invades this area in those who are paralyzed by polio. See also the *attentional network*.

sadism. An affliction that involves receiving pleasure from inflicting cruelty on others. Very little research has been done on this crucially important disorder, in part because psychologists realize that such research might be used by lawyers to excuse the behavior of extremely dangerous individuals. There appears to be a genetic component to sadism.

schizoid personality disorder. A person with this personality disorder is indifferent to social relationships, attempts to avoid interpersonal interactions, lacks *empathy*, and has difficulty with emotional expression. There is disagreement about whether the disorder is related to *schizophrenia*.

schizophrenia. A severe mental disorder characterized by delusions, hallucinations, *paranoia* and other intellectual disturbances. It is associated with *dopamine* imbalances in the brain. Certain *alleles* place a person at risk for schizophrenia.

schizotypal personality disorder. A personality disorder characterized by perceptual dysfunction, depersonalization, interpersonal aloofness, and suspiciousness. The individual shows "magical thinking" with a belief in special powers. It is thought by many to be a mild form of *schizophrenia.*

schizotypy. A condition where someone seems to have some of the traits of *schizophrenia*, such as a paranoid suspiciousness of others.

secondary psychopath. *Psychopaths* are generally known for their lack of remorse or anxiety. However, so-called secondary psychopaths do at least experience anxiety. These individuals show many of the characteristics of *borderline personality disorder.*

serotonin. An important molecule that serves as a messenger to help send signals from one area of the brain to the other. Serotonin plays a critical role in disorders such as depression, *borderline personality disorder*, *bipolar personality disorder*, and anxiety; it is also thought to be involved in sexuality and appetite.

SERT. Short for "*serotonin transporter*." SERT is a protein conveyer belt that helps scoop excess *serotonin* out of the cleft between *neurons* and carry it back into the trigger *neuron*.

short/short. Shortened versions of certain *genes* don't allow for production of as many *serotonin transporter* (*SERT*) molecules. Short/short means that a person inherited a double dose, from both the mother and father, of these short form *genes*. This in turn can lead to nasty traits like anxiety and impulsivity. See also *long/long*.

sociopath. See *psychopath*.

splitting. A coping process where a *borderline* swings between seeing people as either all good or all bad. On a larger scale, splitting behavior is shown when the *borderline* pits people against one another, making one group the "white hats" and the other the "black hats"—although who is considered as good or bad can shift from day to day, or even hour to hour. Also known as *black-and-white thinking*.

stable sinister system. A tightly interlocked, corrupt social group controlled by Machiavellians. Such systems may not be efficient or effective, but the Machiavellians work behind the scenes to eliminate competition with external groups and to keep naysayers in check through a mixture of intimidation and bribery.

subclinical. Showing some aspects of a personality disorder, but not to the extent that a trained clinician would feel comfortable making a diagnosis.

successfully sinister. A synonym for *Machiavellian*.

synapse. The gap between two *neurons*. *Neurotransmitters* such as *serotonin* help bridge the gap when signals are being sent from one part of the brain to another.

thalamus. An egg-shaped gray mass deep in the center of the brain that receives sensory inputs such as sight and sound and routes them on to both the *limbic system* (for emotional processing) and the *prefrontal cortex* (for rational processing).

theta rhythms. Brain structures that are working together have neurons that "blink" (fire) together—rather like harmonized fireflies. Out-of-synch brain waves may be related to mood disorders and diseases such as *schizophrenia*. The confusion and depression that results from *gaslighting* (denial of reality) may be due to disruption of theta rhythms.

transporter. A protein conveyer belt that helps scoop molecules into a cell.

val. Short for *valine*—an amino acid that is crucial in making proteins. *Genes* are sometimes referred to as being a val version, because the protein made by that *gene* has valine in it. Val/val means that the *genes* from both the mother and the father are each val versions.

ventromedial prefrontal cortex. The ventromedial prefrontal cortex helps us link subconscious to conscious thought—it plays a role in emotional cognition, as opposed to the rational cognition affiliated with the *dorsolateral prefrontal cortex*. The ventromedial prefrontal cortex is also involved in the processing of risk and fear, and gives meaning to our perceptions. Damage to this area can result in antisocial behavior and cognitive-perceptual impairment.

white matter. The parts of neural tissue that are covered with a white insulating substance called myelin. This contrasts with *gray matter*, which is neural tissue that is not sheathed with myelin. Pathological liars have more white matter in their brain, while autistics, who have trouble lying, have more *gray matter*.

Williams syndrome. Perhaps the most endearing of all diseases. Those afflicted are very polite and sociable, show great *empathy*, and are completely unafraid of strangers. The syndrome is caused by the loss of a tiny snippet of roughly 21 *genes* on *chromosome* 7.

INDEX

5-HT_{1B} role in aggression, 71
5-HT_{2A} role in self-mutilation, anorexia, and suicide, 71
5-HT_{3A} influence on amygdala and reaction time, 71–72

abandonment, frantic attempts to avoid, as DSM-IV trait of borderline personality disorder, 135, 159–60
abstract reasoning
  dorsolateral prefrontal cortex and, 203
  in psychopaths and dysfunction in right anterior superior temporal gyrus, 97–98
  relation to love, empathy, guilt, and remorse, 100
Abu Ghraib, questionable nature of Philip Zimbardo's conclusions regarding, as described in *The Lucifer Effect*, 303–304n
abuse. *See* child abuse
academia
  difficulty of eradicating older beliefs about nurture in, 38, 38n
  ease with which Machiavellian students can manipulate altruistic—or narcissistic—professors, 338n
  importance of remembering names, 312
  and its sinister aspects compared with other groups, 30
  monolithic nature of American public education system provides ready cover for Machiavellians, 336–37
  neurologically speaking, teaching ethics really *is* just preaching to the choir, 322
  students serve as indentured servants for Machiavellians, 320n
  Texas Southern University as example of "stable sinister system," 278–80
acting out, 37
actors and throwaway quip about mirror neurons, 104n
Adams, Abigail, on George Washington: if he wasn't the best intentioned man in the world he'd be very dangerous, 300
addictive behavior
  alcoholism
    of author's mother, uncles, grandfathers, 26, 122–28

borderline personality disorder and, 80, 140, 205
of Carolyn, 141–42, 324, 341–42
chromosome 2 region and genetic tendency toward, 160
MAO-A, "Cluster B" disorders and, 80
Milosevic, drinking and stress, 161
related to drug and sex addiction, 233
Russian propensity for, 177–78
serotonin transporters and, 73
stress, genetics, and, 66
drug addiction
borderline personality disorder and, 140, 205
COMT gene and, 79
MAO-A "Cluster B" disorders and, 80
Mao's, 217, 232–33, 245, 248
relation to alcoholism and sex addiction, 233
sexual addiction
general discussion of, 233–34
Mao's, 232–34
substance abuse
borderline personality disorder and, 140
DARPP-32 and, 820
MAO-A and, 82–83
produces prefrontal cortex dysfunction in antisocial personality disorder, 204–205
Adorno, Theodor W. (*The Authoritarian Personality*), 46
Adult Temperament Questionnaire, 201
affective instability (moodiness). *See also* anger; irritability
borderline personality disorder
DSM-IV diagnostic criteria for, 158
in-depth discussion of how it feels, 230
limbic system dysfunction, 193–95
mentioned as trait in, 135, 137, 140
overview related to neuroscience results, 205–206, 232
"poorly regulated emotions" as dimensional trait of, 164
Posner's studies related to affectivity and executive control, 200–202
as shared trait with bipolar personality disorder, 142n
coexisting borderline-schizotypal types show gray matter reduction in anterior and posterior cingulates, 227
heritability of, 85
in people
Callas, Maria, 300
Fastow, Andrew, Enron CFO, 295–98
Hitler, Adolf, 299–300, 305
Lauren, Ralph, 300
Mao. *See under* Mao, Chairman, personality traits and disorders
Milosevic. *See under* Milosevic, Slobodan, borderline-like and psychopathic traits
Nixon, Richard, 300
Princess Diana, 277
Shockley, William, 290
Skilling, Jeffrey, Enron CEO, 295–96
possibility of intervention to help those with, 329
prefrontal cortex dysfunction and, 180
provides for successful manipulation and control, 251–52, 297–98
serotonin transporters and, 73
"splitting" and, 143–44
theta rhythms and, 148
aggressiveness. *See also* affective instability; anger
5-HT_{1B} role in controlling offensive, 71
antisocial personality disorder and, 50
MAO-A and, 80–81
orbitofrontal cortex dysfunction releases brakes on, 94
psychopaths' difference in corpus callosum as source of, 92
psychopaths' unresponsive amygdala produces, 93, 97n
Wrangham's theory that humans are "taming" themselves, 264–65
agriculture and Machiavellians, 264–71
Ahmadinejad, Mahmoud, ability to dupe journalist Mike Wallace, 317
Akhtar, Salman, research related to identity disturbance, 156
alcoholism. *See under* addictive behavior

alleles. *See* genes *and also* individual allele names
Allende, Salvador (president of Chile), 285–86
altruism. *See also* naivete
evolutionary aspects
evolutionary model for emotional behavior of evolutionary psychiatrist Randolph Nesse, 258–59
"human cultural and behavioral diversity can be understood in the same way as biological diversity," David Sloan Wilson, 16
neurological apparatus related to conscience played role in development of, 256
reciprocal altruism, 256–57
"tit for tat" as strategy that may have helped lead to development of, 256–60
Wilson, David Sloan, work involving, 17
neuroscience related to
anterior insula activated when we feel disgust for cheaters, caudate activated when we punish them, 260
associated with increased activity in posterior superior temporal cortex, 100
trust, cooperation, and the caudate, 20–22
vasopressin, oxytocin—and gullibility, 83
social aspects
altruistic individuals whose beneficent acts serve others well
Barry Marshall and the discovery of the cause of ulcers, 307n
Bill and Melinda Gates, 319
Linda Mealey and her father's bestowal of Award for Young Investigators in her name, 253, 283
ordinary people's heroic acts to save others, 285
Socrates, Joan of Arc, the rebellious students of Tianenmen square and their kind, 305–307
a true Communist, 213
Warren Buffett, 319
altruistic individuals whose kindness allows them to be taken advantage of or suffer abuse
"congenital" cooperators, 259
Hitler's belief that humbler people responded more readily to his call, 334
kindheartedness of Laci Peterson, 255
little Adupa judged insane and killed by family in starving Ik tribe, 383n
Mao's wife Kaihui killed rather than denounce Mao—even though Mao had abandoned her, 221–22
Williams syndrome, 98–99
altruists often surprised at how ethics policies are used by Machiavellians for nefarious purposes (e.g., "bad whistle-blowers" and "moral entrepreneurs"), 335
Christian pacifists and Muslim terrorists act as altruists—so what *is* a Machiavellian, 285
intimidation by Machiavellians can keep even the most altruistic silent out of fear for their loved ones, 333–34
social aspects
altruistic individuals whose beneficent acts serve others well
Barry Marshall and the discovery of the cause of ulcers, 307n
Bill and Melinda Gates, 319
Linda Mealey and her father's bestowal of Award for Young Investigators in her name, 253, 283
ordinary people's heroic acts to save others, 285
Socrates, Joan of Arc, the rebellious students of Tianenmen square and their kind, 305–307

a true Communist, 213
Warren Buffett, 319
altruistic individuals whose kindness allows them to be taken advantage of or suffer abuse
"congenital" cooperators, 259
Hitler's belief that humbler people responded more readily to his call, 334
kindheartedness of Laci Peterson, 255
little Adupa judged insane and killed by family in starving Ik tribe, 383n
Mao's wife Kaihui killed rather than denounce Mao—even though Mao had abandoned her, 221–22
Williams syndrome, 98–99
altruists often surprised at how ethics policies are used by Machiavellians for nefarious purposes (e.g. "bad whistleblowers" and "moral entrepreneurs"), 335
Christian pacifists and Muslim terrorists act as altruists—so what *is* a Machiavellian, 285
intimidation by Machiavellians can keep even the most altruistic silent out of fear for their loved ones, 333–34
Alzheimer's disease
of author's father, 114, 324–25, 343
complex genetics of, 64
and connection to genetics of polio susceptibility, 327
copy number variants and, 63
ambition. *See* control, desire for
Amin, Idi (Ugandan dictator), brief overview compared to other dictators, 28
amnesia of author's mother, 113
amygdala
activation in intentional versus accidental transgression of social norms, 257–58
in borderline personality disorder
dysfunction causes emotional overreaction, 194–95, 206, 209
role in executive control
discussion, 199
flowchart, 196
smaller according to imaging results, 194, 205
genetic aspects
5-HT_{3A} influence on, 71
MAO-A alleles can produce smaller, 80–81
serotonin transporters' influence on signal to and from cingulate cortex, 74–75
illustrations of, 73, 93, 101, 183
murderers have heightened activity in, 97, 97n
psychopaths have unresponsive, 93, 97n
role in determining emotional significance of information
discussion, 186
flowcharts, 186, 196
Williams syndrome differences in function of, 99
amyotrophic lateral sclerosis: Mao's fatal illness, 248
anger. *See also* affective instability; aggressiveness; irritability
in borderline personality disorder
in context with overall mood swings, 230
as DSM-IV diagnostic criteria, 158
intense rage at those who hurt them, 142
as manipulative tool, 140
prone to abrupt in workplace, 149
related to clinical versus subclinical borderlines, 201
shared trait with bipolar personality disorder, 142n
in child with psychopathic-like traits, 102
in people *without* apparent signs of affective instability (otherwise *see under* affective instability)
Gandhi, Mahatma: "violent nature," 298–99

general discussion, 298–300
Washington, George: capable of terrible wrath, 298–300
anorexia. *See* eating disorders, anorexia
anterior cingulate cortex
ability to resolve conflicting information as part of executive attentional network, 202, 206, 207
in borderline personality disorder
right is smaller, 205
role in executive control
discussion, 199
flowchart (as "ACC"), 196
borderline-schizotypal types show gray matter reduction in anterior and posterior cingulates, 227
Crick, Francis, free will and the, 328
illustrations of, 101, 181
impulsivity and, 193, 199
linkage with limbic (emotional) systems, 186
role in helping focus attention, empathy, memory storage, 182
anterior insula activated when we feel disgust for cheaters, 260
antisocial behavior. *See also* antisocial personality disorder; individuals by name
complete review of neurological findings related to, 370n25
COMT alleles and, 79
fear conditioning reduces, 95
MAO-A alleles and, 81–82
PDQ-4+ and dimensional approaches to quantifying, 133
prefrontal cortex damage can produce, 94
as viable traits flowing from forces of evolution, 254–56
as winning social strategy, 260–61
without callous, emotionless features of psychopathy, 56
antisocial personality disorder
DSM-IV definition, 50–51
general definition, 135
heritability is spotty, 86
MAO-A and, 54, 80
Medline's number of studies about, 33
no single gene causes, 68
overlap with borderline personality disorder, 208–209
psychopathy in relation to, 51
sadism and, 52
anxiety. *See also* affective instability; neurasthenia, Mao's; neuroticism
BDNF alleles and, 77–78
borderline personality disorder and, 140, 158
COMT gene and, 79
limbic system dysfunction in borderlines and, 193–95
of Mendel, Gregor, 288
polio and, 116
psychopaths' lack of, 92–93
serotonin transporters and, 73
sexual addiction commonly found with, 233
shared trait of borderline and bipolar personality disorders, 142n
APOE4 allele
Alzheimer's disease, other good and bad features, 76
connection to genetics of susceptibility for polio, 327
Appearance of Impropriety, The (Peter Morgan and Glenn Reynolds), 335
Arafat, Yasir, duping of journalists and policians, 316–17
Arendt, Hannah: "banality of evil," 303n
aristocracy
attraction of certain personality types toward, 35, 277
enjoyed enormous reproductive benefits in many societies through history, 266–70
Aristotle's distinction between knowledge and moral virtue, 52
Armstrong, Edwin, invention of FM radio (hijacked by unsavory Lee de Forest), 291
arrogance. *See also* narcissism; narcissistic personality disorder
as DSM-IV trait for narcissistic personality disorder, 244
in Mao's companion women, 225

Milosevic's, 153, 171
psychopaths, narcissists, and Machiavellians share trait, 132
in relation to dimensional trait of psychopathy, 167
Shockley, William, of, 290
ASPM gene (cognition), 262
Asquith, Christina, reporter: corruption at Texas Southern University, 279–80
Assad, Hafez, Syrian dictator, ability to dupe journalist Mike Wallace, 317
Ataturk, Mustafa Kemal
brief overview compared to other dictators, 286, 301
mental flexibility despite presence of mood disorders, 314
attention. *See* executive attentional network
attentional network. *See* executive attentional network
attention deficit disorder
illustration showing overlap with borderline personality disorder, 208
MAO-A and, 80–81
polio and, 116
Auerbach, "Red," mental flexibility of, 301
Authoritarian Personality, The (Theodor Adorno), 46
autism
gray matter and, 106
mirror neurons and, 105
avoidant personality disorder, defined, 135
Axelrod, Robert, and game "Prisoner's Dilemma," 257–58
Axis I and Axis II of DSM-IV, defined, 133–34

Babiak, Paul, and Robert Hare (*Snakes in Suits: Why Psychopaths Go to Work*), 107
Bafti, adopted son of author and her husband, 151–52, 153, 169–70
Baldwin, Mark, 264
Baldwinian evolution, 263–64
Ballas, Jerry, on "Chainsaw" Al Dunlap: "it's terrorizing working for the man," 294
"banality of evil," Hannah Arendt, 303n
Bardeen, John, Nobel prize winner blocked by William Shockley, 290
BDNF (brain-derived neurotrophic factor) gene, 77–78
Beatles, The: assistance by manic-depressive "drama queen" Brian Epstein, 291
behaviorism, restrictive effect on research, 41, 174
Behr, Edward (Ceausescu biographer), 270–71
Betzig, Laura, *Despotism and Differential Reproduction*, 268
Bing, Stanley (*Rome, Inc.*), 276n
bioengineering, description of, 31
bipolar personality disorder. *See also* hypomania
BDNF alleles and, 78
chromosomes and genes specifically related to, 54, 160
placement in Axis I of DSM-IV, 134
serotonin receptor alleles, effect on, 72
ventromedial cortex, overactive in bipolars who find meaning in everything they do, 182
birth control programs, 191
Black, Conrad, indicted Hollinger CEO, 313, 338
black-and-white thinking. *See* relationships, unstable personal, "splitting"
"black Norwegians," 118
Blagojevich, Rod, US representative, 165
blame shifting. *See* projection
Blank Slate, The (Steven Pinker): helped shift researchers away from "people are naturally good" idea, 37, 175
Blink (Malcolm Gladwell): "emote control" as dark side of *Blink*, 379n20
Bohus, Martin, neurobiology of borderline personality disorder, 232
borderline. *See* borderline personality disorder
borderline personality disorder (BPD)
attraction of females with borderline personality disorder toward those with power, 277

"borderpath," 168
effect of immune system on, 207
evolutionary perspective on the disorder by psychiatrist Regina Pally, 282
general definition of, 135
heritability of, 85, 85n
intermediate phenotype and, 66
litigation and unnecessary problems created by those with, 149
Medline's number of studies about, 33
neuroscience behind
cognitive-perceptual impairment: dorsolateral, ventromedial, and orbitofrontal systems, 203–205
impulsivity: anterior cingulate and orbitomedial prefrontal systems, 195–202
anterior cingulate cortex dysfunction inhibits ability to focus on what they don't want to hear, 186
limbic system dysfunction, 193–95
MAO-A effects in "Cluster B" personality disorders (includes borderline personality disorder), 80
N-acetylaspartate compounds found in dorsolateral prefrontal cortex, 205
overview of, 193, 205–207, 232
parietal lobe abnormalities may contribute to identity disturbance in borderlines, 198–99
psychotic symptoms increase with decrease in size of right parietal lobe, 199
reduction in hippocampus size increases borderline symptoms, 199
overlap with other disorders
antisocial personality disorder, 137
bipolar personality disorder, 142n
narcissistic personality disorder, 146
a wide variety of disorders, 168, 208–209, 255n
possibility of intervention to help those with, 329
remorse felt by those with, 139
serotonin in mood disorders such as, 69–75, 184
teddy bears, Binky, and, 384n16
traits and characteristics of
affective instability (moodiness). *See* affective instability; anger
cognitive dysfunction and delusions. *See* cognitive dysfunction
dimensional description, 163–64, 166
dissociative symptoms, 158
DSM-IV description, 157–59
gaslighting. *See* gaslighting
identity disturbance. *See* identity disturbance
impulsivity. *See* impulsivity
lack of object constancy, 221
overview of traits and behaviors of those with, 136–40, 142–50
projection. *See* projection
relationships, unstable personal. *See* relationships, unstable personal
situational competence, 145–46
"splitting." *See* relationships, unstable personal, "splitting"
suicidal or self-mutilating behavior, 159
various individuals with borderline-like traits. *See* Diana, Princess; Hitler, Adolf; Mao, Chairman; Milosevic, Slobodan
Borderline Personality Disorder, Demystified (Robert O. Friedel), 136
"borderpath"
defined, 168
efficacy in leading others toward sinister activities, 303n–304n
as emergenic type with constellation of unfortunate personality characteristics, 315
Mao as, 216–18
Milosevic as, 167–68
Bowen, Ray, on Enron's intimidating CFO, Andrew Fastow, 295–96
BPD. *See* borderline personality disorder
brahmacharya (Hindu ascetic practice of restraint and control), 299
brain-derived neurotrophic factor. *See* BDNF

brain stem nuclei and hypothalamus
  portrayed in flowchart form (as "Hypo/BSN"), 196
  role in producing automatic emotional response, 185
Brando, Christian (Marlon Brando's son), 96
Branson, Louise (Milosevic biographer), 154, 158
Brass, Paul, "riot specialists," 303n
Brown, Oliver, a "TSU 3" hero, 279–80
Brown, Tina (*The Diana Chronicles*), 391n54
Bruno, Richard, observations on polio, 111, 114–16
Buffett, Warren, ethical, emergenic genius, 317–19
bulimia. *See* eating disorders, bulimia
Buller, David J., modular brain theory controversy, 388n12
Bush, George W.
  initial naivete regarding Putin, 315–16
  in relation to neuroimaging study on partisanship, 189–90
business. *See also* Black, Conrad; Buffett, Warren; Carnegie, Andrew; Dunlap, "Chainsaw" Al; Enron; Gates, Bill; Stewart, Martha
  advantages and disadvantages of Machiavellians in business, 336
  attraction of females with borderline personality disorder toward those with power, 277
  Babiak and Hare's *Snakes in Suits: When Psychopaths Go to Work*, 107
  corruption in business reaches awe-inspiring proportions before concrete action is taken, 333
  Machiavellians, with their distorted, self-serving cognitions, can savage their companies, 331
  rise of Machiavellians in a growing business means others redirect themselves toward newer, less corrupt businesses, which in turn become corrupt as Machiavellians are attracted to new nexus of power, 336–37
busing programs to help integrate school systems, 191
"Butcher of the Balkans." *See* Milosevic, Slobodan

Caesar, Julius, could not resist temptation to stay in power, 298
Caligula, chameleon-like behavior of, 276
Callas, Maria, temper of, 300
capitalism. *See* politics
Carnahan, Thomas: research on self-selection of personality types for positions that suit disposition, 303n–304n
Carnegie, Andrew: brilliant Machiavellian, 336
Carnes, Patrick, on sex addiction, 233
Carolyn
  author's first adult memories of, 119–20
  early years, 110–14, 124–27
  final years, 323–25, 331
  final years—the conclusion, 340–43
  last words, 23, 107, 343
  love letters, 120–22, 127–29
  missing decade, 117–19
  personality-related traits and characteristics
    alcoholism, 140–42, 340, 341–43
    anorexia, 140–42, 340
    bipolar personality disorder, 142n
    borderline personality disorder, 142, 340
    charisma, 128
    intelligence, 128
    shame, feelings of, 140
  photographs of, 117, 118, 143, 324, 340, 342
  polio and her attentional network, 114–16, 326–27, 340
  stealing mother's boyfriend, 26–27
Carter, Jimmy, duped by Ceausescu, Arafat, Kim Il Sung, and other Machiavellians, 316
Caspi, Avshalom, childhood maltreatment and MAO-A, 80–81
Castro, Fidel
  hypomanic qualities, 314
  remarkable memory, 312

Catherine the Great, enlightened despot, 286
Catholic Church
compared with sinister aspects of other groups, 30
pedophilia, 35, 107, 332
caudate
activated—and we feel satisfaction—when we punish cheaters, 260
cooperation and trust, 20
Ceausescu, Nicolae (Romanian dictator)
brief overview compared to other dictators, 28
in context with other poorly schooled leaders, 308
hypomanic qualities, 314
malevolent emotional contagion seen in ordinary Romanians, 270–71
Center for Advanced Study in the Behavioral Sciences, 40
Centurion Ministries and Jim McCloskey, 188
cerebral cortex
emotion and, 181–82, 186
illustration of, 93
illustration of four key components of, 181
in-depth discussion of various components of, 179–82
CFO Magazine: names Enron's CFO Andrew Fastow CFO of the Year, 295
Chang, Jung (Mao biographer), 218–19, 239–40
chameleon-like behavior. *See* identity disturbance, chameleon-like behavior
charisma
as advantage for Machiavellians, 282, 297
Carolyn's, 128
of Enron's CEO, Jeffrey Skilling, 296
powerful men attracted to charisma of troubled, sometimes deeply sinister women, 277
of Princess Diana, 277, 391n54
role of memory in charm and, 312–13
of Texas Southern University's corrupt president, Priscilla Slade, 280
"cheaters"
caudate activated (and we feel satisfaction) when we punish, 260
does the percentage of "cheaters" influence culture, 270–71
have led to evolutionary arms race, 258
as Machiavellians, 255–56
Cheng, Nien, suffering during Cultural Revolution, 215–16
Chhang, Youk, haunted by memories of heckling couple being buried alive, 303n
"chicken," game of, exemplifies benefit of seemingly irrational emotional strategies, 260–61
child abuse
interference with development of executive control can cause subclinical to descend into clinical borderline, 202
in Mao, Stalin, Hitler, and Abraham Lincoln, 219, 219n
MAO-A alleles and, 54, 81–82
psychosocial versus neurobiological "push," 95
Chinese speakers versus English speakers, neurological differences of, 175–76
Chirot, Daniel, competition for power rarely won by faint of heart, 314
Chomsky, Noam, 174–75
Chou En-lai (Zhou Enlai, premier of China), 239
Christie, Richard, 40–48, 132, 133, 231, 268, 303n
chromosomes
explanation and illustration of human, 60, 61
illustration of illnesses associated with chromosome seventeen, 64
Churchill, Randolph, talentless, egotistical son of Winston, 293n
Churchill, Winston
alcoholism and depression, 307
benefits of his impassioned "emote control," 188, 293
"I am so conceited . . ." [the point being, he really was], 293, 293n
intelligence, 293
mental flexibility, 301, 314

remarkable memory, 313
Stalin's ability to fool, 29–30
cingulate cortex
illustration, 73
MAO-A alleles and decreased reaction in, 80–81
serotonin transporters' influence on signal to amygdala, 74–75
cingulate gyrus, MAO-A alleles can produce smaller, 80
Cixi, Empress, 27
Clark, Wesley, General, NATO commander, 171
clinically significant
inherent flaw in DSM-IV use of concept, 375–76n32
in relation to borderline personality disorder, 162–63
Clinton, Bill
excellent memory, 313
gullibility regarding Saddam Hussein, 316–17
temper, 300
clock gene, 233
Cluster A, B, and C personality disorders
general description, 133–34
MAO-A and Cluster B personality disorders, 80
Cochran, Gregory
argues against historical theory that only social forces matter, 267
Ashkenazi genetic mutations and intelligence, 87
cognitive dissonance, neuroimaging study reveals processes underlying, 190
cognitive dysfunction
anorexia and, 142n
borderline personality disorder
anterior cingulate cortex dysfunction and inability to focus on something undesirable, 182
as dimensional trait of, 164
as heritable trait in, 85
irrationality under effect of strong emotions, 204
overview related to neuroscience results, 205–206
paranoid thinking (a form of cognitive dysfunction) as trait to define personality disorder used by DSM-IV, 164
delusional thinking
outright, 165, 302–307
in schizophrenia and schizotypal personality disorder, 135, 227
effect of stress on, 202
"end justifies the means" behavior, 204
irrationality provides successful strategy for manipulation and control, 260–61
in Machiavellians, 209
in Machiavellians as part of precise definition used in this book, 281
neuroscience behind
anterior cingulate cortex role in focus and attention, 182
cognitive dissonance, neuroimaging study reveals processes underlying, 190
dorsolateral prefrontal cortex and, 181–82, 203
lack of common sense in those with damage to dorsolateral and ventromedial areas, 203
prefrontal cortex dysfunction and, 180
role in irrationally inflexible behavior, 204
ventromedial cortex and, 182, 203
paranoia (a form of cognitive dysfunction)
provides for success in dangerous social structures, 250
seen in individuals with subclinical symptoms of borderline personality disorder, 201
in people
Diana, Princess, 277
general discussion of good and bad effects, with different examples, 300–307, 306n–307n, 314–15
Lay, Ken, Chairman of Enron, 296–98
Mao. *See under* Mao, Chairman, personality traits and disorders

Milosevic. *See under* Milosevic, Slobodan, borderline-like and psychopathic traits
Skilling, Jeffrey, Enron CEO, 295–98
possibility of intervention to help those with, 329
somatic-marker hypothesis, 203
suicide and, 166
cognitive therapy, resulting changes in brain chemistry, 68
Coleman, Roger, smooth ability to lie about rape and murder, 188–89
Collier, Norma, on Martha Stewart: "She's a sociopath . . . ," 293
Commodus, putative son of Marcus Aurelius, 275
common sense
lack of in those with damage to dorsolateral and ventromedial areas, 203
and Wiener's theories of Jewish family structure, 86
communism. *See under* politics
compassion. *See* empathy
"*Complicity with Evil*" (Adam LeBor), describes corruption at the United Nations, 332
COMT (catechol-O-methyltransferase) "warrior or worrier" gene
general effects of, 78–80
shades of intellect-emotion trade-off in Wiener's analysis of Jewish family structure, 86
conduct disorder in children, 102
confirmation bias. *See also* projection bias
explanation of, 179
neuroimaging study reveals processes underlying, 190
in Soviet perspectives about communism versus capitalism, 179
Conquest, Robert, and *The Great Terror* (or, *I Told You So, You Fucking Fools*), 28n
conscience
ability of Machiavellians to seduce others into ignoring their, 281
fear plays role in development of, 93, 95
imaging studies related to, 100–102
of a narcissist is flexible, dominated by self-interest, 247
neurological apparatus related to conscience played role in development of altruism, 256
psychopaths lack, 29n, 93, 95
conspiracy theorists and delusional thinking, 302, 306n–307n
control, desire for. *See also* manipulation
as coping characteristic in borderline personality disorder, 137, 140, 145
historical examples of good and bad aspects of people with desire for control, 298–99, 308–10
obsessive-compulsive personality disorder and, 135
power, aristocracy, and wealth as magnets for, 277, 337–38
social dominance and neurotransmitter levels: self-medication, those with desire for, 238
Cooke, David, studies of psychopathy across cultures, 265–66
dimensional description, 167
cooperation
caudate, 20
evolutionary aspect, 259–60
copy number variants, 63
"corporate" psychopaths, 106–107, 108
corpus callosum
illustration of, 93
psychopaths' differences in, 92
corruption
in democracies compared to corruption under dictators, 251
in the Ottoman empire, 271–72, 274
rise of Machiavellians means others redirect themselves toward less corrupt systems, which in turn become corrupt as Machiavellians are attracted to new nexus of power, 336–37
in the Roman empire, 275–76, 276n
at Texas Southern University, 278–80
at the United Nations, as described in Adam LeBor's "*Complicity with Evil*," 332

Cosmides, Leda, 175, 388n12
countertransference, 37
credulousness. *See* naivete
Crick, Francis: free will and the anterior cingulate cortex, 328
Crnobrnja, Mihailo, 155
Cromwell, Oliver, could not resist temptation to stay in power, 298
cultural relativism. *See under* politics
Cultural Revolution, 215–16, 235, 237, 239, 249
culture and its relation to morality and Machiavellians. *See also* emotional contagion
  basic features of morality appear to be hard-wired—not product of culture, 322
  can culture create Machiavellians—or Machiavellians create culture?, 264–65, 268–71
  difference between the terms "sociopath" and "psychopath," 51
  Enron's mandated top-down culture of greed, 294
  "feel good" politics, 187–92
  framing lenses, 174–79
  "human cultural and behavioral diversity can be understood in the same way as biological diversity" (David Sloan Wilson), 16
  in the Ottoman empire, 271–72, 274
  "projection bias": the danger of assuming those from other cultures think as we do, 378n8
  psychopathy across cultures, 265–66
  in the Roman empire, 274–76, 276n

D4DR and novelty seeking, extroversion, 82
DARPP-32, intelligence and schizophrenia, 82–83
Darwin, Charles, illustrates advantages of narcissism, 288–89, 292
death penalty, 188, 191
De Beauvoir, Simone, as Mao's dupe, 241
Debunking 9/11 Myths (David Dunbar and Brad Reagan), 302
deceitfulness. *See also* gaslighting; lying; manipulation
  antisocial personality disorder and, 50, 135
  as dimensional trait of psychopathy, 167
  problem of detecting deceitful Machiavellians, 332–39
  as strategy in "tit for tat," 258
defense, psychological: neuroimaging study reveals processes underlying, 190
defense mechanisms, 37
De Forest, Lee, hijacked Edwin Armstrong's invention of FM radio, 291
De Gaulle, Charles (president of France), temper and remarkable memory of, 313
deinstitutionalization and resultant increased prison populations, 330
delusional thinking. *See under* cognitive dysfunction
democracy. *See* politics, capitalism
Democratic party. *See also* politics
  in relation to neuroimaging study of political partisanship, 189–90
Deng Xiaoping (leader of China), 248, 309n
denial. *See* gaslighting; lying; manipulation
dependent personality disorder, defined, 136
depersonalization, a trait of schizotypal personality disorder, 135
depression
  borderline personality disorder and, 140, 149
  genetic effects
    BDNF alleles, 77–78
    chromosome 2 region, 160
    DARPP-32, 82–83
    MAO-A, 80–81
    serotonin transporters, 73–75
    stress in relation to, 66
  neural characteristics
    limbic system dysfunction in borderlines, 193–95

negative moods generated by right hemisphere, 92
theta rhythms, 148
ventromedial cortex, inactive in depressed people who find no meaning in what they do, 182
in people
Lincoln, Abraham, 219n
Mao, 224, 229–32
Milosevic, 161
Wiener, Norbert, 86
polio and, 116
Derby, Lord, on Winston Churchill: "He is absolutely untrustworthy . . . ," 285
despotism, as discussed by researcher Laura Betzig, 268–70
Despotism and Differential Reproduction (Laura Betzig), 268
despots. *See* dictators
devaluation and idealization, alternating between. *See* relationships, unstable personal, "splitting"
Diagnostic and Statistical Manual of Mental Disorders. *See* DSM-IV
dialectical materialism, 231n
dialectic-behavioral therapy to help borderlines (developed by Marsha Linehan), 329
Diana, Princess
attraction toward aristocracy, 35
mood swings, 277
Tina Brown's *The Diana Chronicles*: Diana's bulimia; eating, mood disorders in Diana's family, 391n54
Diana Chronicles, The (Tina Brown), 391n54
DiCaprio, Leonardo, mimicking abilities, 104n
dictators. *See also* individual dictators by name
ability to take advantage of naivete and narcissism of others, 315–17, 321
brief comparative overview, 28–30, 285–86, 321–22
delusions of, 304–305
desire for control, 298, 308–309
with dictators, one often has to chose which son of a bitch to support, 339
dysfunctional traits used as manipulative tools by, 315
emergenic qualities, 314–15
Glad, Betty: article "Why Tyrants Go Too Far," 34
inflexibility of, 301–302
intelligence, memory, hypomania of, 310–14
narcissism of, 287, 297–98
personality in relation to ideology of, 307–308
neuroscience, not mentioned in relation to, 34
temper of, 299–300
would-be dictators always poised at the wings, ready to take power, 332–35
Dilas, Milovan, "The hardest thing about being a communist is trying to predict the past," 211
dimensional approach to
borderline personality disorder, 163–64, 166
personality disorders in general, 131–32
Dingshan, home of Yixing teapots, 212
Discourses, The (Niccolo Machiavelli), 46
dissociative symptoms, as DSM-IV diagnostic criteria for borderline personality disorder, 158
distrust. *See also under* cognitive dysfunction paranoia; paranoid ideation
Milosevic's, 157–59
as a trait of paranoid personality disorder, 134
as a trait of schizotypal personality disorder, 135
Ditka, "Iron" Mike, temper of, 300
Djindjic, Zoran, mayor of Belgrade, 165
DNA
explanation of, 61–62, 64
in Roger Coleman death penalty case, 188
so-called junk, 62, 65, 263
Doder, Dusko (Milosevic biographer), 154, 158

dopamine
COMT gene and relation to intelligence, 78–80
in context with other neurotransmitters, reward system, and movement, 184
and delusional thinking, 304
imbalance in borderlines, 78–80
reticular activating system, poliovirus, attention, and, 114–16
social dominance and neurotransmitter levels, 238
ventral tegmental area role in production of, 196
Dorpat, Theodore (*Gaslighting . . .*), 147
dorsal raphe nucleus: role in production of serotonin, 196
dorsolateral prefrontal cortex
antisocial behavior produced by damage to, 94
dysfunction and inability to learn from punishment, 94
function of and dysfunction in, 181–82
illustrations, 94, 101, 181
linkage with limbic (emotional) systems, 186
N-acetylaspartate compounds found in borderlines, 205
role in
cognitive-perceptual impairment in borderlines, 203–207
commonsense decision making, 203
rational cognition, 186
Dreelin, Elizabeth, borderline behavior in a hospital setting, 143–44
Drudge Report, The, role in reporting on Machiavellians, 251, 338
drug addiction. *See* under addictive behavior
DSM-IV (*Diagnostic and Statistical Manual of Mental Disorders*)
defined, 47
diagnostic criteria for personality disorders
antisocial, 50–51
borderline, 158–59
narcissistic, 244
distinction between Axis I and Axis II, 133–34
exclusion of sadism from, 52
explanatory listing of the ten Axis II personality disorders, divided into clusters A, B, and C, 134–36
lack of connection with Machiavellianism, 47
problems with "clinically significant" strategy for diagnosis, 162–63
Duesberg, Peter (*Inventing the AIDS Virus*), denies standard treatment necessary for AIDS, 306n–307n
Djukic, Slavoljub (*Milosevic and Marković: A Lust for Power*), 155
Dunlap, "Chainsaw" Al
presented Sunbeam as having stunning profitability when it was going broke, 294
repulsive personality of, 30
duplicity. *See* deceitfulness; gaslighting; lying; manipulation
Durkheim, Emile, 50
Duvalier, François "Papa Doc" (Haitian dictator), 286
hypomanic qualities of, 314
intelligence, enjoyment of torture, 311
dyslexia, 57
dysphoria in borderline personality disorder, 158

eating disorders. *See also* Diana, Princess
anorexia
$5\text{-}HT_{2A}$ and, 71
BDNF alleles and, 78
borderline personality disorder and, 140
Carolyn's, 142, 142n
cognitive dysfunction, spiritual states, attempts to self-medicate and, 142n
bulimia
BDNF alleles and, 78
borderline personality disorder and, 140
serotonin transporters and, 73
stress, genetics, and, 66

Eden, Anthony (British prime minister), duped by Hitler, 315
Effective, Efficient Professor, The (Philip Wankat), the importance of remembering names, 312
effortful control, its heritability and relation to negative affectivity, 201
Egan, Michael, work at NIH, 66–67, 363n9
Einstein, Albert, Theory of Relativity was nearly named Invariance Theory, 322n
Einzatgruppen, Nazi, 303n
emergenesis (emergenic phenomena)
 general explanation of, 83–85
 of Machiavellian leadership, 314
"emote control." *See also* emotion
 borderline personality disorder and defaults in "emote control" system
  limbic system, 193–95
  overview, 193
 comprehensive discussion of various aspects of, 187–97
 defined, 187
 McCloskey, Jim, as an example of decision making using "emote control," 188–89
 political partisanship and, 189–90
 role in abetting Machiavellians, 333
emotion. *See also* "emote control"
 cerebral cortex and effect of, 181–82, 186
 delusional thinking and, 304
 evolutionary model for emotional behavior, 258–59
 executive control and, 199–202
 role in shaping "rational" thinking, 187–92
 role of limbic system in, 182–83, 186
emotional contagion
 in Ceausescu's Romania, 270–71
 Enron's top-down mandated culture of greed, 295
 in Hitler's underlings, 305
 an increase in Machiavellians in a population could increase Machiavellian behavior in everyone through, 270–71
 in Stalin's underlings, 309
 worship of Mao, 242, 252
emotional experience, deficient, as dimensional trait of psychopathy, 167
emotionality. *See* affective instability
empathy
 abstract reasoning and anterior cingulate cortex, relation to, 97–98, 100, 182
 lack of in personality disorders
  antisocial, 51
  borderline, 142, 149, 228
  narcissistic, 135, 244
  people
   Mao, 225–27, 227n
   Milosevic, 162, 166
   Stalin, 30
  psychopathy, 51, 56
  schizoid personality disorder, 135
 medial prefrontal cortex, role in, 99
 mirror neurons and, 104–105
 Williams syndrome's abundance of, 98–99
emptiness, as DSM-IV diagnostic criteria for borderline personality disorder, 158
"end justifies means" behavior, 204
endophenotype. *See* phenotype, intermediate
Engineering in Medicine and Biology Society, 211–12
English speakers versus Chinese speakers, neurological differences of, 175–76
Enron
 compared with sinister aspects of other groups, 30
 corrupt rise and fall of, 294–98
 difficulty in stopping Enron's illegal activities despite flagrant nature, 332
environment, effect on personality. *See* nurture
envy, 269
 a DSM-IV trait of narcissistic personality disorder, 244
 is possibly a useful, genetically linked trait, 269

lack of in Alfred Russel Wallace, 292
in Mao's China—creates danger and induces people not to be creative, 215
Epstein, Brian, manic-depressive "drama queen" who launched Beatles, 291
ethics
imaging studies related to, 100
systems of ethics can be ineffectual, even counterproductive, 335
"Evening of Russian Poetry, An," excerpt from Vladimir Nabokov's poem, 174
evil
allusion to possible benefits of, 32
defined, 321–22
and free will, 328–29
genes (in that these genes can contribute to personality disorders as well as positive personality traits), 69–83, 85–87, 314–15
older beliefs about personality and, 37–38
questionable nature of Philip Zimbardo's conclusions about evil, described in *The Lucifer Effect*, 303n–304n
some children born with marked (genetic) tendency toward, 57–58
evolution
Baldwinian, 263–64
evolutionary benefits of psychopathic traits, 254–56, 387–88n2
evolutionary model for emotional behavior, 258–59
evolutionary perspective on borderline personality disorder, "psychiatrist Regina Pally," 282
general definition and discussion of, 65
gradual development of rational as well as emotional capacities in humans, 187
"human cultural and behavioral diversity can be understood in the same way as biological diversity" (David Sloan Wilson), 16
Laura Betzig's *Despotism and Differential Reproduction*: preferential reproductive opportunities of Machiavellians, 268–70
occurs much more quickly than had been supposed, 261–63
executive attentional network
anterior cingulate cortex and the, 182, 202
Carolyn and her, 326–27
Joseph Newman's work on psychopathy as a form of learning disability, 103, 326–27
reticular activating system, poliovirus, and the, 114–16
what of those who don't have ability to focus attention on self and change, 328–29
executive control
in clinical versus subclinical borderline personality disorder, 199–202
coexisting borderline-schizotypal types show gray matter reduction in anterior and posterior cingulates, 227
flowchart describing pathways of, 196
role of DARPP-32 gene in, 82–83
executive dysfunction in
antisocial children and adults, 93–94
borderline personality disorder, 232
exhibitionism
Mao's, 223–24
narcissistic personality disorder and, 135
of Shockley, William, Nobel Prize winner, 290
extroversion and D4DR, 82
Eysenck, Hans J. (*The Psychology of Politics*), 46

Fairbank, John, Harvard professor, as Mao's dupe, 241
Fastow, Andrew, Machiavellianism trumps incompetence in CFO Fastow's rise and Enron's fall, 295–97
father of author
Alzheimer's disease of, 25, 114
love for Carolyn, 122, 126, 323–25
photographs of, 25, 123, 324

Faustina, wife of Marcus Aurelius, 275, 275n
fear
and development of conscience, 95
lack of in psychopaths and relation to genetics, 93
serotonin transporters and inability to suppress, 74–75
Williams syndrome phobias, 99
Feynman, Richard: "you are the easiest person to fool," 307n
Fischer, Louis (Gandhi biographer), 298–299
Fleishman, Jeffrey (journalist), description of defiant, arrogant Milosevic, 153
forgiveness, affiliated with activation of specific areas of brain, 102
framing lens, 176
Frank, Anne, diary of, 27
Frankfurter, Felix, Supreme Court Justice, on the Holocaust: "I cannot believe you," 321
Franklin, Rosalind, codiscoverer of structure of DNA, 290, 392n13
free will and evil, 328–29, 330n
"frequency dependence"
allusion to (without use of term), 256
discussion related to, 259–60
Friedel, Robert O.
Borderline Personality Disorder, Demystified, 136
dimensional description of borderline personality disorder, 164
discussion of borderline sister, Denise, 136
Friedman, Russell, "You can't love someone into mental health," 320
Fukuyama, Francis (*Trust*), 219
functional magnetic resonance imaging (fMRI) defined, 90

Galileo, 306
gambling, as addictive behavior, 233
Gandhi, Mahatma
"had a violent nature," took pains to control, 298–99, 328
hypomanic qualities and remarkable memory, 313
as unsuccessful role model for Kosovar resistance against Serbs, 169
gaslighting. *See also* lying; manipulation
defined, 146–47
as demonstrated by
Hitler and his henchmen, 305
Milosevic, 164–65
dorsolateral prefrontal cortex and, 182
etiology of term, 146n
Gaslighting . . . (Theodore Dorpat), 147
theta rhythms may be thrown off by, 148
used as coping charactistic in
antisocial personality disorder, 209
borderline personality disorder, 137, 209
used for successful manipulation and control, 147, 250–52
Gates, Bill
ability to argue competently with him is prized, 296n
philanthropy (with Melinda Gates), 319
temper, 300
Gates, Melinda, philanthropy of (with Bill Gates), 319
genes (alleles of genes are also listed here)
alleles, explanation of, 63
chromosome 17, illnesses associated with genes on, 64
copy number variants, 63
explanation of basics about, 60–65
future reengineering of, 329–30
and personality
"evil" (in that these genes can contribute to personality disorders as well as to positive personality traits), 69–83, 85–87
no single gene causes a personality disorder, 68
older beliefs about personality and genes, 37–38, 38n
overview of effect on personality, 59
religious predisposition, 59, 59n
poliovirus infects only those with certain genes, 111
Quantitative Trait Loci (QTL) model, 71

specific genes and alleles
5-HT_{1B} role in aggression, 71
5-HT_{2A} role in self-mutilation, anorexia, and suicide, 71
5-HT_{3A} influence on amygdala and reaction time, 71–72
ASPM (cognition), 262
BDNF, related to neural growth, 77–78
clock, related to sleep cycles, 233
COMT, "warrior" or "worrier," 78–80
D4DR and novelty seeking, extroversion, 82
DARPP-32, intelligence and schizophrenia, 82–83
MAO-A, helps break down neurotransmitters, 54, 80–82
microcephalin (cognition), 262
related to the enzyme tryptophan hydroxylase (TPH), 82
related to the hormones vasopressin, oxytocin that help produce feelings of love, 83
serotonin-related, 69–75
stress's effect on, 65–66
genetics. *See also* genes
family discord flowing from, 60
fundamental overview of, 60–65
genetic drift, 65
"genetic stamping," large effect of certain historical individuals, 267–70
"imaging genetics," 67–69
intermediate phenotype and, 66–69
loss of olfactory and detoxification-related genes, 262, 262n
of personality traits and disorders
affective instability, heritability, 86
impulsivity, heritability, 86
lack of fear, 93
narcissism (very strongly heritable), 244, 287
negative affectivity and effortful control, 201
psychopathy, 55–58
sadism, 52
schizophrenia and schizotypal personality disorder, 227
separate heritability of borderline traits for mood disorders, impulsivity, and cognitive dysfunction, 206
Williams syndrome, 99
of susceptibility to polio, 111, 327
various populations have genetics adapted to cold or to high altitudes, 262
vitamin D and reshaping of genetics in northern climes, 263
Wiener's theory of Jewish versus Christian traditions and effect on, 86–87
genetic stamping, 267–68
Genghis Khan
and his roadies spread their genes, 267
surprisingly benevolent and visionary leader, 286
Genghis Khan and the Making of the Modern World (Jack Weatherford), a highpoint of compulsive biography inhalation, 286n
genocide
complicity of the United Nations in, 332
do well-intentioned policies of appeasement lead to, 307
Milosevic's role in, 156n, 162, 169–71
occurs in part because people can't believe Machiavellians, with their neurologically based sinister underpinnings, exist, 321–22
questionable nature of Philip Zimbardo's conclusions related to genocide, described in *The Lucifer Effect*, 303n–304n
Samantha Power's *"A Problem from Hell,"* how and why genocide happens, 321
genome
defined and discussed, 62, 67, 68, 69n, 84
human genome project, 34, 38, 290
genotype, explanation of, 64
Geoghan, John, pedophile priest, 107
George, Bill, on Warren Buffet's profound effect on ethics in business, 319

Giocangga, progenitor of Qing dynasty, and genetic stamping, 267
Glad, Betty, "Why Tyrants Go Too Far," 34
Gladwell, Malcolm
dark side of *Blink* and relation to "emote control," 379n20
strengths and weaknesses of imaging, 367–68n2
"gold diggers," 271–78
Goldhagen, Daniel Jonah (*Hitler's Willing Executioners*), 303n
gossip
importance of "gossip" magazines in reporting on Machiavellians, 251
"office gossip" and its role in alerting about Machiavellians, 337–38
gray matter
in autistics is increased, in pathological liars is reduced, 106
differences in psychopaths', 106
loss with one form of schizophrenia, 205
Great Leap Forward, 226, 234–36
"Great Society" legislation, 258
Great Terror, The (Robert Conquest), or, *I Told You So, You Fucking Fools*, 28n
guilt. *See also* empathy
abstract reasoning, relation to, 100
affiliated with activation of specific areas of brain, 102
Guinness Book of World Records and Ismail the Bloodthirsty, 268
gullibility. *See* naivete

Halifax, Lord: "His virtues have done more harm than the vices of hundreds of others," 292
Hall, Matthew Henry: cartoon "You're the only peson in this department I trust," 44–45
Halliday, Jon (Mao biographer), 218–19, 239–40
Hanks, Tom, in relation to neuroimaging study on partisanship, 189–90
Hardy, Jason, Ashkenazi genetic mutations and intelligence, 87
Hare, Robert
comorbidity of psychopathy with other disorders, 168
"corporate psychopaths," 106, 108
sadism's wider context, 53–54
Harpending, Henry: Ashkenazi genetic mutations and intelligence, 87
Harris, Judith Rich
Harris's *Nurture Assumption* laid profound case for influence of genes on personality, 37, 175
higher-status members of group have more influence, 270
No Two Alike, 56n
Hauser, Marc, studies of neurological features of morality, 322
Heller, Leland, comments on mood swings in borderlines, 230
heritability. *See* genetics *and also* individual personality disorders
He Zizhen, Mao's third wife: difficult life ending in madness, 221–22, 226–27, 227n
Hill, Christopher, Ambassador, 165
Hilton, Paris, and family, 273, 337
hippocampus
in borderline personality disorder
differences may explain hypersensitivity to emotional cues, 194, 206
smaller, 194, 205
differences in unsuccessful versus successful psychopaths, 105
enlarged in London taxi drivers, 176
illustrations of, 93, 101, 183
murderers have "turbocharged," 97
portrayed in flowchart form (as "Hippocampal Formation"), 185, 196
Historia Augusta, 274–75
histrionic personality disorder
defined, 135
illustration showing overlap with borderline personality disorder, 208
MAO-A in "cluster B" personality disorders, 80
Hitchens, Christopher (*The Missionary Position*), 285

Hitler, Adolf
ability to dupe credulous, 315
affective instability (rage), 299–300, 305
brief overview compared to other dictators, 28, 298, 315
control, desire for, 308–309
extraordinary memory of, 311–12
father's abuse, 218
hypomanic qualities, 314
identity disturbance
delusional and magical thinking, "gaslighting," 304–305
inflexibility, 204, 204n
many different diagnoses of, 33
mystery of his success, 107
narcissism of, 287
photograph of, with children, 316
was not totally evil, 322
Hoffa, Jimmy, 30
hostility and BDNF alleles, 77
House of Hilton (Jerry Oppenheimer), 273
Hsu Chen, Nationalist captain, warned of dangers of communism, 241–42
HT serotonin receptor alleles. *See* 5-HT
Hudson, William, a "TSU 3" hero, 279–80
Huettel, Scott, imaging studies of altruism, 100
human nature, Philip Zimbardo's questionable conclusions about, described in *The Lucifer Effect*, 303n–304n
humiliation
Mao, 239
in relation to antisocial personality disorder, 135
Sabin, Alfred, of Jonas Salk, 291
Huns, 303n
hunter-gatherer versus nomadic societies, trade-offs in, 266
Hussein, Saddam
Bill Clinton's gullibility regarding, 316–17
brief overview compared to other dictators, 28
Dan Rather's gullibility regarding, 317
mystery of his success, 35, 107
hypersensitivity
borderline personality disorder
differences in hippocampi and amygdalae may explain, 194
as trait of, 137, 145
narcissistic personality disorder and, 135
of various individuals
Hitler, 299
Mao, 246
Milosevic, 153–54
Princess Diana, 277
hypertension, primary pulmonary, and serotonin transporters, 75–77
hypocrisy (inconsistency). *See* identity disturbance, inconsistency
hypomania. *See also* bipolar personality disorder
examples of good and bad leaders with hypomanic qualities, 313–14
Stalin's apparent (through ability to work prodigiously), 30
hypothalamus and brain stem nuclei
portrayed in flowchart form (as "Hypo/BSN"), 185, 196
role in producing automatic emotional response, 196

Iacoboni, Marco, and research on mirror neurons, 104–105
ice cream, superb Russian, child wrestling, 286n
idealization and devaluation, alternating between. *See* relationships, unstable personal, "splitting"
identity diffusion. *See* identity disturbance
identity disturbance
affective instability and relation to, 230
chameleon-like behavior
as aspects of "painful incoherence" and "lack of authenticity," 156
as characteristic of borderlines, 137, 149, 207
defined, 144–46
in relation to executive control network, 199–202
seen in those attracted to aristocracy and wealth, 277

sometimes impossible to believe someone is as chameleon-like as others say, 337–38
as defining DSM-IV trait for borderline personality disorder, 135, 157, 158
general definition, 155–57
inconsistency (hypocrisy), 156–57, 246–47
inflexibility and rigidity
as feature of identity disturbance along with *over*-flexibility, 155–56, 207, 232
new data difficult to assimilate due to emotionally based biases, 189–90
obsessive-compulsive personality disorder and, 135
paranoid personality disorder, a trait of, 134
role of orbitofrontal cortex in suppressing emotional memories that affect decision making, 204
Machiavellians and, 156–57, 297–98
overview related to neuroscience results, 206–207, 232
parietal lobe abnormalities may contribute to identity disturbance in borderlines, 198–99
in people
beneficial visionaries, delusions of, 305–307, 307n
Caligula and Nero, 276
Diana, Princess, 277
Duesberg, Peter, denies standard treatment necessary for AIDS, 306n–307n
Fastow, Andrew, Enron CFO (chameleon-like behavior), 295
general discussion of inflexible tyrants, 301–302
Hilton, Kathy (Paris Hilton's mother), "she nailed him with her fake personality," 273
Hitler. *See* Hitler, identity disturbance
lack of chameleon-like behavior in people
Ataturk, Mustafa Kemal, 309
Thatcher, Margaret, 309
Maggiore, Christine, denies standard treatment necessary for AIDS even after death of daughter from AIDS, 306n–307n
Mao. *See under* Mao, Chairman, personality traits and disorders
Milosevic. *See under* Milosevic, Slobodan, borderline-like and psychopathic traits
Roxalena, wife of Suleyman the Magnificent, 271–72
role absorption, 156
stress and identity diffusion, 202
ideological commitment, lack of
in Machiavellians, 42
Mao's, 231–32, 231n
Milosevic's, 154
a symptom of narcissistic personality disorder, 247–48
imaging. *See also* functional magnetic resonance imaging (fMRI); magnetic resonance imaging (MRI); PET
"imaging genetics," 67–69
strengths and weaknesses of imaging, 367–68n2
immune system and link with personality disorders, 207
impaired perception and reasoning. *See* cognitive dysfunction
impulsivity. *See also* affective instability; anger
anterior cingulate cortex and, 193
antisocial personality disorder and, 50, 135
borderline personality disorder
as defining DSM-IV trait, 159
dimensional trait of, 164
executive control of impulses as defining difference for clinical versus subclinical, 199–202
neuroimaging results related to, 195–201, 205–206
as shared feature with bipolar personality disorder, 142n
as symptom of, 137, 140, 232
in children with conduct disorder, 102–104

heritability of, 85
MAO-A and violent, 80–81
orbitofrontal cortex dysfunction releases brakes on, 94, 180, 197–98
in people. *See under* affective instability; anger
in relation to dimensional trait of psychopathy, 167
serotonin transporters and, 73
Incan aristocracy, extraordinary reproductive opportunities of, 269–70
inconsistency (hypocrisy). *See* identity disturbance, inconsistency
inferior parietal cortex, difference in Chinese versus English language use of, 175–76
inflexibility. *See* identity disturbance, inflexibility
Innocence Project, 188
insula, left, and "emote control" in neuroimaging study of political partisanship, 189–90
intelligence. *See also* common sense
as advantage for Machiavellians, 297, 310–12
Ashkenazi genetic mutations, Tay-Sachs disease, and, 87
association with memory, 312
borderline personality disorder and, 142
"cheaters" may be behind dramatic leap in human, 259
COMT alleles and, 78–80
coupled with subtly impaired reasoning skills (ventromedial dysfunction), 95–96
DARPP-32 and relation to, 82–83
polio leaves intact, 116
of various individuals
Carolyn, 128, 142, 327
Churchill, Winston, 293
Duvalier, "Papa Doc" (Haitian dictator), 311
lack of in Andrew Fastow, Enron's CFO (compensated for with Machiavellian traits), 295
Mao, 250
Shockley, William (also his disparaging comment about his children's), 290–91
Stalin, Joseph, 312
Stroessner, Alfredo (Paraguayan dictator), 311
intermediate phenotype. *See also* subclinical personality disorders
cautions regarding this approach, 363n9
defined, 66–69, 255n
in relation to "Cluster B" disorders and MAO-A gene, 238
International Society for Human Ethology, 253, 283
intimidation, *See under* manipulation
Inventing the AIDS Virus (Peter Duesberg), 307n
Irfan, adopted son of author and her husband, 151–52, 153, 169–70
iron lung, photo of ward, 125
irrationality. *See* cognitive dysfunction
irresponsibility
antisocial personality disorder and, 51, 135
as dimensional trait of psychopathy, 167
irritability, 142n. *See also* affective instability; aggressiveness; anger
antisocial personality disorder and, 50
borderline personality disorder and, 158
Islam, fundamentalist, Machiavellians who have found purchase in, 307
Ismail the Bloodthirsty and his (somewhat exaggerated) prolific nature, 268
isolation, social
borderline personality disorder and, 137, 140
Carolyn's, 124–26, 141, 327
Mao's, 225

Jackson, Andrew, temper of, 300
Jefferson, Thomas (one of most "double-faced" politicans in America), 298, 301
Joan of Arc, 306
Johnson, Lyndon, and "the Johnson treatment," 258

Jordan, Justin, a "TSU 3" hero, 279–80
journalism. *See* media, press, and journalists
Justinian, Byzantine emperor who gave "despotism" bad name, 268–70

Kaihui, Mao's second wife: disillusionment, marriage, and undying love, 220–22, 228
Kang Sheng, Mao's torturer extraordinaire, 239
Kernberg, Otto, and malignant narcissism, 33
Kerry, John, in relation to neuroimaging study on partisanship, 189–90
Kershaw, Ian (Hitler biographer), 300
Kiehl, Kent, imaging studies of psychopaths, 90–92
Kim Il Sung and Kim Jong Il
  Jimmy Carter duped by Kim Il Sung, 316
  their "Happy Corps" of women, 269n
kindness. *See* altruism
Kinner, Stuart, psychopaths can't project onto others feelings they themselves don't have, 105
Knudson, Dean, problems with borderlines in the workplace, 149
Kozlowski, Dennis, Tyco CEO, ice statue urinating Stolichnaya vodka, 294
Krauthammer, Charles: "blanket anti-son-of-a-bitch policy is . . . righteous self-delusion," 339
Kreger, Randi (*Stop Walking on Eggshells*), 139–40, 146, 147
Kreisman, Jerold
  borderlines and empathy, 142
  refers to borderlines by diagnosis, 139n
  Sometimes I Act Crazy: "typically, the borderline seeks partners who are in a position of power," 277
Kretschmer, Ernst, psychiatrist: "In normal times we diagnose them; in disturbed times they govern us," 332, 334–35
Kroc, Roy, McDonald's founder, "rat eat rat" industry, 292
Kuklinski, Richard (hired killer), perfunctory attitude toward work, 96

labor unions, 310
lack of authenticity. *See also under* identity disturbance, chameleon-like behavior
  defined, 156
lack of object constancy, 221
Langer, Walter (psychoanalyst and Hitler biographer), 299
language: learning and theories related to, 174–77, 261–62
lateral frontal cortex and "emote control" in neuroimaging study of political partisanship, 189–90
Lauren, Ralph, temper of, 300
law and free will, 330, 330n
Lay, Ken, Chairman of Enron, 296–98
learning, new data difficult to assimilate due to political and other emotional biases, 189–90
learning disability, psychopathy as, 103
LeBor, Adam
  corruption at the United Nations, as described in *"Complicity with Evil*, 332
  observations in *Milosevic: A Biography*, 154, 164–65
Leibniz and Newton, feud over invention of calculus, 291
Leichter, Steven, borderline behavior in a hospital setting, 143–44
Lenin, Vladimir, could not resist temptation to stay in power, 298
Lesser, Guy, on Milosevic's war crimes tribunal at The Hague, 156
liars. *See* lying
Lifton, Robert Jay, Mao's policies as "psychism," 235
limbic system. *See also* amygdala; hippocampus; nucleus accumbens; thalamus
  in borderline personality disorder
    dual pathology in limbic and prefrontal systems, 166, 208
    PET scans show lower glucose metabolism, 197

emotional processing of signals in, 182–86, 187
illustration of key components of, 183
MAO-A in relation to smaller organs in, 80
murderers have "turbocharged," 97
in psychopaths', 90–91, 209
ventromedial cortex, intimate connection with, 182
Lincoln, Abraham, abuse by father and depression, 219n
Linehan, Marsha
dislike of word "manipulative" when applied to borderline patients, 138
her dialectic-behavioral therapy has been shown effective in helping borderlines, 329
litigation more likely from borderlines, 149
Li Zhisui, Mao's doctor, observations of Chairman Mao. *See also* Mao, Chairman
addiction to sleeping pills, sex, 233
devoid of human feeling, ignores injured child acrobat, willing to lose half population of China, sad interactions with ex-wife, 225–27
a few of programs probably conceived with vague decency, 249
"greatest manipulator of all," 237
"I felt no sorrow at his passing," 248
Mao's reaction to devastating effects of his policies was to pretend weren't happening, 234–35
Mao states "Getting upset is one of my weapons," 229
methods for gaining loyalty from others, 241
paranoia begins to tighten its grip, 245
"ruthless though he was, I believe Mao launched the Great Leap Forward to bring good to China," 236
terms Mao's condition "neurasthenia," 231
Lobel, Jules, "emote control," 187
Loewenstein, George, "emote control," 187
Lombroso, Cesar, 50
Long, Huey, Louisiana's political "Kingfish," 251
"long" in relation to serotonin transporters, 72–75, 78
Lou Gehrig's disease: Mao's fatal illness, 248
love
abstract reasoning, relation to, 100
alternating loving and hateful behavior in borderlines, 149
Carolyn's diaries record her true feelings related to love for father, 341–43
Milosevic's unquestioned love for his wife, 155
vasopressin and oxytocin hormones help produce feelings of, 83
Lowenstein, Roger (biographer of Warren Buffett), 318
Lucifer Effect, The (Philip Zimbardo), questionable nature of its conclusions, 303–304n
lying. *See especially* manipulation; *and also* gaslighting; deceitfulness
antisocial personality disorder and, 50, 135
borderline personality disorder and, 140
easy ability to lie serves as advantage for Machiavellians, 297
gaslighting and intentional, 147
narcissism as motivation for, 162
pathological, 106
by people. *See* manipulation by people
as trait defined by Christie for Machiavellians, 42

Mach (high and low), 46–47, 47n
Mach-IV test for Machiavellianism
development of, 46–47
Mao illustrates why a Machiavellian would not achieve perfect score, 243–44
online test to determine your level of Machiavellianism, 47n
used in conjunction with the PDQ-4+, 133–36

Machiavelli, Niccolo
as early psychological pioneer, 41–42
ideas used to help found discipline, 32, 46
Machiavellianism. *See also* Machiavellians
Christie and study of, 41–48
evolutionary benefits of Machiavellian behavior, 254–56, 387–88n2
psychopathy and, 42, 131–32
test for. *See* Mach-IV test for Machiavellianism
Machiavellians. *See also* Machiavellianism; Mao, Chairman; Milosevic, Slobodan; *and also* draw your own conclusions related to others mentioned in text
borderlines as, 137, 145
broad societal and historical implications
ability to discern Machiavellians is important defensive tool, 318–22
advantages of having at least a few Machiavellian traits, 336
blanket anti-son-of-a-bitch policy is righteous self-delusion (Charles Krauthammer), 339
does percentage of Machiavellians influence culture, 270–71
Machiavellians more frequently found in positions of power and control, 333, 335
Machiavellians have ability to seduce others into ignoring conscience, 281
Machiavellians take advantage of natural altruism and kind-hearted naivete, 255
Machiavellians use our own neural quirks to fool us into working against our dearest ideals, 192
in Ottoman empire, 271–72, 274
preferentially selected for in more densely populated agricultural societies, 264–71
religion and, 35, 250–51, 307, 331. *See also* Mao, Chairman, religious cult of personality
in Roman empire, 275–76, 276n
ultimate effects on society, 331–39
why people support Machiavellians in politics, 334, 339
Christie's proposed set of traits for, 42–45
definition of a Machiavellian
Christian pacifists and Muslim terrorists act as altruists—so what *is* a Machiavellian, 285–87
general, 32, 35
precise and detailed, 255n, 280–83
emergenic constellation of borderpathic traits in Machiavellian leaders, 314–15
at Enron, 294–98. *See also* business
as "gold diggers," 271–78
identity disturbance and, 156–57
ideological commitment, lack of. *See* ideological commitment, lack of
inexactitude of term "Machiavellian"—as with term "antisocial personality disorder," 255n, 281
no single gene causes, 68
problem of detecting, 332–39
Madonna, narcissism as tool to reach superstar status, 292
Maggiore, Christine, denies standard treatment necessary for AIDS, 306n–307n
magical thinking
fusiform gyrus, associated with unusual features in, 235
Hitler's, 305
Mao's, 235
schizotypal personality disorder, a trait of, 135
magnetic resonance imaging (MRI), defined, 89–90
malaria and Baldwinian evolution, 264
malignant narcissism, Medline's lack of citations on, 33
manipulation. *See also* control, desire for; gaslighting; lying
advantage of "dysfunctional" personality traits for, 250–52

aristocracy, wealth, power, and control attract those with manipulative traits, 277, 333–35
caused by neural quirks, 331
as defining characteristic for Machiavellians
in Christie's original work, 42, 45, 46, 47
as part of precise definition used in this book, 281
dictators' use to hear only what they want to hear, 315
murderers and, 97
narcissism as motivation for, 162
our best traits used as levers for our, 192
by people
Carolyn, 141, 142, 327
Coleman, Roger, smooth ability to lie about rape and murder, 188–89
Fastow, Andrew, Enron CFO, 295–98
Gandhi, Mahatma, 169, 298
Gordy, Berry ("Motown" creator), 291–92
Hilton, Kathy (Paris Hilton's grandmother and similarly named mother), tactics of intimidation, 273
Hitler, Adolf, 299–300, 315
Jefferson, Thomas, one of the most "double-faced politicians in America," 301
Mao. *See* under Mao, Chairman, personality traits and disorders
Milosevic. *See under* Milosovic, Slobodan, borderline-like and psychopathic traits, identity disturbance *and* lying
Skilling, Jeffrey, Enron CEO, 296–98
Snyder, Solomon, and his usurpation of credit for discovery of opiate receptors, 291
Stalin, Joseph, 315
Stewart, Martha, 293–94, 315
Washington, George: if he wasn't the best intentioned man in the world, he'd be very dangerous, 300
predatory murderers and, 97
of the press. *See under* media, press, and journalists
psychopathy and, 29n, 107, 132
as seen in various personality disorders
antisocial, 135
borderline, 137, 140, 145, 148–49, 200n
histrionic, 135
narcissistic, 135
pathological lying and, 106
subclinical borderline, 200
tools, *for more comprehensive explanations and examples see under main headings* gaslighting; memory; projection; identity disturbance; theta rhythms; *and* control, desire for
gaslighting, 146–48
intimidation, 29n, 53, 273, 297, 299–300, 315, 333–34, 336
irrationality, 260–61
manipulation of press, intellectuals, and politicians, 241–42, 315–17
memory prowess, 311–12
projection (blame shifting), "splitting," chameleon-like, and controlling behavior, 145, 162, 234–35, 250
sadism, 250, 274
silent treatment, threats, no-win situations, anger, 140
theta rhythm disruption, 148
Mao, Chairman. *See also* Li Zhisui, Mao's doctor, observations of Chairman Mao
brief overview compared to other dictators, 28–29, 308, 315, 332
Cultural Revolution, 215–16, 235, 237, 239, 249
death from amytrophic lateral sclerosis, 248
family
earliest years and antisocial tendencies, 218–19
father's abuse of Mao—and Mao's abuse of his father, 218–19, 219n

suffering and mental disturbances of Mao's children, 222–23, 227
wives
He Zizhen, Mao's third wife, a difficult life ending in madness, 221–22, 226–27, 227n
Jiang Qing, Mao's hellish fourth wife, 223–24
Kaihui, Mao's second wife, disillusionment, marriage, and undying love, 220–22, 228
Mao's evanescent first wife, 220
Great Leap Forward, 226, 234, 236
ideological commitment, lack of, 231–32, 231n, 247–48
insomnia, 232–33
personality traits and disorders
addictions
drug abuse, 217, 232–33, 245
sexual, 233–34
advantages of Mao's "dysfunctional" traits, 250–52
affective instability (mood dysfunction, "neurasthenia"), 217, 224, 229–32, 235
antisocial behavior, early, 218–19
attention, strangely warped, 249–50
borderline personality disorder and, 217
as "borderpath," 216–18
charm and chameleon-like behavior, 229, 241–42
cognitive dysfunction and magical thinking
general discussion, 234–39
paranoia, 217, 245–46
control, desire for, 225, 229, 236–39, 298
empathy
lack of, 225–26, 227n
rare displays of, 226–28
exhibitionism, 223–24
good qualities, rare but real, 226–28, 248–49
humiliation of others and feelings of self-humiliation, 246
hypersensitivity, 246
hypomanic qualities, 314
identity disturbance
inconsistency (hypocrisy), 222, 233–36, 246–47
Mao's personalization of yin and yang, and Marxism, sneaks perilously close to, 230–31, 231n
impulsivity, 228–29
inflexibility ("very hard for me to change," difficulty handling criticism), 235
insomnia, 225, 232–33
intelligence, 250
lack of object constancy, 221
manipulation of others
general discussion of techniques, 241, 246–47, 250
of journalists and prominent individuals, 241–42
Mao was the "greatest manipulator of all," 237
use of mood to frightening effect, 229
memory, excellent, 218
narcissism, 225, 238–39, 242–43, 245, 248
projection, 234–35
psychopathic traits, 217, 224–25
relationships, unstable personal, 217–25
religious cult of personality, 242–43, 252, 332
sadism, 239–40
vindictiveness, 246
photograph (with wife Jiang Qing), 224
photograph of as idolators wave their *Little Red Books*, 243
Pye, Lucian, explains reticence about declaring Mao a narcissist with a borderline personality, 217
results of his policies, 216–17, 248, 251–52
schizophrenia of son, 222–23, 227
MAO-A (monoamine oxidase A)
alleles produce different intermediate phenotypes, 80

childhood stress and predisposition for antisocial behavior, 54, 80–82
Marcus Aurelius, 275, 275n
Markovic, Mira (Slobodan Milosevic's wife and virtual Svengali), 155, 166–67
marriages, troubled, in relation to family discord and genetics, 60
Marshall, Barry, "crackpot" researcher on ulcers, is ultimately proved correct, 307n
Martha Stewart—Just Desserts (Jerry Oppenheimer), 293–94
Martin, Bradley K. (*Under the Loving Care of the Fatherly Leader*), 269n
Marx, Karl: disproving Marx's theory that only social forces matter, 267, 308
Marx's concept of dialectical materialism in relation to identity disturbance, 231n
Mason, Paul (*Stop Walking on Eggshells*), 139–40, 146, 147
Massey, Douglas, observations on emotion from evolutionary perspective, 187
May, Richard, presiding judge at Milosevic's trial, 162
McCloskey, Jim, and "Centurion Ministries," 188
McFarland, Sam, self-selection of personality types for positions that suit disposition, 303n–304n
McHoskey, John
  study relating Machiavellianism to general personality disorders, 131–36
  study relating Machiavellianism to psychopathy, 49–50
Mealey, George, establishment of in-perpetuity fund in daughter's name, 283
Mealey, Linda
  "Linda Mealey Award for Young Investigators," 283
  mystery of her disappearance, 253–54, 283
  psychopaths can't project onto others feelings they themselves don't have, 105
media, press, and journalists
  bamboozling by the successfully sinister of the, 316–17
  Mao's manipulation of the, 241–42
  role in reporting on Machiavellians, 251, 338
medial prefrontal cortex
  association with empathy and social interactions; Williams syndrome, 99
  illustration of, 101
Medline, explanation of, 32
Meloy, J. Reid
  good parents sometimes still have psychopath-like child, 102–103
  observation about psychopath as predator, 52
  remarks on paranoid personality disorder in relation to Mao, 245–46
memory
  anterior cingulate cortex and role in making permanent, 182
  conscious ability to suppress, 165
  current and historical examples of good and bad leaders with remarkable memories
    general discussion of, 310–13, 313n
    Mao, 218
    Milosevic, 154
    Warren Buffett, 318
  effect of genes and alleles on memory
    APOE4, 76
    BDNF, 77
    COMT, 79
  Parker, Robert (wine connoisseur) and taste memory, 262n
  role of memory in charm, charisma, and manipulation, 312
  "splitting," mood, and fragmentation in relation to storage of, 144
"*met*"
  BDNF allele, poorer memory, less anxiety, 77
  COMT allele, slower dopamine metabolization, "worrier," 79
Meyer, Edith, pioneering studies of psychological effects of polio, 114–16

Meyer, Michael Leverson, described Milosevic, 171
microcephalin gene and cognition, 262
midbrain
  murderers have "turbocharged," 97
  poliovirus invasion of the, 114–16
Milgram, Stanley, research on blind tendency to obey authority, 333n
Milosevic, Borislav (Slobodan's brother), 160, 166–67
Milosevic, Slobodan
  borderline-like and psychopathic traits
    affective instability (mood changes), 155, 161
    arrogance and defiance, 153, 171
    borderline personality disorder
      overview of Milosevic's many borderline-like characteristics, 153–68
      why Milosevic did NOT have borderline personality disorder, 162–63
    as "borderpath," 167–68
    cognitive dysfunction and delusions
      general discussion, 164–66
      seeming normalcy masked, 209
    control, desire for, 161
    empathy, lack of, 162, 166
    identity disturbance
      chameleon-like behavior, 154–57
      distrust, 157–59
      general discussion of, 155–57
      inflexibility, 156, 156n
    impulsivity and anger, 159, 161
    lying, 153–54, 162, 164–66, 171
    narcissism, 162
    projection (by attacking other side at trial), 162
    rudeness, 154
    sadism, 161
    "splitting," 157
    suicide and, 160–61, 166
    vindictiveness, 161
  brief overview compared to other dictators, 28, 315
  compartmentalization of activities, 158–59
  diabetes, type II, 161
  ideological commitment, lack of, 154
  mimicking abilities, 154
  only way to influence was through threat of force, 160
  photograph of Slobodan with wife Mira, 167
  positive relationships with others
    childhood: he was good, if unctuous, 218
    love for wife and family, 155, 166–67
    loyalty to friends if they were strictly loyal, 166
  results of his policies, 153, 169–71
  role in genocide, 156n, 162, 169–71
Milosevic and Marković: A Lust for Power (Slavoljub Djukic), 155
mimicking abilities. *See also* mirror neurons
  DiCaprio, Leonardo, 104n
  Milosevic, 154
mirror neurons, 104–105. *See also* mimicking abilities
Missionary Position, The (Christopher Hitchens), 285
Mitchie, Christine, and dimensional trait description of psychopathy, 167
Mitevic, Dusan, Milosevic's friend and propaganda chief, 161
mitochondria in American Indians adapted to cold, 262
modular brain theory, 175, 388n12
Moll, Jorge
  description of psychopaths, 51–52
  imaging studies related to morality, 100
monoamine oxidase A. *See* MAO-A
Montefiore, Simon Sebag: Stalin's real genius was his charm, 30
moodiness. *See* affective instability
"moral entrepreneurs," 335
morality
  basic features hard-wired, not product of culture, 322
  imaging studies related to, 100–102, 322
  Wilson, David Sloan, work involving, 17

Morgan, Peter (*The Appearance of Impropriety*), 335
Morse, Stephen, free will and criminality, 330n
Morris, Gouverneur: eulogy of George Washington, 298
mother of the author
  amnesia of, 113
  love for Carolyn, 122, 126
  photograph of, 124
Mother Teresa's buttering up of despotism, 285
motor cortex
  learning to play a musical instrument produces changes in, 176
  Mao's fatal amyotrophic lateral sclerosis killed cells in, 248
  polio kills cells in, 111, 114–16, 326
  portrayed in flowchart form, 185, 196
Mugabe, Robert
  in context with other poorly schooled leaders, 308
  control, desire for, 308
  hypomanic qualities, 314
Mullis, Kary, foreword to *Inventing the AIDS Virus*, 307n
murderers
  affective versus predatory, 96–97
  psychosocial versus neurobiological "push," 95
Murphy, Jane, 265
Mussolini, Benito
  brief overview compared to other dictators, 28, 286
  hypomanic qualities, 314
  prodigious memory of, 312
mutations, genetic, 63, 264
myelin, definition and relation to psychopathy, 92

Nabokov, Vladimir, excerpt from "An Evening of Russian Poetry," 174
N-acetylaspartate (NAA) compounds, borderline personality disorder, and schizophrenia, 205
naivete. *See also* altruism
  general discussion of
    examples of credulity by many well-known individuals, 241, 315–17
    experiences of ordinary people—consequences of their credulity, particularly regarding genocide, 255, 319–22
  vasopressin, oxytocin—and gullibility, 83
  Williams syndrome, 98–99
Nandhra, Harpal, coinage of word "borderpath," 168
Napoleon, Bonaparte, "They wanted me to be another Washington." But he wasn't, 298
narcissism. *See also* arrogance; narcissistic personality disorder
  acquired situational narcissism, 248n
  apology for loosely interchangeable use of narcissism, ego, self-esteem, self-importance, conceit, and arrogance, 288n
  borderline personality disorder, as coping characteristic in, 137
  can motivate lying, 162
  in children with conduct disorder, 102–104
  in conjunction with paranoid personality disorder, 246
  and connection with "self" as opposed to "other" neural circuitry, 302–303
  as a crucial asset in art, science, business, and politics, 297–98
  DSM-IV definition of, 244
  as *the* key trait behind Machiavellianism, 281, 283
  in people
    Ceausescu, Nicolae, 287
    Churchill, Winston, 292–93
    Darwin, Charles, and the advantages of a bit of narcissism, 288–89, 292
    Fastow, Andrew, CFO Enron, has himself named CFO of year, 295–97
    Hilton, Paris, and family, 273
    Hitler, 287

Madonna, 292
Mao. *See under* Mao, Chairman, personality traits and disorders
Mendel, Gregor's lack of narcissism, and resulting problems, 288
Milosevic. *See under* Milosevic, Slobodan, borderline-like and psychopathic traits
Niyazov, Saparmurat (dictator of Turkmenistan), 287
Shockley, William, inventor of junction transistor, 290
Stalin, Joseph, 309
Stewart, Martha, 293
Thatcher, Margaret, 309
Wallace, Alfred Russel, and the disadvantages of being free of narcissism, 288–89, 292
Watson, James, sublimely arrogant, misogynistic codiscoverer of DNA, 290, 392n13
psychopathy and, 49
social dominance and neurotransmitter levels, 238
narcissistic personality disorder. *See also* narcissism
borderline personality disorder, overlap with, 146, 208
defined, 135
MAO-A in "cluster B" personality disorders, 80
placement in Axis II of DSM-IV, 134
sadism and, 52
National Enquirer
Martha Stewart sues after being characterized as borderline by, 293–94, 315
reasons behind surreptitious enjoyment of, 37
reputation for investigative reporting, 338
National Institutes of Health. *See* NIH
Nazi Einzatgruppen, 303n
negative affectivity, its heritability and relation to executive control, 201
Nero, chameleon-like behavior of, 276
Nesse, Randolph, evolutionary model for emotional behavior, 258–59
neurasthenia, Mao's, 217, 231
neuromodulators, 196
neurons
general explanation, 183–85
illustrations of, 70, 184
neuroscience, not mentioned in relation to research about dictators, 34
neuroticism. *See also* affective instability; anxiety
aristocracy, wealth attract those with semi-, 277
BDNF alleles and, 77–78
COMT gene and, 78–80
Mendel's, 288
neurotransmitters. *See also* dopamine; serotonin
defined, 69, 184
and delusional thinking, 304
imbalances in borderlines, 195–201
reticular activating system, poliovirus, attention, and, 114–16, 326
social dominance and, 238
Newman, Joseph
psychopathy as learning disability, 103
theories tie neatly to Posner's on attentional networks in borderlines, 202, 326
Newton and Leibniz, feud over invention of calculus, 291
Niall of the Nine Hostages and genetic stamping among Irish, 268
Nielsen, Nete Munk: long-term psychiatric effects of polio, 325–26
NIH (National Institutes of Health)
overview of work mapping genetic bases of psychiatric illnesses, 68
research related to intermediate phenotype, 66–67
Nisbett, Richard E., codirector of University of Michigan's Culture and Cognition Program, 175
Nixon, Richard, temper of, 300
Niyazov, Saparmurat, dictator of Turkmenistan, narcissism of, 287
Noble, Ralph: Siamese fighting fish, 39
nomadic versus hunter-gatherer societies, trade-offs in, 266
Norwegians, "black," 118

No Two Alike (Judith Rich Harris), 56n
novelty seeking and D4DR, 82
nucleotides, explanation of, 62
nucleus accumbens
in borderline personality disorder
role in executive control
discussion, 199
flowchart, 196
role in determining motivation and reward, action site for drugs that produce euphoria
discussion, 186
flowchart, 185
nurture
older beliefs about influence on personality of, 37–38
psychopathy and effect of, 51, 55–58
Nurture Assumption, The (Judith Rich Harris), laid profound case for influence of genes on personality, 37, 175

obsessive-compulsive disorder
COMT alleles and, 79
defined, 135
Oppenheimer, Jerry (*Martha Stewart—Just Desserts* and *House of Hilton*), 293–94
Oppenheimer, Robert, mental flexibility of, probable polio survivor, 301
orbitofrontal cortex
in borderline personality disorder, 206
problems with impulsivity similar to those with damage to, 196–98, 206
role in cognitive-perceptual impairment, 203–207
role in executive control
discussion, 199
flowchart (as "Orbital PFC"), 196
role in suppressing emotional memories that affect decision making and can cause inflexibility, 204
dysfunction releases brakes on impulses, 94, 180–81
illustration of, 94, 181
linkage with limbic (emotional) systems, 186
MAO-A alleles and decreased reaction in, 80–81
Ottoman empire—the Sick Man of Europe, 271–72, 274
Ovid: "All things may corrupt when minds are prone to evil," 334
oxytocin hormones, 83

painful incoherence, defined, 156
Pally, Regina, description of borderline personality disorder from evolutionary perspective, 282
Palmer, Mark, Enron's head of corporate communications, 298
Panic, Milan: Milosevic's attempted suicide, 160–61
paranoia. *See* cognitive dysfunction, paranoia
paranoid ideation (defining DSM-IV trait for borderline personality disorder), 157, 158
paranoid personality disorder. *See also* cognitive dysfunction, paranoia
defined, 134
shame, narcissism, vindictiveness, and hypersensitivity in relation to; Mao's features of, 246
parietal lobes in borderline personality disorder
right lobe appears to be smaller, 198
smaller right lobe produces psychotic symptoms, 199
Paris, Joel
relation between antisocial and borderline personality disorders, 137
roller coaster emotional life of those with borderline personality disorder, 138
Parker, Robert (wine connoisseur) and taste memory, 262n
partisanship, political. *See* politics
Paul, Annie, 41–42
PDQ-4+ (Personality Diagnostic Questionnaire), 132–33
Perisi, General (Milosevic's chief of general staff), 165
Perry, Rick, Texas state governor: his inef-

fectual handling of Texas Southern University's corruption, 279–80
personality
effect of immune system on personality disorders, 207
older beliefs about influence of nurture on, 37–38, 38n
overview of how genes affect, 59
personality types sort themselves into positions that suit personalities, 277, 303–304n
personal relationships, unstable. *See* relationships, unstable personal
Pert, Candace, credit for her discovery of opiate receptors usurped by advisor, 291
PET (positron emission tomography), defined, 197
Peterson, Laci, murdered by husband, previously dated violent man, 255
Phelps, Jim
discussion of BDNF, 77–78
preface by, 19–22
phenotype
explanation of, 64
intermediate. *See* intermediate phenotype
Pinker, Steven: his *The Blank Slate* helped shift researchers away from "people are naturally good" idea, 37, 175
Pinochet, Augusto (president of Chile), 286
pleiotropy, 75–77
Plomin, Robert, 59, 59n, 363n9
Polio Paradox, The (Richard Bruno), 114–16
poliovirus
how it infects and affects people, 111, 114–16
long-term psychiatric effect on, 325–27
stress related to treatment of those with polio, 125–26
political partisanship, neuroimaging study of, 189–90
Political Psychology (journal), 34
politicians, advantage of "dysfunctional" personality traits for, 250–52
politics
"emote control"
confirmation bias, seeing and hearing only confirmation of our expectations, 179
neuroimaging study of partisanship of Democrats and Republicans, 189–92
"projection bias," the danger of assuming those from other cultures think as we do, 378n8
general observations about political gamesmanship
blanket anti-son-of-a-bitch policy is righteous self-delusion (Charles Krauthammer), 339
gerrymandering and blocking of transparency rules typify protections Machiavellians provide themselves, 336
at high levels, the game of "find the Machiavellian" becomes a house of mirrors—because each party is rife with Machiavellians, 338
Machiavellians, with their distorted, self-serving cognitions, can twist good intentions to ill and set entire populations aflame with hatred, 331
people with grudges can twist "good" laws to bad purposes, 335
rise of Machiavellians means others redirect themselves toward less corrupt systems, which in turn become corrupt as Machiavellians are attracted to new nexus of power, 336–37
self-select, so that higher political levels have higher percentages of Machiavellians, 333, 335
intellectuals and journalists provide kinder treatment to brutal foreign dictators than home politicians, 241, 317
political ideologies
are seized by people of certain tem-

peraments, for good or for ill, 308, 322, 333
capitalism
capitalists as cultural chumps, 177–79
freedom of speech in capitalist society meant capitalist atrocities magnified, while communist atrocities remained hidden, 242
communism
purges: "everybody makes mistakes," 28, 173–74, 179, 237, 240
true Communist traditions of helpfulness, 212–13
unabashed idealism provides perfect cover for Machiavellians, 250
fascism, 311, 333
Mao's fascist ideas, 217
right-wing dictators wave anticommunist flag to maintain power, 286
sham democracy of right-wing dictators, 250
opaque ideologies and politically oriented religions with few checks and balances provide perfect cover for Machiavellians, 334
relativism
basic features of morality hardwired, not a product of culture, 322
cultural policies of appeasement lead to genocide, 307
"projection bias," 378n8
power attracts females with borderline personality disorder, 277
Pol Pot
brief overview compared to other dictators, 28
lack of psychological literature related to, 33
polygeny (genetics), definition, 76–77
polygyny (having more than one wife): role in increasing reproduction opportunities for Machiavellians, 268–70
population density affects number of psychopaths and Machiavellians, 264–71
Posner, Michael, studies of
attentional networks and their relevance to dysfunctional people, 301, 326, 333
executive control as defining difference for clinical versus subclinical borderlines, 200–202
Post, Jerrold
conscience of narcissist is flexible, dominated by self-interest, 247
malignant narcissism, 33
posterior cingulate cortex, coexisting borderline-schizotypal types show gray matter reduction in, 227
posterior superior temporal cortex, increased activity in relation to altruism, 100
posttraumatic stress disorder, illustration of overlap with borderline personality disorder, 208
power, attraction of those with borderline personality disorder toward those with, 277
Power, Samantha (*"A Problem from Hell"*): how and why genocide happens, 321
prefrontal cortex. *See also* dorsolateral prefrontal cortex; orbitofrontal cortex; ventromedial cortex
antisocial behavior produced by damage to, 94
borderline and antisocial patients both show differences in function of, 208–209
in borderline personality disorder
boosted signal from amygdala causes excess activity, 195
disrupted connection to other regions of brain cause many borderline features, 206
PET scans show higher glucose metabolism, 197
PET scans show lower serotonin levels, 197

role in executive control
discussion, 199
flowchart, 196
emotion and, 186
illustration of three major components of, 94
portrayed in flowchart form, 185
press. *See* media, press, and journalists
Prince, The (Niccolo Machiavelli), 41–42, 46
prison, deinstitutionalization and trade-offs with increased prison populations, 330
"Prisoner's Dilemma," 257
"Problem from Hell, A" (Samantha Power): how and why genocide happens, 321
professors and Machiavellianism. *See* academia
projection. *See also* manipulation
borderline personality disorder
as coping characteristic in, 137
defined, 145
connection with delusional thinking, 302–303
dorsolateral prefrontal cortex and, 182
paranoid personality disorder, trait of, 134
provides for successful manipulation and control, 250
use of by different people
Carolyn, 141, 145
Mao, 234–35, 250
Milosevic, 162
"projection bias," the danger of assuming those from other cultures think as we do, 378n8
"projects," to eliminate bad housing for the poor, 191
pruning, strange neural in psychopaths, 92
pseudopsychopaths, subtly impaired reasoning skills of, 96
"Psychism," Mao's policies dubbed as by Robert Jay Lifton, 235
Psychology of Politics, The (Hans J. Eysenck), 46
psychopaths. *See also* psychopathy
"borderpath," 168
"corporate," 106–107, 108
defined, 29n, 51–52, 255n
good parents can still have psychopath-like child, 38n, 89, 102–104
neuroscience behind studies of psychopaths
amygdala, unresponsive, 93, 209
corpus collosum, neural pruning and myelin sheaths of, 92
differences in reactions to emotionally charged words, imaging studies of, 90–91
gray, white matter and, 106
image of psychopathic dysfunction and moral reasoning overlap, 101
mirror neurons, role in, 105
neural differences between borderlines and, 209
"pseudopsychopaths" created by brain damage, 96
right anterior superior temporal gyrus dysfunction and problems with abstract reasoning, 97–98
one in six supervisors in Britain is thought by subordinates to be psychopath, 335
overlap in terminology with borderlines, Machiavellians, and the "successfully sinister," 255n
psychopaths take advantage of natural altruism and kindhearted naivete, 255
"successful," 105–108, 322
terrifying feeling evoked by, 52
psychopathy. *See also* psychopaths
across cultures, 265–66
dimensional traits of, 167
evolutionary benefits of psychopathic traits, 254–56, 387–88n2
genetics
general discussion of, 55–58
heritability spotty for, 86
no single gene creates, 68
hypotheses on origin of, 102–104
as learning disability and attention (focusing) issue, 103, 202, 326

Machiavellianism and, 42, 131–32
manipulation and, 107, 132
overlap with other personality disorders, 167, 208–209
sadism and, 52, 239
secondary, 168
psychosis
defined, 413
dopamine system dysfunction and, 184
increases with decrease in size of right parietal lobe in borderlines, 199
Pulsudski, Jozef, brief overview compared to other dictators, 286
Punch cartoon "Doubtful friends," 44
punishment, neural, activated when thinking about partisan information, 189
Putin, Vladimir
ability to dupe credulous, 316
plagiarism as evidence of underlying Machiavellian tendencies, 338
Putnam, Katherine, research related to identity disturbance, 207
Pye, Lucian, explains reticence about declaring Mao a narcissist with a borderline personality, 217

Quantitative Trait Loci (QTL) model, 69n, 71

Racak, massacre of, 156n, 169–71
Raine, Adrian
background of, 94–95
cheating strategy *must* involve a gene machine that lacks a core moral sense, 259
early reference to evolutionary benefits of psychopathic traits, 387–88n2
good parents can still have psychopath-like child, 89
image of overlap between psychopathic dysfunction and moral reasoning areas, 101
problem of free will, 328–29
review of neurological findings on antisocial behavior, 370n25
"successful" psychopaths, 105–108
Rape of Nanking, 303n
Rather, Dan, gullibility regarding Saddam Hussein, 317
rationality (rational cognition), *as opposed to* cognitive dysfunction
borderlines have difficulty with, 195–201
evolutionary development of rational as well as emotional capacities in humans, 187
no simple algorithm for teasing rationality from emotion, 192
not active when engaged in partisan thinking, 189–90
in relation to the legal system, 330n
"rational" approaches by politicians, 307, 309–10
Rauschning, Hermann: Hitler could entertain most incompatible ideas, 305
Reagan, Ronald, remarkable memory of, 313, 313n
receptors. *See also* individual receptor names
BDNF, 78
poliovirus, 327, 413
serotonin, 69–72
reciprocal altruism. *See* altruism
Red Guards, Chinese, 215–16
relationships, unstable personal. *See also* individuals by name
in antisocial personality disorder, 219
borderline personality disorder
as defining DSM-IV trait, 159
as dimensional trait of, 164
as symptom complex of, 135, 137, 219
"splitting" and black-and-white thinking
as an aspect of DSM-IV criteria for borderline personality disorder, alternating between extremes of idealization and devaluation, 157
as borderline personality disorder coping characteristic, 137, 143–44
overactivation of emotional pro-

cessing (ventromedial prefrontal cortex) and, 144
provides for successful manipulation and control, 250
relativism. *See under* politics
religion
established religions
Catholic Church and pedophilia, 30, 35, 107, 332
Christian pacifists and Muslim terrorists act as altruists—so what *is* a Machiavellian, 285–87
Islam, fundamentalists who have found purchase in, 307
neuroscience and genetics
fasting, anorexia, and spirituality, 142n
genes that predispose toward, 59, 59n
sociological aspects
corruption in religion must often reach awe-inspiring proportions before people take concrete action, 333
Machiavellians self-select for positions of power and control, 333
Machiavellians, with their distorted, self-serving cognitions, can twist good intentions to ill and set entire populations aflame with hatred, 331
Mao's religious cult of personality, 242–43, 252, 332
may have served to help promote altruistic behavior, 256
opaque religious ideologies with little by way of checks and balances provide perfect cover for Machiavellians, 333
religious ideologies are seized by people of certain temperaments, whether for good or for ill, 308
there are some things religion cannot easily instill, 322
Wiener's theory of Jewish versus Christian traditions and genetics, 86–87
Wilson, David Sloan, work involving, 17
remorse, 100. *See also* empathy
remorselessness. *See* empathy, lack of
Republican party. *See also* politics
in relation to neuroimaging study on political partisanship, 189–90
resentment
affiliated with activation of specific areas of brain, 102
paranoid personality disorder, a trait of, 134
reticular activating system
illustration of, 115
poliovirus invasion of the, 114–16
reticular formation. *See also* reticular activating system
illustration of, 115
reward system, neural, activated when thinking about partisan information, 189
Reynolds, Glenn (*The Appearance of Impropriety*), 335
rigidity. *See* identity disturbance, inflexibility
Riina, Toto, dreaded *capo di tutti i capi*, remarkable memory of, 313
Ripa, Kelly, "Pin Thin" at 105 pounds, 142n
"riot specialist" Paul Brass, 333n
role absorption, defined, 156
Roman empire, 274–76, 276n
Rome, Inc. (Stanley Bing), 276n
Romulus, Count, 27
Roosevelt, Franklin Delano
"cunning of a schemer and the ambitions of a genuine altruist," 310
remarkable memory, 313
Roxalena, wife of Suleyman the Magnificent, 271–72
Rugova, Ibrahim, leader of Kosovar Albanians, 169
Rummel, R. J., statistics related to those killed by Mao, 216–17

Sabin, Alfred, continual public humiliation of Jonas Salk, 291
sadism
general description of, 52–54

in people
child with psychopath-like traits, 102–103
Hilton, Kathy (Paris Hilton's grandmother and eponymous mother), 273
Ismail the Bloodthirsty, 268
Mao, 239–40
Milosevic, 161
provides for successful manipulation and control, 250–52
research hampered by fears of use in denying culpability of sadists, 52, 330
Sahakian, Barbara: beware possibilities in criminals reengineering personalities, 330
Salk, Jonas, continually humiliated by mean-spirited Alfred Sabin, 291
Sartre, Jean-Paul, as Mao's dupe, 241
Schelling, Thomas (*The Strategy of Conflict*), 260
schizoid personality disorder
decrease in right parietal lobe size in borderlines causes increased symptoms of, 199
defined, 135
schizophrenia
association with borderline personality disorder, 205, 227
chromosomes and genes specifically related to, 54
complex genetics of, 64, 66–67
DARPP-32 and, 82–83
dopamine system and, 184
general discussion of, 227
gray matter loss with one type of, 205
intermediate phenotype and, 66–67
of Mao's son, 223, 227
placement in Axis I of DSM-IV, 134
theta rhythms and, 148
toxoplasma gondii and, 85
schizotypal personality disorder
borderline personality disorder
association with, 205
illustration showing overlap with, 208
defined, 135
traits can provide for success in dangerous social structures, 250
Scott, Robert Falcon, dislike of Ernest Shackleton, 313
self-selection of personality types for positions that suit disposition, 303n–304n, 333
Selim the Sot, 272
sensitivity, personal. *See* hypersensitivity
Sequim, Washington, 23–24, 26, 323–24
serotonin
in context with other neurotransmitters and general mood disorders, 184
dorsal raphe nucleus role in production of, 196
lower levels in prefrontal regions of borderlines, 197
receptors, 69–72
reticular activating system, poliovirus, attention, and, 114–16
sexual addiction, plays key role in, 234
social dominance and neurotransmitter levels, 238
transporters, 72–75
tryptophan hydroxylase (TPH), an enzyme that processes, 82
SERT. *See* transporters, serotonin
Severan dynasty, 276
sexual addiction. *See under* addictive behavior
Shackleton, Ernest, Antarctic explorer, remarkable memory of, 313
shame
affiliated with activation of specific areas of brain, 102
felt by Carolyn, 140
Shang, The Book of Lord, 46
Shermer, Michael: Wallace was "agreeable to a fault," 289
Sherron Watkins, Enron whistle-blower, 297
Shockley, William, intelligence, narcissism, and rages of, 290–91
Short, Philip (Mao biographer), 238, 247
"short" in relation to serotonin transporters, 72–75, 78

sickle-cell anemia and Baldwinian evolution, 264
Siever, Larry J., borderlines elicit responses to soothe themselves, 140
Silk, Ken
  compassionate understanding of his borderline patients, 200, 200n
  research related to identity disturbance, 207
situational competence as coping characteristic of borderline personality disorder, 137, 145–46
  executive control network and, 199–200
Skilling, Jeffrey, Enron CEO, 295–98, 313
Skinner, B. F. *See* behaviorism, restrictive effect on research
Slade, Priscilla, ex-president of Texas Southern University, 278–80
Slotten, Ross, on Darwin's delay in publication, 289
Snakes in Suits: When Psychopaths Go to Work (Paul Babiak and Robert Hare), 107
Snow, Edgar, as Mao's dupe, 241
Snyder, Solomon's usurpation of credit for discovery of opiate receptors, 291
social dominance. *See* control, desire for
sociopath. *See also* psychopath
  definition of, 51
Sociopath Next Door, The (Martha Stout), 322, 398n12
Socrates, 305–306
Solomon, George, research linking personality disorders and the immune system, 207
somatic-marker hypothesis, 203
Sometimes I Act Crazy (Jerold Kreisman), 277
Somoza, Anastasio, brief overview compared to other dictators, 28
Soviet perspective on their system, 177–79, 186
Speer, Albert, observations on Hitler, 300
Spence, Jonathan, noting Mao's charm, 240–41
"splitting." *See under* relationships, unstable personal
"stable sinister systems"
  definition, 278
  discussion, 334, 336–37
  Enron, 294–98
  Ottoman empire—the Sick Man of Europe, 271–72, 274
  Roman empire, 274–76, 276n
  Texas Southern University as prototypical example, 278–80
Stalin, Joseph
  author's discussions with Soviets about, 173–74
  brief overview compared to other dictators, 28, 298, 315
  charm and charisma of, 29–30, 241
  in context with other poorly schooled leaders, 308
  control, desire for, 309
  empathy, lack of, 30
  father's abuse, 218
  illustration: "It's not who votes that counts . . . ," 178
  intelligence and extraordinary memory of, 312
  intimidation techniques, 28
  manipulative duplicity of, 29–30
  narcissism, 309
  temper, 30
Stewart, Jon: "obsolete power corrupts obsoletely," 323
Stewart, Martha, dark business genius, 293–94, 315
Stop Walking on Eggshells (Paul Mason and Randi Kreger), 139
Stout, Martha (*The Sociopath Next Door*), 322, 398n12
Strategy of Conflict, The (Thomas Schelling), 260
stress. *See also* child abuse
  borderline personality disorder and, 85
  effect on genes and personality, 65–66
  family discord and genetics, 60
  malnutrition switching on of APOE4 allele, 76

MAO-A alleles and, 81–82
polio and, 125–26
Stroessner, Alfredo (Paraguayan dictator)
intelligence of, 311
subclinical personality disorders
borderline personality disorder
anterior cingulate cortex dysfunction and inability to focus on something undesirable, 182
concordance of subclinical borderline personality disorder in twins, 85n
control, desire for, 236
executive control as defining difference for clinical versus subclinical, 199–202
inflexibility because emotional memories poorly suppressed by orbitofrontal cortex, 204
larger numbers of subclinical than clinical, 137
inherent flaw in DSM-IV use of "clinical significance" in discriminating between clinical and subclinical personality disorders, 375n3, 376n32
MAO-A and, 80
narcissistic personality disorder and its mixed benefits in subclinical form, 244
overlap in terminology with psychopaths, borderlines, Machiavellians, and the "successfully sinister," 255n
PDQ-4+ and dimensional approaches to quantifying, 132–33
polio and, 326
stress can push a person with propensity for disorder from subclinical to clinical, 65–66, 85. *See also* stress
"tip of the iceberg" understanding of serotonin receptor alleles and, 72
subconscious thought, 182, 186
substance abuse. *See under* addictive behavior
"successful psychopaths," 105–107
successfully sinister (synonym for Machiavellian). *See* Machiavellians
suicide
5-HT_{2A} and, 71
chromosome 2 region and genetic tendency toward, 160
Milosevic and, 160–61
serotonin transporters and, 73
suicidal behavior as defining DSM-IV trait of borderline personality disorder, 159
Suleyman the Magnificent, 271–72
Sumo wrestlers and "tit for tat" strategy, 258
superior temporal gyrus, right anterior and abstract reasoning deficits in psychopaths, 97–98
Suroweicki, James (*The Wisdom of Crowds*), 176, 243, 245, 301, 334
suspiciousness. *See* distrust
Swift, Nicholas: coinage of word "borderpath," 168
Sykes, Bryan: comments on "Mini-Genghises," 267–68
synapse, explanation and illustrations of, 70, 184
systems biology, definition of, 34
Szent-Gyorgyi, Albert: " Research is to see what everybody else has seen . . . ," 40

taxi drivers, London, 176
Tay-Sachs disease and intelligence, 87
teaching and Machiavellianism. *See* academia
TEDS (Twins Early Development Study), 55
temper. *See* anger
temporo-occipital cortex, boosted signal from amygdala causes excess activity in borderlines, 194, 195
Terrill, Ross (Mao biographer)
early Mao had genuine good aspects, 248–49
evidence that Mao was borderline mounted by the year, 217
imperial-plus-Leninist system acted as a magnifying glass for Mao's personality quirks, 250
"Mao *did not* believe a lot of what he proclaimed in public," 235–36

Texas Southern University as example of "stable sinister system," 278–80
thalamus
in borderline personality disorder
role in executive control
discussion, 199
flowchart, 196
illustration of, 183
murderers have "turbocharged," 97
portrayed in flowchart form, 185, 196
as router for neural signals
discussion, 185–86
flowchart, 185
thalassaemia and Baldwinian evolution, 264
Thatcher, Margaret
desire for control and egotism, 309
hypomanic qualities, 314
remarkable memory of, 313, 313n
theta rhythms, 148
Tiananmen Square, photo of unknown man in front of tank, 306
"tit for tat" as a strategy that may have helped lead to altruism, role of amygdala in reinforcing, 256–60
Tolstoy, Leo: disproving his theory that only social forces matter, 267
Tonya, The Tonya Show, ". . . and he's refreshingly free of emotional baggage," 49
Tooby, John, 175, 388 n12
torture
by a child with psychopath-like traits, 102–103
by Ismail the Bloodthirsty, 268
in Mao's regime, 216, 225, 239–40
in Milosevic's regime, 156
by "Papa Doc" Duvalier, 311
in relation to sadism, 52–54
in Stalin's regime, 28, 239
toxoplasma gondii and schizophrenia, 85
Toynbee, Arnold, duped by Hitler, 315
transporters, serotonin, 72–75
Trebjesanin, Zarko: Milosevic is "cold narcissus," 161–62
Trevisan, Dessa, *London Times* Balkan correspondent, confronts Milosevic, 153
Trivers, Robert, and the concept of reciprocal altruism, 256–57
Trujillo, Rafael, brief overview compared to other dictators, 28
trust
caudate and, 20–22
Williams syndrome's abundance of, 98–99
Trust (Francis Fukuyama), 219
tryptophan hydroxylase (TPH), 82
Turnbull, Colin, studies of the Ugandan Ik tribe, 283n
Turner, Ted, duped by Putin, 316
tyrants. *See* dictators

United Nations and *"Complicity with Evil"* as in Adam LeBor's book of that name, 332
universities and Machiavellianism. *See* academia
unstable personal relationships. *See* relationships, unstable personal
unstable sense of self, shared trait of borderline and bipolar personality disorders, 142n

"val"
BDNF allele, improved memory, more anxiety, 77
COMT allele, faster dopamine metabolization, "warrior," 79
vasopressin hormones, 83
venereal disease
Mao's, 233
modern despots don't have as many children because of, 233, 269n
vengeance. *See* vindictiveness
Venter, Craig, speeds up decoding of human genome, 290
ventral tegmental area, role in production of dopamine, 196
ventromedial cortex
damage creates "pseudopsychopaths" with impaired reasoning skills, 95–96
and "emote control" in neuroimaging study of political partisanship, 189–90

executive dysfunction, antisocial behavior and, 94
function of and dysfunction in, 182
illustrations of, 94, 181
limbic system, intimate connection with, 182, 186
role in cognitive-perceptual impairment in borderlines, 203–207
role in commonsense decision making, 203
role in emotional cognition, 186
Viding, Essi, studies of genetic risk for psychopathy, 55–58
vindictiveness
Hilton, Kathy (Paris Hilton's grandmother), 273
Mao's, 239, 250
Milosevic's, 161
paranoid personality disorder, trait of, 134, 246
provides for successful manipulation and control, 250, 260–61
violence. *See* aggressiveness; impulsivity
vitamin D and the reshaping of genetics in northern climates, 263
Vogel, Carl: mixed blessings of mild narcissism, 244
Volkan, Vamik: "the narcissist in power can restructure reality by eliminating those who threaten his self-esteem," 247

Wade, Nicholas, writes on the emergence of the modern mind, 266, 389n25
Waite, Robert (Hitler biographer), 308–309, 311–12, 315
Walker, William, ambassador, description of massacre at Racak, 170
Wallace, Alfred Russel, and the disadvantages of being free of narcissism, 288–89, 292
Wallace, Mike, journalist, gullibility regarding Middle Eastern terrorists and tyrants, 317
Wankat, Philip (*The Effective, Efficient Professor*), the importance of remembering names, 312
"warrior" (*val*) COMT allele, 79
Washington, George: a man of "tumultuous passions" whose "genius was his judgment," 298–300, 307–308, 328
Watson, James, sublimely arrogant, misogynistic codiscoverer of DNA, 290, 392n13
wealth, power, aristocracy, attraction of borderlines toward those with, 277
Weatherford, Jack, *Genghis Khan and the Making of the Modern World*, a high-point of compulsive biography inhalation, 286n
Weinberger, Daniel
BDNF, research on, 77–78
comments on genetics and personality research, 67–68
DARPP-32, research on, 82–83
Welch, Jack, mental flexibility of, 301
Westen, Drew, work related to
emotional reasoning, 189–90
identity disturbance, 156
whistle-blowers, "bad," 335
white matter, in pathological liars is increased, in autistics is reduced, 106
Wiener, Norbert, emergenic prodigy, 86–87
Wilkinson-Ryan, Tess, work related to identity disturbance, 156
Williams syndrome, 98–99
Wilson, David Sloan
emphasizes the speed of human evolution, 263
foreword by, 15–17
Wilson, Matthew, research on theta rhythms, 148
Wisdom of Crowds, The (James Surowiecki), 176, 192, 243, 245, 301, 334
"worrier" (*met*) COMT allele, 79
Wrangham, Richard: theory that humans are "taming" themselves, 264–65

Yang, Yaling
image of overlap between psychopathic dysfunction and moral reasoning areas, 101

pathological liars and increased white matter, 106
Yixing teapots, 212–15, 382n1

Zhou Enlai (Chou En-lai), 239
Zimbardo, Philip (*The Lucifer Effect*), questionable nature of its conclusions, 303–304n
Zimmermann, Warren, Ambassador
calls Milosevic "con man," 153
left speechless as gaslighting Milosevic pleads, "I'm not so bad, am I," 164–65
notes Milosevic fooled himself as well as others, 154
zisha purple clay, 212–15, 382n1

Photo by John Meiu Photosite